Springer Series on

ATOMIC, OPTICAL, AND PLASMA PHYSICS 57

Springer Series on

ATOMIC, OPTICAL, AND PLASMA PHYSICS

The Springer Series on Atomic, Optical, and Plasma Physics covers in a comprehensive manner theory and experiment in the entire field of atoms and molecules and their interaction with electromagnetic radiation. Books in the series provide a rich source of new ideas and techniques with wide applications in fields such as chemistry, materials science, astrophysics, surface science, plasma technology, advanced optics, aeronomy, and engineering. Laser physics is a particular connecting theme that has provided much of the continuing impetus for new developments in the field. The purpose of the series is to cover the gap between standard undergraduate textbooks and the research literature with emphasis on the fundamental ideas, methods, techniques, and results in the field.

Please view available titles in *Springer Series on Atomic, Optical, and Plasma Physics* on series homepage http://www.springer.com/series/411

Preface

Thermodynamics and statistical physics study the physical properties (mechanical, thermal, magnetic, optical, electrical, etc.) of the macroscopic system. The tasks and objects of study in thermodynamics and statistical physics are identical. However, the methods of investigation into macroscopic systems are different.

Thermodynamics is a *phenomenological theory*. It studies the properties of bodies, without going into the mechanism of phenomena, i.e., not taking into consideration the relation between the internal structure of substance and phenomena, it generalizes experimental results. As a result of such a generalization, postulates and laws of thermodynamics made their appearance. These laws make it possible to find general relations between the different properties of macroscopic systems and the physical events occurring in them.

Statistical physics is *a microscopic theory*. On the basis of the knowledge of the type of particles a system consists of, the nature of their interaction, and the laws of motion of these particles issuing from the construction of substance, it explains the properties being observed on experiment, and predicts the new properties of systems. Using the laws of classical or quantum mechanics, and also the theory of probability, it establishes qualitatively new *statistical appropriatenesses* of the physical properties of macroscopic systems, substantiates the laws of thermodynamics, determines the limits of their applicability, gives the statistical interpretation of thermodynamic parameters, and also works out methods of calculations of their means. The Gibbs method is based on statistical physics. This method is the most canonical. Therefore, in this book, the exposition of the Gibbs method takes an important place.

Results, stemming from phenomenological thermodynamics, bear the general character and can be applied to any macroscopic systems; however, the internal mechanism of physical phenomena and properties, being observed in the experiments, is not disclosed. In other words, thermodynamics only describes the phenomena and establishes the relation between them, but does not answer the question why it happens just so.

Statistical physics relates the properties of bodies to their internal construction, creates the microscopic theory of physical phenomena, and answers the question why it happens just so. The disadvantage of this method resides in the fact that results, obtained here, bear a particular character and are right only in frames of the considered model of the structure of substance.

Thermodynamics and statistical physics study not only equilibrium systems, but also systems in which specified currents and flows (the electric current, flow of energy and substance) exist. In this case, the theory is called *thermodynamics of non-equilibrium systems* or *kinetics*. Kinetics originates from the Boltzmann equation (1872) and has continued developing up to the present time.

The development of phenomenological thermodynamics started in the first half of the nineteenth century.

The first law of thermodynamics was discovered by the German physiologist Julius Robert von Mayer (1842) and the English physicist James Prescott Joule (1843). They showed the equivalence of heat and mechanical work. The first law of thermodynamics is a law of conservation of energy for closed processes. In 1847, the German physicist and physiologist Hermann von Helmholtz generalized this law for any non-closed thermodynamic processes.

The second law of thermodynamics was discovered independently by both the German physicist Rudolf Clausius (1850) and the English physicist William Thomson (Lord Kelvin). They introduced in the theory a new function of state – *entropy*, in the statistical sense, and discovered *the law of increasing entropy.*

The third law of thermodynamics was discovered in 1906 by the German physicist–chemist Walther Nernst. According to this law, *entropy of all systems independently of external parameters tends to the identical value (zero) as temperature approaches the absolute zero.*

Note that the first law of thermodynamics is a law about energy, and the second and the third ones are about entropy.

The founders of thermodynamics are J.R. von Mayer, J.P. Joule, H. von Helmholtz, R. Clausius, W. Kelvin, and W. Nernst.

Statistical physics received its development only in the last quarter of the nineteenth century. The founders of classical statistical physics are R. Clausius, J.C. Maxwell, L. Boltzmann, and J.W. Gibbs. The height of development of classical statistical physics is the method of Josiah Willard Gibbs (1902).

The application of classical statistics to many problems provided results, though not coinciding with the experimental facts of that time. Black radiation (thermodynamics of a photon gas), heat capacity of metals, Pauli paramagnetism, etc. can serve as examples. These difficulties of classical statistics were circumvented only after the rise of quantum mechanics (L. de Broglie, W. Heisenberg, E. Schrödinger, and P. Dirac) and quantum statistics, created on its basis (E. Fermi, P. Dirac, S.N. Bose, A. Einstein) during 1924–1926.

The method of thermodynamic functions and potentials, and also the Gibbs statistical method or the methods of free energy, being the key-note of the book, occupy an important place. It is shown that of all the thermodynamic functions, the most important are the function of free energy and grand thermodynamic potential, which are determined from the Gibbs canonical distribution. It is expalined that the basic postulate of statistical physics – the *microcanonical distribution* of isolated systems – is based on the statistical theory of the macroscopic properties of a system, from which all canonical distribution stems.

Understanding free energy and grand thermodynamic potential, it is easy to determine entropy, thermal and caloric equations of state, and also all thermodynamic coefficients, measured by testing. To do this in the case of classical systems, it is sufficient to know the Hamilton function – energy as a function of coordinates and impulses of particles of the system, forming it, and for quantum systems, it is the energy spectrum, i.å., the dependence of energy on quantum numbers. It is also an essence of the Gibbs method, which is applied to ideal and non-ideal gases, and also to a crystalline solid. The exposition of the Fermi-Dirac and Bose–Einstein quantum statistics and its application to different quantum gases occupy a large place. It is shown how the difficulties of classical statistics, associated with its application to an electron gas in metals, are circumvented. The statistics of the electron gases are considered in detail in this book.

A separate chapter is devoted to the statistical theory of thermodynamic properties of an electron gas in a quantizing magnetic field. Note that the investigation of properties of an electron gas in extremal conditions, in particular, at ultra-low temperatures and in strong quantizing magnetic fields, is one of the actual tasks of contemporary physics.

In the last chapter, on the basis of the Boltzmann kinetic equation, the electron gas in metals and semiconductors is considered in a nonequilibrium state. Nonequilibrium processes are associated with charge carrier motion in a crystal under external disturbances such as the electric field and the temperature gradient in the magnetic field. They include electric conductivity, thermoelectric, galvanomagnetic, and thermomagnetic effects.

Baku,
November 2009

Bahram M. Askerov
Sophia R. Figarova

Contents

1

Basic Concepts of Thermodynamics and Statistical Physics

Summary. The basic concepts and postulates of thermodynamics and statistical physics are expounded in this chapter. Different ways of description of the state of macroscopic systems, consisting of a very large number of particles such as atoms, molecules, ions, electrons, photons, phonons, etc., are adduced. Such concepts as the distribution function over microstates, statistical weight of the preassigned macroscopic state of a system, absolute temperature, and pressure are also introduced.

1.1 Macroscopic Description of State of Systems: Postulates of Thermodynamics

As noted in the Foreword, thermodynamics and statistical physics study physical properties of macroscopic systems with a large number of degrees of freedom. The lifetime of these systems ought to be sufficiently long to conduct experiments on them. A usual gas, consisting of atoms or molecules, photon gas, plasma, liquid, crystal, and so on, can serve as an example of such systems. A small but macroscopic part of the considered system is called *a subsystem.*

Macroscopic systems can interact between themselves or with the surrounding medium by the following channels:

1. An interaction when the considered system performs work on other systems, or vice versa, is called *mechanical interaction* ($\Delta A \neq 0$). In this case, the volume of the system changes.
2. An interaction in which the energy of the system changes only at the expense of heat transfer (without performing work) is called *thermal interaction* ($\Delta Q \neq 0$).
3. An interaction leading to exchange of particles between systems or between the system and the surrounding medium is called *material interaction* ($\Delta N \neq 0$).

Depending on which of the above-indicated channels is open or closed, different types of macroscopic systems exist in nature.

A system is called *isolated* if energy and material exchange with the surrounding medium is absent ($\Delta A = 0,\ \Delta Q = 0,\ \Delta N = 0$). For such systems all channels of interaction are closed.

If a system is surrounded by a heat-insulated shell, the system is called *an adiabatically isolated system* ($\Delta Q = 0$).

If a system does not exchange particles with the surrounding medium ($\Delta N = 0$), such a system is called *closed*, and, on the other hand, if exchange of particles ($\Delta N \neq 0$) takes place, the system is called *open*.

If the considered system is a small but macroscopic part of a large system, physical processes occurring in it will hardly influence the thermodynamic state of the large system. In this case, the large system is called *a thermostat*, and the system interacting with it is *a system in the thermostat*.

The thermodynamic state of each system at pre-assigned external conditions can be described by a restricted number of physical quantities which can be measured on test. Such quantities are called *thermodynamic parameters*. The number of particles in a system N, its volume V, pressure P, absolute temperature T, dielectric $\mathscr{P}$ and magnetic $\boldsymbol{M}$ polarization vectors, electric $\boldsymbol{\mathcal{E}}$ and magnetic $\boldsymbol{H}$ field strengths are examples of thermodynamic parameters. These parameters characterize the system itself and also the external conditions in which it is found.

Parameters that are determined by coordinates of external bodies interacting with a system are called *external parameters*: volume, the electric and the magnetic field strength, etc. Parameters that, apart from coordinates of external bodies, depend also on coordinates and impulses of particles entering into the system are called *internal parameters*: pressure, temperature, the internal energy, dielectric and magnetic polarizations.

Internal parameters can be *intensive* or *extensive*. Parameters not depending on the number of particles in a system are called *intensive*: pressure, temperature, etc. Parameters that are proportional to the number of particles or the amount of substance are called *extensive* or *additive*: entropy, energy and other thermodynamic potentials.

The state of a system determined by the totality of the above-enumerated, measured-on-test thermodynamic parameters is called *a macroscopic state* of the system:

$$\text{Macroscopic state} \Rightarrow (N,\ V,\ P,\ T,\ \mathscr{P},\ \boldsymbol{M},\ \boldsymbol{\mathcal{E}}, \boldsymbol{H}, \ldots)\,.$$

It is evident that these macroscopic parameters determine the averaged state of a system, in that the details and the nature of the complex motion of particles composing the system are disregarded. Such a description of a system bears the phenomenological, i.e. the descriptive, character.

If thermodynamic parameters determining the state of a system do not depend on time, such a state is called *stationary*. Moreover, if stationary flows and currents are absent in a system, such a state is called a *thermodynamic equilibrium*. This state is the simplest macroscopic state of a system. It is to be noted that even in this state, inside the system particles perform complex chaotic motion; however, this motion is not of interest in thermodynamics.

After introducing the primary basic thermodynamic concepts, we pass on to the exposition of two postulates comprising the basis of thermodynamics. These postulates were established from generalizations of experimental data.

The first postulate of thermodynamics states that each isolated system has only one intrinsic state, that of thermodynamic equilibrium. If a system is not in the equilibrium state, it tends toward its equilibrium state over a period of time, and once it has achieved that state, can never come out of it spontaneously without an external force being exerted on it.

Also called *the general principle of thermodynamics*, this postulate defines the thermodynamic equilibrium state. This principle is demonstrated by an example of the macroscopic parameter L (Fig. 1.1). The time τ during which the parameter $L(t)$ passes to the equilibrium state L_0 is called *the relaxation time*. The quantity τ depends on the nature of interaction and intensity of the motion of the particles composing the system.

The first postulate of thermodynamics determines the limit of applicability of the laws of thermodynamics. Indeed, inasmuch as each system consists of chaotically moving particles, the parameter $L(t)$ can deviate from its mean value, i.e. a fluctuation occurs. These deviations are schematically shown in Fig. 1.1. Thermodynamics disregards these fluctuations and takes only the mean values measured on test into consideration. Therefore the laws of thermodynamics are applicable only to systems in which deviations from the mean values are much smaller than the mean values themselves. But this is possible only in systems with a sufficiently large number of particles.

If a system consists of a small number of particles, the relative fluctuation can be large and the system itself can move away from the equilibrium state. In this case, the concept of "the equilibrium state" loses its sense, and the first postulate of thermodynamics is violated. It can be demonstrated by a simple example. Let us assume that a gas contained in a rectangular vessel contains N particles. Mentally divide the vessel into two equal parts. In the equilibrium state, in each half $N/2$ molecules ought to be found. If $N = 4$,

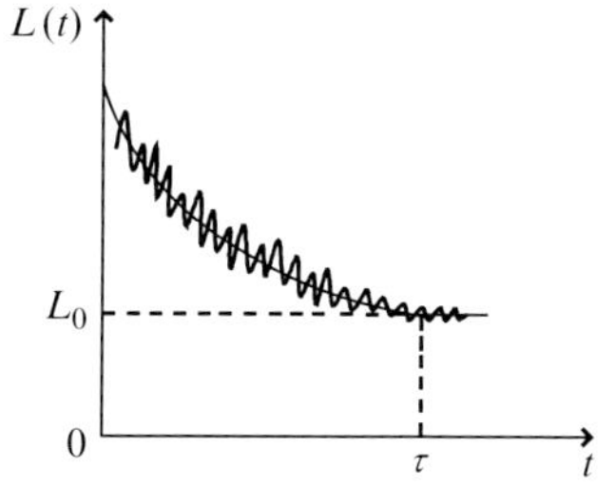

Fig. 1.1. The fluctuation of the thermodynamical parameter

often the following picture can be observed: in the first part there are three molecules, and in the second there is one molecule; or, in the first part there are four molecules, and the second part is empty. Such a situation means that the system itself is not in the equilibrium state, in which $N/2 = 2$ molecules should be present in each part of the vessel.

Thus, from the first postulate the following conclusion can be made: *Laws of thermodynamics are not applicable to systems consisting of a small number of particles.*

The second postulate of thermodynamics states that if two systems A and B are separately found in thermodynamic equilibrium with the third system C, A and B are also found in thermodynamic equilibrium between themselves, i.e.

$$\left.\begin{array}{l} A \sim C \\ B \sim C \end{array}\right\} \Rightarrow A \sim B. \tag{1.1}$$

This postulate is also called *the zeroth law of thermodynamics* and, as we will see below, defines the concept of *absolute temperature.*

The second postulate determines the upper boundary of applicability of the laws of thermodynamics. As seen from this postulate, when bringing into thermal contact subsystems A, B, C, or while disconnecting them, the state of equilibrium is not violated, i.e. the energy of interaction of the subsystems is negligibly small and the energy of the whole system is an additive quantity[1]

$$E = \sum_{\alpha} E_{\alpha}, \tag{1.2}$$

where E_{α} is the energy of the subsystem α.

Thus, the laws of thermodynamics are applicable only to systems for which the condition of additivity of energy (1.2) is fulfilled. Naturally, the condition of additivity is not fulfilled for large systems, individual parts of which interact through a gravitational field.[2] Therefore, the laws of thermodynamics are not applicable to very large, complex systems, e.g. to the universe as a whole.

From the second postulate, besides the principle of additivity of energy, the second, not less important, conclusion stems. Indeed, from this postulate it follows that if A, B, C are subsystems of a large system in the equilibrium state, their state, besides the external parameters, ought to be characterized also by the general internal parameter. This internal intensive parameter is called *temperature* and is identical in all parts of the large system, which is in the thermodynamic equilibrium state. Temperature is determined by the intensity of the thermal motion of the particles in the system. Thus, according to the second postulate, the thermodynamic equilibrium state of a system is determined by the totality of external parameters and temperature. Consequently,

[1] Equality (1.2) supposes that the energy of interaction between subsystems is negligible small compared with the internal energy of a subsystem.

[2] In this case, the gravitational energy of interaction between parts cannot be neglected.

according to the second postulate, *each internal parameter is a function of external parameters and temperature.* This conclusion relates any internal parameter A_i to temperature T and external parameters $a_1, a_2, \ldots, a_n$:

$$A_i = A_i(a_1, a_2, \ldots, a_n; T); i = 1, 2, \ldots \kappa, \tag{1.3}$$

where k is the number of internal parameters. This equation, written in the symbolic form, is called *the equation of state.* The number of such equations, naturally, equals the number of internal parameters k.

If in the capacity of an internal parameter we accept the internal energy of a system $A_i \equiv E$, (1.3) can be presented in the form:

$$E = E(a_1, a_2, \ldots, a_n; T). \tag{1.4}$$

This equation is called *the caloric equation of the state of a system.*

If in the capacity of an internal parameter we accept pressure $A_i \equiv P$, from (1.3) we get *the thermal equation of state*:

$$P = P(a_1, a_2, \ldots, a_n; T). \tag{1.5}$$

Thus, from the set of equations (1.3) it is seen that the thermodynamic state of a system is single-valuedly determined by $(n+1)$ independent number of parameters. Therefore, the number $(n+1)$ is called *the thermodynamic degree of freedom* of a system. Depending on the complexity of a system, n takes on values $n = 1, 2, 3, \ldots$.

In the simplest case of closed systems,[3] if in the capacity of an independent external parameter volume V is accepted, the internal parameter pressure P and internal energy E, conforming to (1.3), can be expressed as follows:

$$P = P(V; T); \quad E = E(V; T). \tag{1.6}$$

The explicit form of these equations for ideal gases is experimentally determined and theoretically substantiated by statistical methods:

$$P = \frac{N}{V} k_0 T; \quad E = \frac{3}{2} k_0 TN, \tag{1.7}$$

where N is the number of particles of an ideal gas, and k_0 is the Boltzmann constant.

If from (1.4) we determine temperature and substitute it into (1.3), all internal parameters can be expressed by E and external parameters $a_1, a_2, \ldots, a_n$. Thus, the second postulate of thermodynamics can be also expressed as follows: *All internal parameters of a system found in thermodynamic equilibrium are functions of external parameters and energy*:

$$A_i = A_i(a_1, a_2, \ldots, a_n; E). \tag{1.8}$$

[3] A gas consisting of the pre-assigned number of neutral atoms or molecules can be considered as an example.

Systems satisfying this condition are called *ergodic*. Consequently, thermodynamics is applicable only to ergodic systems.

For an ideal gas, using (1.7), the equation of type (1.8) takes the following explicit form:

$$P = \frac{2}{3}\frac{E}{V}. \tag{1.9}$$

Finally, note once more that the first postulate of thermodynamics defines the concept of *thermodynamic equilibrium* and the second one defines the concept of *absolute temperature.*

1.2 Mechanical Description of Systems: Microscopic State: Phase Space: Quantum States

It is known that any macroscopic system consists of a colossal but finite number of particles, and also that each particle can have a complex internal structure. Here the structure of a particles does not interest us, and we will regard that the considered system as consisting of N number of chaotically moving material points interacting among themselves. Thus, the number of degrees of freedom of the considered system is $3N$. Note that under normal conditions $1\,\mathrm{cm}^3$ of air contains 3×10^{19} molecules. The linear dimension of each molecule is $10^{-8}\,\mathrm{cm}$. In order to have a notion of the number of particles and their dimensions, we quote a known example by Kelvin, according to which the number of H_2O molecules in a glass of water is 100 times the number of glasses of water available in all oceans and seas of the world.

Naturally, it is impossible to describe in detail the state of such a macroscopic system with a small number of thermodynamic parameters, since these parameters disregard the internal structure of the system. For the complete description of a system, it is necessary to know which particles it consists of, what nature of their interaction is and by which equations their motion is described, i.e. whether the motion of particles obeys classical or quantum mechanical laws. In conformity with this, in nature two types of systems exist: classical and quantum systems. We consider these cases separately.

Classical systems. The motion of particles forming such systems obeys the laws of classical mechanics, and the state of each of them is determined by three generalized coordinates $q_i(t)$ and by three corresponding generalized impulses $p_i(t)$,[4] where i takes on the value $i = 1, 2, 3$. Consequently, the general state of a classical system consisting of N particles at the instant of time t is determined by $6N$ quantities:

$$\textit{microstate} \Rightarrow (q, p) \equiv (q_1, q_2, \ldots, q_{3N}; p_1, p_2, \ldots, p_{3N}). \tag{1.10}$$

[4] In classical statistical physics the motion of particles is characterized not by velocity, but by impulse, as the Liouville theorem (Sect. 1.3) is just not in the space of coordinates and velocity $(q, \dot{q})$, but in the phase space (q, p).

The state of a system being determined by $3N$ generalized coordinates and $3N$ generalized impulses is called *the microscopic state* or, briefly, *the microstate* of the classical system. The quantities $q_i(t)$ and $p_i(t)$, i.e. the microstates of the system, are found from a solution of the system of Hamilton canonical equations:

$$\dot{q}_i = \frac{\partial \mathscr{H}}{\partial p_i}, \dot{p}_i = -\frac{\partial \mathscr{H}}{\partial q_i}; i = 1, 2, 3, \ldots, 3N, \tag{1.11}$$

where $3N$ is the number of degrees of freedom of the system, the points over q_i and p_i mean the time derivative and $\mathscr{H}$ is the Hamilton function of the system. For conservative systems, $\mathscr{H}$ coincides with the total energy $E(q, p)$ of the system:

$$\mathscr{H} \equiv E(q, p) = \sum_{i=1}^{3N} \frac{p_i^2}{2m} + U(q_1, q_2, \ldots, q_{3N}), \tag{1.12}$$

where m is the mass of a particle and $U(q)$ is the potential energy of interaction of particles; it is supposed that an external field is absent.

To describe a microstate of classical systems, it is convenient to introduce the concept of *phase space* or the Γ-*space*. Each system has its intrinsic phase space. For instance, the phase space of a classical system consisting of N particles represents an abstract $6N$-dimensional space. The position of each "point" in this space is determined by $3N$ generalized coordinates q_i and $3N$ generalized impulses p_i, i.e. by $6N$ quantities.

Thus, a microstate of a classical system consisting of N particles has a corresponding "point" – *a phase point* – in the $6N$-dimensional phase space.

Henceforth, for an elementary volume of the phase space $\mathrm{d}\Gamma$ we will use the following symbolic notation:

$$\mathrm{d}\Gamma = \mathrm{d}q\,\mathrm{d}p \equiv \prod_{i=1}^{3N} \mathrm{d}q_i\,\mathrm{d}p_i. \tag{1.13}$$

Hence it is seen that the dimensionality of an element of "volume" of the phase space of a classical system consisting of N particles is $(action)^{3N}$.

The phase space can be subdivided into two subspaces: coordinates and impulses. Then, for an element of volume of the phase space one can write $\mathrm{d}\Gamma = \mathrm{d}\Gamma_q \cdot \mathrm{d}\Gamma_p$.

For some systems (e.g. an ideal gas) instead of the Γ-space, we may introduce the concept of the μ-space. The μ-space is a six-dimensional space, each point of which is determined by a coordinate and an impulse of one particle (x, y, z, p_x, p_y, p_z). It is evident that in the μ-space a microstate of an ideal gas consisting of N particles is described by a multitude of N points.

Coordinates $q_i(t)$ and impulses $p_i(t)$ of particles forming a system continually change in conformity with equations of motion (1.11); therefore, the microstate of the system changes with time. As a result, the position of phase points changes, describing a specified "curve" in the $6N$-dimensional phase space. This curve is called *the phase trajectory.*[5] The equation of the phase trajectory, in principle, can be found from the solution of the system of equations (1.11). These solutions can be written symbolically in the following form:

$$\begin{cases} q_i = q_i(t; q_{01}, q_{02}, \ldots, q_{03N}; p_{01}, p_{02}, \ldots, p_{03N}) \\ p_i = p_i(t; q_{01}, q_{02}, \ldots, q_{03N}; p_{01}, p_{02}, \ldots, p_{03N}), \end{cases} \tag{1.14}$$

where q_{0i} and p_{0i} are initial coordinates and impulses, respectively, of a particle.

Note that the phase trajectory can be closed, but it cannot intersect or touch itself. This result follows from the principle of determinism of classical mechanics, i.e. the single-valuedness of the solution (1.14) of the equation of motion (1.11).[6]

The phase trajectory of even simple systems cannot be described graphically. It is sufficient to remember that the phase space of a system consisting of one particle is six-dimensional. The phase trajectory can be graphically described only for one particle moving in a one-dimensional space. In this case, the phase space is two-dimensional. In Fig. 1.2, the phase trajectory of a freely moving particle with a mass m and a pre-assigned energy $p^2/2m = \varepsilon_0 = \text{const}$ in "a one-dimensional box" with dimensions $-L/2 \le q \le L/2$ is presented.

As seen from Fig. 1.3, the phase trajectory of a linear harmonic oscillator with a pre-assigned energy $p^2/2m + m\omega^2 q^2/2 = \varepsilon_0 = \text{const}$ represents

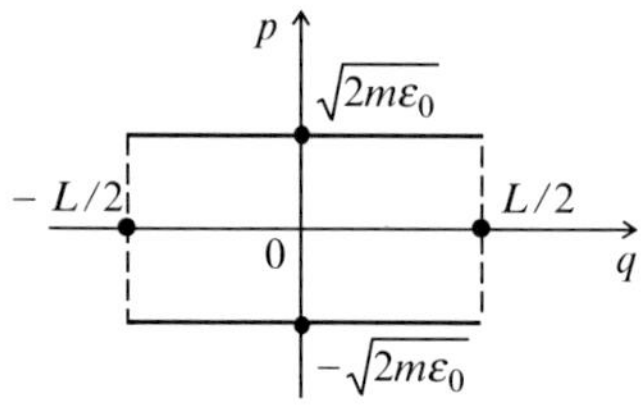

Fig. 1.2. The phase trajectory of the one-dimensional free moving

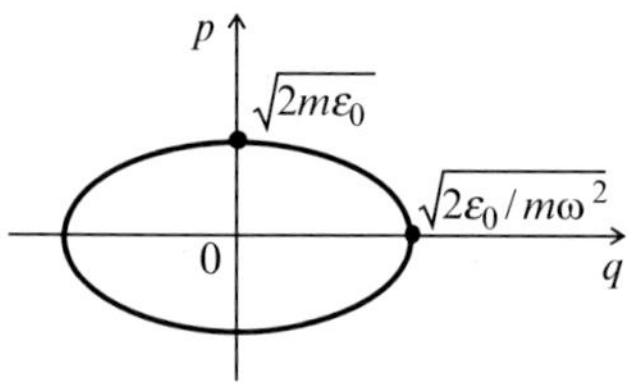

Fig. 1.3. The phase trajectory of a linear harmonic oscillator

---

[5] This curve should not be confused with the trajectory of a particle's motion in the usual three-dimensional space.

[6] Indeed, if the phase trajectory would intersect itself, the intersection point could be accepted as an initial one, and the further change in the state of the system would not be single-valued.

an ellipse with semi-axes $\sqrt{2\varepsilon_0/m\omega^2}$ and $\sqrt{2m\varepsilon_0}$, where ω is the circular frequency of the linear harmonic oscillator with a mass m.

Quantum systems. For quantum systems, i.e. systems in which the motion of particles is described by equations of quantum mechanics, concepts of the phase space, phase point and phase trajectory have no sense. Indeed, according to the Heisenberg uncertainty principle, the coordinate q and impulse p of a particle cannot be single-valuedly determined simultaneously. Principal errors Δq and Δp are related by the following relationship:

$$\overline{(\Delta q)^2} \cdot \overline{(\Delta p)^2} \geq \frac{\hbar^2}{4}, \tag{1.15}$$

where

$$\hbar = h/2\pi = 1.05 \times 10^{-27}\,\mathrm{erg\,s},$$

h being the Planck constant.

The state of a system in non-relativistic quantum mechanics is described by the stationary Schrödinger equation

$$\hat{\mathscr{H}}\Psi_n = E_n\Psi_n. \tag{1.16}$$

Here Ψ_n and E_n are an eigenfunction and eigenvalue of energy, respectively. The Hamilton operator $\hat{\mathscr{H}}$ entering into (1.16) can be obtained from expression (1.12) by replacing $p_i \Rightarrow \hat{p}_i = -i\hbar\frac{\partial}{\partial q_i}$. If in place of the generalized coordinate q_i we take the Cartesian coordinate $\boldsymbol{r}_i = \boldsymbol{r}(x_i, y_i, z_i)$, the Hamilton operator $\hat{\mathscr{H}}$ takes the following form:

$$\hat{\mathscr{H}} = \sum_{i=1}^{N}\left(-\frac{\hbar^2}{2m}\nabla_i^2\right) + U(\boldsymbol{r}_1, \boldsymbol{r}_2, \ldots, \boldsymbol{r}_N), \tag{1.17}$$

where ∇_i^2 is the Laplace operator. In (1.16) n is the totality of quantum numbers, determining one quantum state of the system. The wave function of the system Ψ_n depends on the coordinates of all particles, i.e. $\Psi_n = \Psi_n(\boldsymbol{r}_1, \boldsymbol{r}_2, \ldots, \boldsymbol{r}_N)$.

Thus, the microscopic state of a quantum system is determined by the totality of all quantum numbers describing one quantum state Ψ_n. One value of the energy of the system E_n can correspond to each microstate n, or several microstates can correspond to one value of energy, i.e. it is possible to have quantum degeneracy.

For real systems, we cannot exactly solve the Schrödinger equation (1.16) and determine the microstate of a system. Here, we consider the simplest ideal systems, allowing an exact solution.

Ideal gas in a rectangular box. Assume that in a rectangular box with dimensions L_x, L_y, L_z, N non-interacting particles with a mass m are found. It is known that the total energy of an ideal gas is the sum of the energies of the individual particles:

$$\mathrm{E}_n = \sum_{i=1}^{N} \varepsilon_{n_i}, \tag{1.18}$$

where

$$\varepsilon_{n_i} = \frac{\hbar^2}{2m}\left(k_{ix}^2 + k_{iy}^2 + k_{iz}^2\right) \tag{1.19}$$

is the quantum mechanical energy of the i *th* particle;

$$k_{ix} = \left(\frac{\pi}{L_x}\right) n_{ix}, \quad k_{iy} = \left(\frac{\pi}{L_y}\right) n_{iy}, \quad k_{iz} = \left(\frac{\pi}{L_z}\right) n_{iz} \tag{1.20}$$

are components of the wave vector, where

$$n_{ix}, n_{iy}, n_{iz} = 1, 2, 3, \ldots \tag{1.21}$$

are quantum numbers of the ith particle, taking on any positive integer value; and $n \Rightarrow (n_{1x}, n_{1y}, n_{1z}; n_{2x}, n_{2y}, n_{2z}; \ldots; n_{Nx}, ; n_{Ny}, n_{Nz})$ is the totality of quantum numbers determining the state of a system.

If we substitute the value of the wave vector (1.20) into (1.19), the energy of the ith particle ε_{n_i} can be expressed by the quantum numbers

$$\varepsilon_{n_i} = \frac{\pi^2\hbar^2}{2m}\left(\frac{n_{ix}^2}{L_x^2} + \frac{n_{iy}^2}{L_y^2} + \frac{n_{iz}^2}{L_z^2}\right). \tag{1.22}$$

As can be seen, the energy of the ith particle of an ideal gas is determined by three quantum numbers n_{ix}, n_{iy}, n_{iz} (the spin of a particle is disregarded). From quantum mechanics, it is known that in such a case each totality of quantum numbers has only one corresponding wave function, i.e. one quantum state, and the degeneracy is absent.

If the energy of a particle (1.22) is taken into account in (1.18), it can be said that one microstate of an ideal gas in the box is determined by assigning $3N$ quantum numbers, i.e. for the considered system

$$\begin{aligned} &\text{microscopic state} \quad \Rightarrow \\ &\Rightarrow \qquad (n_{1x}, n_{1y}, n_{1z};\ n_{2x}, n_{2y}, n_{2z};\ \ldots;\ n_{Nx}, n_{Ny}, n_{Nz}). \end{aligned} \tag{1.23}$$

System consisting of harmonic oscillators. Assume that the considered system consists of N non-interacting linear harmonic oscillators with frequency ω. Owing to the absence of the interaction between oscillators, expression (1.18) for the energy of the system remains in force; only in the given case, energy of the ith particle of an oscillator has the form

$$\varepsilon_i = (n_i + 1/2)\,\hbar\omega, \tag{1.24}$$

where $n_i = 0, 1, 2, \ldots$ are quantum numbers of an oscillator. Each value n_i has one corresponding wave function, i.e. one quantum state.

Thus, one microstate of an ideal gas consisting of N linear harmonic oscillators is determined by the totality of N quantum numbers:

$$\text{microstate} \Rightarrow (n_1, n_2, \ldots, n_N). \tag{1.25}$$

System consisting of rotators. Consider a system consisting of N non-interacting rotators formed by rigidly bound atoms of masses m' and m'' at a distance r from each other (a diatomic molecule). Assume that the rotators rotate around the axis passing through the fixed centres of the masses. Inasmuch as the rotators do not interact, the total energy of the system can be determined from expression (1.18); however, in the given case, the quantum mechanical energy of each rotator is pre-assigned by the expression

$$\varepsilon_i = \frac{\hbar^2}{2I} l_i(l_i + 1), \tag{1.26}$$

where $l_i = 0, 1, 2, \ldots$ is the azimuthal quantum number, $I = mr^2$ is the moment of inertia of a rotator (a molecule), and $m = m'm''/(m' + m'')$ is the reduced mass.

Note that the wave function of a rotator, apart from the azimuthal quantum number l_i, depends also on the magnetic quantum number m_i, which takes on integer values in the limits

$$-l_i \leq m_i \leq l_i. \tag{1.27}$$

Hence it follows that each quantum state of a rotator is determined by two quantum numbers (l_i and m_i). Inasmuch as the energy of a rotator does not depend on the quantum number m_i, its energy levels are $(2l_i + 1)$-fold degenerate.

Thus, one microstate of a system consisting of N rotators is determined by the totality of $2N$ quantum numbers:

$$\text{microstate} \Rightarrow (l_1, m_1;\, l_2, m_2; \ldots;\, l_N, m_N). \tag{1.28}$$

From the examples given above, it is seen that the total of quantum numbers determining one microstate of a system equals the number of degrees of freedom. In fact, if we consider an ideal gas consisting of N diatomic molecules and take into account that each molecule has three translational, one vibrational and two rotational degrees of freedom, then the system as a whole has $6N$ degrees of freedom. From (1.23), (1.25) and (1.28), it is seen that for the considered system the total of quantum numbers determining one microstate also equals $6N$.

Quasi-classical approximation. It is known that at specified conditions (the quasi-classical approximation) the quantum mechanical description of a system can be replaced with its classical counterpart, i.e. in some particular cases, quantum systems behave as classical ones. Consider the conditions of quasi-classicity. The quasi-classical approximation is applicable when the difference of energies of two adjacent quantum levels is much less than the energy of the particles:

$$[\varepsilon(n+1) - \varepsilon(n)] \ll \varepsilon(n). \tag{1.29}$$

For the case (1.22), this inequality takes the form $(2n+1) \ll n^2$, i.e. the quasi-classical approximation is applicable for very large ($n \gg 1$) quantum numbers.

Condition of quasi-classicity (1.29) obviously can be expressed as follows: if the mean energy of the thermal motion of particles $\overline{\varepsilon(n)} = k_0T$ is much more than the discreteness of the energy spectrum, the motion can be regarded as classical. For free particles in a box with volume $V = L^3$, i.e. for case (1.22), the condition of quasi-classicity (1.29) can be written down in the form

$$\frac{\pi^2\hbar^2}{2mL^2} \ll k_0T, \tag{1.30}$$

where T is the absolute temperature.

Condition of quasi-classicity (1.30) can also be presented as

$$L \gg \lambda, \tag{1.31}$$

where $\lambda = h/p$ is the de Broglie wavelength of a particle, and $p = \sqrt{2mk_0T}$ is the mean impulse.

Thus, *the free motion of particles is classical when the linear dimensions of the space L in which the motion occurs are much more than the de Broglie wavelength of a particle.*

Consider the question of the number of microstates of a system. For quantum systems, the number of microstates in the pre-assigned range of energy equals the number of quantum states. In the quantum case, the number of microstates accounting for the specified range of energy is finite because of the property of discreteness.

For classical systems, one phase point corresponds to each microstate and therefore, formally, the "infinite" number of microstates corresponds to any element of the phase space. If, from the point of view of quantum mechanics, it is taken into account that the least "volume" of the phase space $h = 2\pi\hbar$ (Fig. 1.4) corresponds to one state of a particle with one degree of freedom,

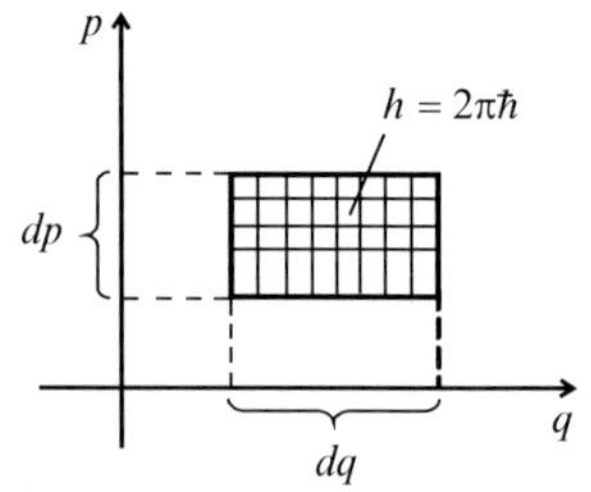

Fig. 1.4. The "volume" of the quantum state of a particle with one degree of freedom

then the "volume" $(2\pi\hbar)^{3N}$ corresponds to a microstate of a system with $3N$ degrees of freedom in the phase space.

Thus, the number of quantum states corresponding to "an element of volume" $\mathrm{d}\Gamma = \mathrm{d}q\,\mathrm{d}p$ equals

$$\mathrm{d}G = \frac{\mathrm{d}\Gamma}{(2\pi\hbar)^{3N}} = \frac{\prod\limits_{i=1}^{3N} \mathrm{d}q_i\,\mathrm{d}p_i}{(2\pi\hbar)^{3N}}. \tag{1.32}$$

Note that we will use the relation (1.32) between "an element of volume" of the phase space and the number of quantum states of a system for the single-valued determination of entropy of a system in the quasi-classical case (Sect. 1.6).

1.3 Statistical Description of Classical Systems: Distribution Function: Liouville Theorem

As noted in Sect. 1.2, to determine a microstate of a classical system consisting of N particles it is necessary to know $6N$ parameters. To do this, it is required to solve a system of $6N$ equations (1.11). Even if the explicit form of the Hamilton function (1.12) and initial conditions are known, it is very difficult or practically impossible to solve the system (1.11) owing to the huge number of particles. Moreover, knowing the coordinates and impulses of all particles gives no complete information about properties of the system as a whole. It is associated with the fact that in the behaviour of macrosystems qualitatively new statistical appropriateness arises. Such appropriateness bears a probabilistic character and is distinct from mechanical laws. Hence it follows that states of macrosystems should be described by statistical methods. In these methods, the idea is not the exact determination of the microstates of a system but their determination with a certain probability.

Assume that the considered system is a small but macroscopic part of a large system – the thermostat. A microstate of a system interacting with the thermostat chaotically changes over a course of time, and we cannot exactly determine the coordinates (q, p) of these states in the phase space. Then the question can be posed in the following form: *What is the probability that microstates of a system may be found in a small element of volume* $(\mathrm{d}q\,\mathrm{d}p)$ *of the phase space*? To determine this probability, mentally trace the change in the microstates of a system in the course of a long time interval T. Assume that over the course of a short time range $\mathrm{d}t$ a microstate of the system is found in the element of volume $(\mathrm{d}q\,\mathrm{d}p)$ taken near the point (q, p). If the time of observation is regarded as very long, the quantity

$$\lim_{T\to\infty}\left(\frac{\mathrm{d}t}{T}\right) = \mathrm{d}W \tag{1.33}$$

can be taken as the probability of the microstate of the system being found in the element of volume $(\mathrm{d}q\,\mathrm{d}p)$.

It is evident that the probability $\mathrm{d}W$ depends on around which point $(q,p) = (q_1, q_2, \ldots, q_{3N}; p_1, p_2, \ldots, p_{3N})$ the element of volume $(\mathrm{d}q\,\mathrm{d}p) = \prod_{i=1}^{3N} \mathrm{d}q_i\,\mathrm{d}p_i$ is taken, and, naturally, the quantity $\mathrm{d}W$ ought to be directly proportional to this element of volume $(\mathrm{d}q\,\mathrm{d}p)$:

$$\mathrm{d}W = \rho(q,p)\mathrm{d}q\,\mathrm{d}p. \tag{1.34}$$

The coefficient of proportionality $\rho(q,p)$ is the probability of the microstate found in the unit volume of the phase space taken near the point (q,p) and is called *the distribution function.* From the geometrical point of view, $\rho(q,p)$ is the density of phase points corresponding to the microstate of the system near the point (q,p) of the phase space, i.e. it characterizes the distribution of microstates of the system in the phase state.

Knowing the distribution function, we can calculate the mean statistical value of any quantity $L(q,p)$, depending on coordinates and impulses of particles of the system:

$$\overline{L}_\rho = \int L(q,p)\rho(q,p)\mathrm{d}q\,\mathrm{d}p. \tag{1.35}$$

Notice that in reality the state of a system is determined by macroscopic quantities measured experimentally. Knowing the internal structure of the system, the mean value of the measured quantities with respect to time can be immediately calculated:

$$\overline{L}_t = \lim_{T\to\infty} \frac{1}{T} \int_0^T L(q(t), p(t))\mathrm{d}t. \tag{1.36}$$

To do this, besides the explicit form of the dependence of the quantity L on q and p, it is necessary to know the dependences of the coordinates and impulses of a particle on time, i.e. functions $q = q(t)$ and $p = p(t)$. And this means a solution of system of equations (1.11). Inasmuch as this system is unsolvable, the immediate calculation of $\overline{L}_t$ by (1.36) is not possible. In order to overcome this difficulty, we assume that the mean value with respect to time (1.36) can be replaced with the statistical mean (1.35):

$$\overline{L}_t \Rightarrow \overline{L}_\rho. \tag{1.37}$$

This supposition is called *the ergodic hypothesis*, and systems satisfying condition (1.37) are called *ergodic systems.* Henceforth, we will consider only ergodic systems and omit the statistical index ρ, i.e. $\overline{L}_\rho = \overline{L}$.

An advantage of calculation of the statistical mean value (1.35) lies in the fact that in this case it is sufficient to know dependences of the quantity L

on q and p, i.e. it is not necessary to know the time dependences of q and p. The explicit form of the function $L(q, p)$ for different systems can be found in classical mechanics. In the particular case, L can be the Hamilton function $H = E(q, p)$. However, as seen from (1.35), in order to find $\overline{L}$, besides $L(q, p)$ it is necessary to know the distribution function $\rho(q, p)$. Finding the explicit form of this function is the basic task of statistical physics. It is necessary to find such a distribution function in order that the mean statistical value, calculated with its aid, would coincide with the mean value with respect to time, i.e. condition (1.37) would be satisfied.

Properties of the distribution function. In order to find the explicit form of the distribution function, note its following properties.

1. The distribution function ought to satisfy *the normalization condition.* It is evident that if we integrate the expression (1.34) over all the phase space, we get the normalization condition

$$\int \rho(q, p) \mathrm{d}q \, \mathrm{d}p = 1, \tag{1.38}$$

   which means that the event of a microstate of a system being found at some point or other of the phase space is real.
2. To define the second property of the distribution function, we introduce the concept of *a statistically independent subsystem.* Assume that the considered system consists of two macroscopic subsystems. It is evident that in the process of interaction of these subsystems, basically particles that are found on the boundary, the number of which is much smaller than the total number of particles in the subsystems, participate. Therefore, in a time range which is less than the relaxation time, these subsystems can be considered as self-dependent, i.e. the change in the state of one subsystem in the given time interval does not influence the state of the other subsystem. Subsystems satisfying this condition are *statistically independent.*

Elements of volume of the phase space of the considered subsystems are denoted by $\mathrm{d}q^{(1)}\mathrm{d}p^{(1)}$ and $\mathrm{d}q^{(2)}\mathrm{d}p^{(2)}$, respectively. Then probabilities of microstates of the subsystems being found in these elements of volume have the form

$$\mathrm{d}W^{(1)} = \rho_1 \, \mathrm{d}q^{(1)}\mathrm{d}p^{(1)}; \quad \mathrm{d}W^{(2)} = \rho_2 \, \mathrm{d}q^{(2)}\mathrm{d}p^{(2)}. \tag{1.39}$$

It is evident that the distribution function ρ_1 depends on coordinates and impulses of particles of the first subsystem, and ρ_2 depends on coordinates and impulses of particles of the second subsystem.

Mathematically, the statistical independence means that the probability $\mathrm{d}W = \rho \, \mathrm{d}q \, \mathrm{d}p$ of the microstate of the system consisting of two subsystems found in the element of volume $\mathrm{d}q \, \mathrm{d}p = \mathrm{d}q^{(1)}\mathrm{d}p^{(1)}\mathrm{d}q^{(2)}\mathrm{d}p^{(2)}$ ought to equal the product of the probability of the microstate of the first subsystem found in the element of volume $\mathrm{d}q^{(1)}\mathrm{d}p^{(1)}$ and the probability of the microstate of the second subsystem found in the element of volume $\mathrm{d}q^{(2)}\mathrm{d}p^{(2)}$, i.e. $\mathrm{d}W = \mathrm{d}W^{(1)} \cdot \mathrm{d}W^{(2)}$. In the explicit form

$$\rho \, \mathrm{d}q \, \mathrm{d}p = \rho_1 \mathrm{d}q^{(1)} \mathrm{d}p^{(1)} \cdot \rho_2 \mathrm{d}q^{(2)} \mathrm{d}p^{(2)}, \tag{1.40}$$

or

$$\rho = \rho_1 \cdot \rho_2. \tag{1.41}$$

In the general case of a large system consisting of n_0 number of subsystems, equality (1.41) takes the form

$$\rho = \rho_1 \cdot \rho_2 \cdot \ldots \cdot \rho_{n_0} = \prod_{\alpha=1}^{n_0} \rho_\alpha. \tag{1.42}$$

Thus, *the distribution function of a large system is the product of the distribution functions of the statistically independent subsystems forming the large system.* The converse is also true: if the distribution function of the whole system can be presented in the form of the product of the distribution functions of individual subsystems, these subsystems are statistically independent.

If we take the logarithm of the equality (1.42), we can obtain the second important property of the distribution function:

$$\ln \rho = \sum_{\alpha=1}^{n_0} \ln \rho_\alpha. \tag{1.43}$$

This means that the logarithm of the distribution function of a large system equals the sum of the logarithms of the distribution functions of the individual subsystems; in other words, *the logarithm of the distribution function of the system is an additive quantity.*

3. *Liouville theorem.* The third property of the distribution function stems from this theorem, according to which *the distribution function is constant along the phase trajectory,* i.e. $\rho(q,p) = \text{const}$. This is one of the definitions of the Liouville theorem. To prove this theorem, mentally keep a watch over microstates of the given subsystem for an extended time and subdivide the time of observation into very small identical ranges. Imagine that phase points $A_1, A_2, A_3, \ldots, A_n$ correspond to microstates of the subsystem at instants of time $t_1, t_2, t_3, \ldots, t_n$ in the phase space.

Now assume that each phase point $A_1, A_2, A_3, \ldots, A_n$ corresponds to a microstate of one of the subsystems at the instant of time t. It is evident that the multitude of subsystems mentally formed in this way is a multitude of states of identical subsystems (with identical Hamilton function) and is called *the Gibbs statistical ensemble.* The number of subsystems n entering into this ensemble ought to be very large.

A microstate of the statistical ensemble, i.e. positions of phase points $A_1, A_2, \ldots, A_n$ in the course of time changes and at the instant of time t', is described by the multitude of the phase points $A'_1, A'_2, \ldots, A'_n$:

$$(A_1, A_2, \ldots, A_n)_t \Rightarrow (A'_1, A'_2, \ldots, A'_n)_{t'}. \tag{1.44}$$

If Δn phase points occupy an element of the phase volume $\Delta\Gamma$, by definition of the distribution function it can be written as

$$\Delta n = \rho(q, p, t)\Delta\Gamma. \tag{1.45}$$

For the pre-assigned ensemble, phase points do not disappear or arise. Therefore, the distribution function $\rho(q, p, t)$ in the phase space ought to satisfy the continuity equation. In order to write the continuity equation in the $6N$-dimensional phase space, first remember its form in the usual three-dimensional space:

$$\frac{\partial \rho}{\partial t} + \mathrm{div}(\rho v) = 0. \tag{1.46}$$

Here $\rho(x, y, z, t)$ and $v(x, y, z, t)$ are the density and velocity of the flow of points at the instant of time t, respectively. Equation (1.46) actually is the law of conservation of substance and can be written down in the form

$$\frac{\partial \rho}{\partial t} + v\,\mathrm{grad}\,\rho + \rho\,\mathrm{div}\,v = 0. \tag{1.47}$$

If we take into account that the sum of the first two terms is a total derivative of the function ρ with respect to time, (1.47) takes the form:

$$\frac{\mathrm{d}\rho}{\mathrm{d}t} + \rho\,\mathrm{div}\,v = 0. \tag{1.48}$$

Then, by analogy with (1.48), for the continuity equation in the $6N$-dimensional phase space we have:

$$\frac{\mathrm{d}\rho(q, p, t)}{\mathrm{d}t} + \rho\,\mathrm{Div}\,\boldsymbol{V} = 0, \tag{1.49}$$

where $\boldsymbol{V}$ is the vector of "velocity" in the $6N$-dimensional phase space with components $\dot{q}_1, \dot{q}_2, \ldots, \dot{q}_{3N}; \dot{p}_1, \dot{p}_2, \ldots, \dot{p}_{3N}$; $\mathrm{d}\rho/\mathrm{d}t$ is the velocity of the change in the density of phase points around the point (q, p); and $\mathrm{Div}V$ is the symbol of divergence in the $6N$-dimensional phase space:

$$\mathrm{Div}\,\boldsymbol{V} = \sum_{i=1}^{3N} \left(\frac{\partial \dot{q}_i}{\partial q_i} + \frac{\partial \dot{p}_i}{\partial p_i} \right). \tag{1.50}$$

Taking into account the Hamilton canonical (1.11), we get:

$$\sum_{i=1}^{3N} \left(\frac{\partial \dot{q}_i}{\partial q_i} + \frac{\partial \dot{p}_i}{\partial p_i} \right) = \sum_{i=1}^{3N} \left(\frac{\partial^2 \mathscr{H}}{\partial q_i \partial p_i} - \frac{\partial^2 \mathscr{H}}{\partial p_i \partial q_i} \right) = 0. \tag{1.51}$$

Thus, for the $6N$-dimensional phase space

$$\mathrm{Div}\,\boldsymbol{V} = 0. \tag{1.52}$$

From (1.49) and (1.52), it is seen that a total derivative of the distribution function equals zero:

$$\frac{\mathrm{d}\rho(q,p,t)}{\mathrm{d}t} = 0, \tag{1.53}$$

i.e. ρ is constant along the phase trajectory:

$$\rho(q,p,t) = \text{const.} \tag{1.54}$$

When proving this property of the distribution function, we have used equation of motion (1.11) describing the phase trajectory. Therefore, in conformity with Liouville theorem (1.54), *the third property of the distribution function* can be formulated as follows: *The distribution function along the phase trajectory does not depend on time*, i.e. it is an integral of the motion. Note that $\rho(q,p,t)$ along the phase trajectory remains constant though coordinates $q(t)$ and impulses $p(t)$ strongly depend on time, and their changes are described by a solution of (1.11).

Besides the above, one more equivalent definition of the Liouville theorem can be given; it follows from expression (1.45) and condition (1.54): an element of volume $\Delta\Gamma$ of the phase space occupied by phase points for the pre-assigned part (Δn = const) of the ensemble does not depend on time, i.e. $\Delta\Gamma$ = const.

Thus, another definition of the Liouville theorem can be formulated thus: *The phase volume occupied by phase points corresponding to microstates of the statistical ensemble is conserved*, i.e.

$$\Delta\Gamma = \Delta\Gamma', \tag{1.55}$$

where $\Delta\Gamma$ and $\Delta\Gamma'$ are elements of volume of the phase space occupied by phase points of the ensemble at the instants of time t and t', respectively.

Consider the general conclusions stemming from the indicated properties of the distribution function. As it is seen from (1.35), to calculate the mean statistical value of the physical quantities it is necessary to know two functions. The explicit form of $L(q,p)$ is found in classical mechanics. The distribution function $\rho(q,p)$ is determined in statistical physics. It is clear that for both $L(q,p)$ and $\rho(q,p)$ the universal form for all systems cannot be found. However, using the above-indicated properties of the distribution function, we can determine the general form applicable for any system.

According to the third property stemming from the Liouville theorem (1.54), the distribution function $\rho(q,p)$ along the trajectory is constant, though its arguments q and p substantially depend on time. This means that the distribution function ρ can depend on coordinates q and impulses p only in terms of mechanical invariants $I(q,p)$ – integrals of the motion:

$$\rho(q,p) = \rho(I(q,p)). \tag{1.56}$$

From classical mechanics, it is known that systems have seven additive integrals of motion: the total internal energy of a system $E(q,p)$; components of the impulse P_x, P_y, P_z; and components of the moment of the impulse M_x, M_y, M_z of the system as a whole. Note that the frame of reference can be so connected with the body (the system) that in the given frame of reference $\boldsymbol{P}$ and $\boldsymbol{M}$ would be equal to zero. Then, in this frame of reference the only mechanical invariant – the total internal energy $E(q,p)$ – remains, and as a result dependence (1.56) can be written down in the form

$$\rho(q,p) = \rho(E(q,p)). \tag{1.57}$$

Such a dependence of the distribution function $\rho(q,p)$ indicates the exclusively important role of the internal energy $E(q,p)$ in statistical physics.

Thus, we get the most general and fundamental property of the distribution function: *The distribution function $\rho(q,p)$ depends only on the total internal energy of a system $E(q,p)$.*

And what is the explicit form of this dependence? To this question, a universal answer for any system does not exist. We consider a concrete system.

Assume that the considered system consists of several subsystems. Then, taking into account the condition of additivity of energy (1.2) and $\ln \rho$ (1.43), we see that the logarithm of the distribution function of any α-subsystem can depend on its energy E_α as follows:

$$\ln \rho_\alpha = A_\alpha + \beta E_\alpha(q,p). \tag{1.58}$$

The constants A_α and β here are found from some specified conditions. It is evident that the coefficient β ought not to depend on the number of the subsystem and is identical for all subsystems. Only in this case, for all the system $\ln \rho(q,p)$ and, consequently, the distribution function $\rho(q,p)$ itself satisfies the condition (1.57): i.e. $\rho(q,p)$ depends only on the total internal energy of the system: $E = \sum\limits_\alpha E_\alpha$.

It should be noted that just the distribution function of a system found in the thermostat, i.e. the Gibbs canonical distribution (see Chap. 4), has such a general appearance (1.58): or

$$\rho_\alpha = \exp(A_\alpha + \beta E_\alpha(q,p)). \tag{1.59}$$

1.4 Microcanonical Distribution: Basic Postulate of Statistical Physics

The microcanonical distribution is concerned with completely isolated systems. Owing to the absence of any interaction with the surrounding medium ($\Delta A = 0,\ \Delta\Theta = 0,\ \Delta N = 0$), its energy remains fixed as $E_0 = \text{const}$, i.e. in whatever microstate the system is found, its total internal energy is

$$E(q,p) = E_0. \tag{1.60}$$

Microstates satisfying this condition are called *possible states.* In the abstract phase space, (1.60) is an equation of a specified "hypersurface". Energy of all microstates corresponding to phase points on this surface is identical, and $E_0 = \text{const}$. Therefore this hypersurface is called *the isoenergetic surface.* It is evident that the system cannot be found in microstates to which phase points outside this surface $E(q,p) \neq E_0$ correspond. These states are called *impossible states.*

The explicit form of the distribution function for completely isolated classical systems that are found in the thermodynamic equilibrium is determined on the basis of the postulate of statistical physics.

The basic postulate of statistical physics is as follows: *If a completely isolated system is found in the thermodynamic equilibrium state, the probability of it being found in any possible microstate is identical, i.e. preference can be given to none of the possible states.*

Mathematically, this postulate can be expressed in the form

$$\rho(q,p) = \begin{cases} C = \text{const}, & \text{at} \quad E(q,p) = E_0, \\ 0, & \text{at} \quad E(q,p) \neq E_0. \end{cases} \tag{1.61}$$

For the distribution function, expression (1.61) has the appearance

$$\rho(q,p) = C\delta(E(q,p) - E_0). \tag{1.62}$$

This distribution function is called *the microcanonical distribution.* The constant C is determined from the normalization condition of the distribution function (1.38):

$$\int \rho(q,p)\mathrm{d}q\,\mathrm{d}p = C\int \delta(E(q,p) - E_0)\mathrm{d}q\,\mathrm{d}p = 1. \tag{1.63}$$

Recall that the δ-function entering into expression (1.62) has the following properties:

$$\begin{aligned} &1. \quad \int \delta(x-a)\mathrm{d}x = 1, \\ &2. \quad \int f(x)\delta(x-a)\mathrm{d}x = f(a), \\ &3. \quad \int f(x)\delta[\phi(x)]\mathrm{d}x = \sum_s \frac{f(x_s)}{|\phi'(x_s)|}, \end{aligned} \tag{1.64}$$

where x_s are roots of the equation $\phi(x_s) = 0$, and a is an arbitrary constant. In order to use the second property of the δ-function (1.64), in (1.63) pass from the integral over $\mathrm{d}q\,\mathrm{d}p \equiv \mathrm{d}\Gamma$ to the integral over $\mathrm{d}E$. Then we get

$$C\int \delta(E(q,p) - E_0)\mathrm{d}q\,\mathrm{d}p = C\int \delta(E(q,p) - E_0)\frac{\mathrm{d}\Gamma}{\mathrm{d}E}\mathrm{d}E = 1.$$

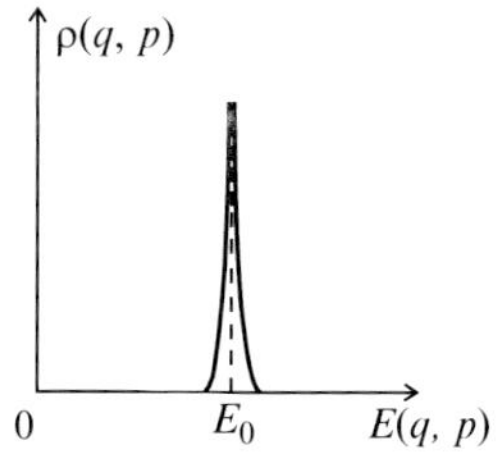

Fig. 1.5. The microcanonical distribution for isolated classical systems

Here

$$C = \frac{1}{(\mathrm{d}\Gamma/\mathrm{d}E)_{E=E_0}} = \frac{1}{\Omega(E_0)} \tag{1.65}$$

is the normalizing constant.

Hence, it is seen that the quantity $\Omega(E_0) = (\mathrm{d}\Gamma/dE)_{E=E_0}$ represents the "volume" of the phase space found between two hypersurfaces, differing from one another by one energetic unit and taken around the isoenergetic hypersurface $E(q, p) = E_0$.

As a result, the distribution function for isolated classical systems – *the microcanonical distribution* – takes the form

$$\rho(q, p) = \frac{1}{\Omega(E_0)}\delta(E(q, p) - E_0). \tag{1.66}$$

The microcanonical distribution is schematically depicted in Fig. 1.5.

Note that (1.66) is a mathematical expression of the basic postulate of statistical physics. The justification of this postulate is corroborated by the coincidence of the results obtained by use of (1.66) with the experimental results.

For isolated systems, with the aid of the microcanonical distribution the mean value of any physical quantity depending on energy $L(q, p) = L(E(q, p))$ can be computed:

$$\overline{L} = \int L(E(q, p))\frac{1}{\Omega(E_0)}\delta(E(q, p) - E_0)\mathrm{d}q\,\mathrm{d}p. \tag{1.67}$$

If with the aid of the replacement

$$\mathrm{d}q\,\mathrm{d}p = \mathrm{d}\Gamma = \frac{\mathrm{d}\Gamma}{\mathrm{d}E}\mathrm{d}E \tag{1.68}$$

we pass from the integral over $\mathrm{d}\Gamma$ to the integral over $\mathrm{d}E$ and take into account (1.65), for the mean value we get

$$\overline{L} = L(E_0). \tag{1.69}$$

In the particular case when $L(E(q, p)) = E(q, p)$, for the mean value of energy of the system we have

$$\overline{E(q, p)} = E_0. \tag{1.70}$$

1.5 Statistical Description of Quantum Systems: Statistical Matrix: Liouville Equation

Imagine that the motion of particles of the considered macroscopic system bears quantum character. Inasmuch as the number of particles in the system is very large, the purely quantum mechanical description of the system as well as for classical systems is practically impossible. In fact, for the quantum mechanical description of a system it is necessary to solve (1.16) and find the wave function Ψ_n, depending on $3N$ number of variables – coordinates of all particles. Furthermore, the wave function Ψ_n can be used to find the quantum mechanical mean value $\overline{L}$ of the physical quantity L. The practical impossibility of working with this problem is evident.

Another principal difficulty is associated with the fact that a system cannot be found in the purely stationary quantum state, inasmuch as its energy spectrum is continuous. In nature, a completely isolated system does not exist. Each system to some extent interacts with the surrounding medium and the energy of this interaction is more of the difference between energy levels. Therefore, the macroscopic system being found in the stationary state is impossible. For this reason, the macroscopic system is found not in the purely quantum state but in "the mixed" state. According to quantum mechanics, in "a mixed" state the system is described not by the stationary wave function, but by *the density matrix*. In statistical physics, it is called *the statistical matrix*.

Impossibility of the stationary state of a macroscopic system also follows from the uncertainty principle for energy. Indeed, the difference of energies of two adjacent levels $\Delta E = (E_{n+1} - E_n)$ ought to be much more than the uncertainty of energy:

$$\delta E \sim \frac{\hbar}{\Delta t}. \tag{1.71}$$

To fulfill this condition, the time of measurement Δt ought to be infinitely large. In reality, however, Δt is finite. Therefore, in the range of the uncertainty of energy δE, several energy levels can be found and, consequently, it cannot be asserted that the system is found in some specified stationary state.

Thus, inasmuch as the quantum mechanical description of the system is impossible, the problem needs to be solved by statistical methods. To do this, we proceed as follows. Separate one subsystem, which is a small part of a large system. Suppose that the subsystem does not interact with the surroundings. Then, we can speak about "the stationary state" of the subsystem. The wave function of the subsystem in these stationary states is denoted by $\psi_n(q)$. Here, q are coordinates of all particles of the system, and n is the totality of quantum numbers determining its stationary state. Let the energy of this stationary state be E_n.

Assume that at a certain instant of time the state of the subsystem is described by the wave function $\Psi(t)$, which can be expanded into a series with respect to orthonormalized wave functions $\psi_n(q)$:

$$\Psi(t) = \sum_n C_n \psi_n(q), \tag{1.72}$$

where the coefficient C_n depends on time

$$C_n \sim \exp\left(-\frac{\mathrm{i}E_n t}{\hbar}\right), \tag{1.73}$$

and satisfies the normalization condition

$$\sum_n |C_n|^2 = 1. \tag{1.74}$$

The mean value of a physical quantity L corresponding to the operator $\hat{L}$ is defined as follows:

$$\overline{L} = \int \Psi^* \hat{L} \Psi \, \mathrm{d}q. \tag{1.75}$$

If we use the expansion (1.72), for $\overline{L}$ we get

$$\overline{L} = \sum_{nm} C_n^* C_m L_{nm}, \tag{1.76}$$

where

$$L_{nm} = \int \psi_n^*(q) \hat{L} \psi_m(q) \mathrm{d}q \tag{1.77}$$

is the matrix element of the physical quantity L corresponding to the operator $\hat{L}$.

If we introduce the notation

$$C_n^* C_m \Rightarrow W_{mn}, \tag{1.78}$$

we get the formula to calculate the mean value:

$$\overline{L} = \sum_{mn} W_{mn} L_{nm}. \tag{1.79}$$

The totality of quantities W_{mn} is called *the statistical matrix*. If we denote the statistical operator corresponding to this matrix by $\hat{W}$, expression (1.79) takes the form

$$\overline{L} = \sum_n (\hat{W}\hat{L})_{nn}. \tag{1.80}$$

Diagonal elements of the statistical matrix $W_{nn} \equiv W_n$ show the probability of the system being found in the stationary state n. Therefore W_n in the quantum statistics corresponds to the distribution function $\rho(q, p)$ in the classical statistics:

$$\rho(q, p) \Rightarrow W_n, \tag{1.81}$$

and the normalization condition (1.38) has the appearance

$$\sum_n W_n = 1. \tag{1.82}$$

Recall that for a classical system the distribution function $\rho(q,p)$ determines the probability of the system that is found in a microstate corresponding to the phase point $(q;p)$. And in quantum systems, W_n means the probability of the system that is found in a microstate with energy E_n, which is determined by the totality of quantum numbers n.

For classical systems, the distribution function $\rho(q,p)$ possesses the property stemming from the Liouville theorem: the distribution function is an integral of the motion and therefore depends only on mechanical invariants (1.56). In quantum systems, for the statistical matrix W_{nm} a theorem analogous to the Liouville theorem can be proved. To do this, using (1.73) write down the derivative of the statistical matrix with respect to time. Then we get

$$\frac{\partial}{\partial t}(C_n^* C_m) = \frac{i}{\hbar}(E_n - E_m) C_n^* C_m, \tag{1.83}$$

or, if we use notations (1.78),

$$\frac{\partial}{\partial t} W_{mn} = \frac{i}{\hbar}(E_n - E_m) W_{mn}. \tag{1.84}$$

The right-hand side of this equation can be presented in the form

$$(E_n - E_m) W_{nm} = \sum_k (W_{mk}\mathscr{H}_{kn} - \mathscr{H}_{mk} W_{kn}). \tag{1.85}$$

Here $\mathscr{H}_{mn}$ is the matrix element of the Hamilton operator $\hat{\mathscr{H}}$. In the energetic representation, $\mathscr{H}_{mn}$ is a diagonal matrix:

$$\mathscr{H}_{mn} = E_n \delta_{mn}. \tag{1.86}$$

If we take this into account, the entry of (1.85) becomes clear.

As a result, (1.84) takes the form

$$\frac{\partial}{\partial t} W_{mn} = \frac{i}{\hbar} \sum_k (W_{mk}\mathscr{H}_{kn} - \mathscr{H}_{mk} W_{kn}). \tag{1.87}$$

Equation (1.87) can be also written down in the matrix form, i.e. for the operator of the density matrix $\hat{W}$,

$$\frac{\partial}{\partial t}\hat{W} = \frac{i}{\hbar}(\hat{W}\hat{\mathscr{H}} - \hat{\mathscr{H}}\hat{W}). \tag{1.88}$$

This equation, as well as (1.87), is called *the Liouville equation.*

As seen from the Liouville equation, to fulfill the stationary condition $\partial\hat{W}/\partial t = 0$, the operator $\hat{W}$ ought to commutate with the Hamilton operator $\hat{\mathscr{H}}$ of the system:

$$\hat{W}\hat{\mathscr{H}} - \hat{\mathscr{H}}\hat{W} = 0. \tag{1.89}$$

Physical quantities corresponding to the operator that commutates with the Hamilton operator are conservation quantities. Therefore, according to (1.89), it can be asserted that *the statistical matrix is an integral of the motion.*

In quantum statistics, this conclusion is an analogue of the Liouville theorem (1.54) in classical statistics.

From the energetic presentation of the statistical matrix, one more of its property follows. Indeed, as seen from the Liouville equation (1.84), to fulfill the stationary condition $(\partial W_{mn}/\partial t = 0)$ the following ought to take place:

$$(E_n - E_m)W_{mn} = 0. \tag{1.90}$$

And to fulfill this condition, the matrix element W_{mn} ought to be diagonal:

$$W_{mn} = W_n \delta_{mn}. \tag{1.91}$$

With regard to (1.91), the formula for the mean value (1.79) takes the form

$$\overline{L} = \sum_n W_n L_{nn}. \tag{1.92}$$

As can be seen, to calculate the mean value of any physical quantity L it is sufficient to know the distribution function W_n and only the diagonal elements of the matrix L_{nn}. For the considered system, L_{nn} is found from quantum mechanics. And the finding of the explicit form of the distribution function W_n is the task of statistical physics. Naturally, a universal expression for W_n applicable to any system does not exist. However, as is known, in quantum statistics the Liouville theorem is fulfilled, too, i.e. W_n is a conservation quantity. And this means that the dependence W_n on the totality of the quantum numbers n is expressed by conservation quantities, namely, by E_n:

$$W_n = W(E_n). \tag{1.93}$$

This property is an analogue of the property of the distribution function (1.61) in classical statistics.

The explicit form of the function $W(E_n)$ is different for different physical systems. Assume that the considered system with energy $E = E_n$ consists of statistically independent subsystems. If we denote the energy of the α-subsystem by E_{n_α}, the energy of the complete system is

$$E_n = \sum_\alpha E_{n_\alpha}, \tag{1.94}$$

where n_α is the totality of quantum numbers determining the state of the α-subsystem, and n is the totality of quantum numbers determining the state of the whole system, i.e. $n \Rightarrow n_1, n_2, \dots n_\alpha$.

Inasmuch as the subsystems are statistically independent, the distribution function W_n ought to possess the property analogous to the property of $\rho(q,p)$ in classical statistics (1.43):

$$\ln W_n = \sum_{\alpha} \ln W_{n_\alpha}, \tag{1.95}$$

i.e. in the quantum case the logarithm of the distribution function is an additive quantity, too. Then, the logarithm of the distribution function can be presented in the form

$$\ln W(E_{n_\alpha}) = A_\alpha + \beta E_{n_\alpha}, \tag{1.96}$$

where E_{n_α} is energy of the α-subsystem; A_α is a constant, which is found from the normalization condition and depends on the number of the subsystem; and the coefficient β ought not to depend on the number of the subsystem, since only in this case conditions (1.95) and (1.94) are fulfilled at the same time.

Notice that the canonical distribution for systems in the thermostat has the same appearance as (1.96).

Consider the microcanonical distribution for isolated quantum systems with the pre-assigned energy $E = E_0 = \text{const}$. As noted above, the energy spectrum of macroscopic systems is continuous. Denote by $\mathrm{d}G$ the number of quantum states in an infinitely small range of energy $\mathrm{d}E$ taken around the given value of energy E.

If it is supposed that the system consists of several subsystems, then

$$\mathrm{d}G = \prod_{\alpha} \mathrm{d}G_\alpha, \tag{1.97}$$

where $\mathrm{d}G_\alpha$ is the number of quantum states in an infinitely small range of energy $\mathrm{d}E$, taken close to the given value of energy E_α of the subsystem with number α.

Notice that (1.97) corresponds to the relationship

$$\mathrm{d}\Gamma = \prod_{\alpha} \mathrm{d}\Gamma_\alpha \tag{1.98}$$

for the classical case, which means that an element of volume of the phase space of the whole system equals the product of elements of volumes of the phase space of individual subsystems.

For an isolated system, quantum states $\mathrm{d}G$ falling in the range of energy $\mathrm{d}E$ can be considered *as possible states*. According to the basic postulate of statistical physics, the probability of the system found in any microstate is identical, i.e. preference can be given to none of them. On the other hand, the probability $\mathrm{d}W$ of the system found in any of the states $\mathrm{d}G$ ought to be proportional to the number $\mathrm{d}G$. Then it can be written as

$$\mathrm{d}W = \text{const}\, \delta(E - E_0) \prod_{\alpha} \mathrm{d}G_\alpha. \tag{1.99}$$

Equation (1.99) is called *the microcanonical distribution* for quantum systems. Here the $\delta(E - E_0)$ function shows the isolatedness of the system, and

$$E = \sum_{\alpha} E_\alpha \tag{1.100}$$

is the total energy of the system.

1.6 Entropy and Statistical Weight

We introduce one of the basic concepts of thermodynamics and statistical physics – *entropy of a system*. Entropy, as well as energy, is a function of state, i.e. it determines the microscopic state of a system.

At first, consider the concept of the statistical weight of a system, which is closely associated with entropy. To do this, suppose that the considered quantum system is found in the thermodynamic equilibrium state. Subdivide this system into a multitude of subsystems. Let n be the totality of quantum numbers determining a microstate of any subsystem with energy E_n, and $W_n = W(E_n)$ be the probability of the system found in the given microstate. Pass from the distribution over microstates $W(E_n)$ to the distribution over energy $w(E)$. It is known that the energy spectrum of a macroscopic system is almost continuous, and therefore a multitude of energy levels corresponding to the quantum states accounts for a sufficiently small range of energies. In order to find the probability $w(E)\mathrm{d}E$ of the system found in the state with energy in the range of E and $E + \mathrm{d}E$ taken close to E, it is necessary to multiply the function $W(E)$ by the number of quantum states (microstates) accounting for the range of energy $\mathrm{d}E$.

If we take into account that the number of these microstates is

$$\mathrm{d}G(E) = \frac{\mathrm{d}G(E)}{\mathrm{d}E}\mathrm{d}E = g(E)\mathrm{d}E, \tag{1.101}$$

the distribution function over energies takes the form

$$w(E) = g(E)W(E). \tag{1.102}$$

Here

$$g(E) = \frac{\mathrm{d}G(E)}{\mathrm{d}E} \tag{1.103}$$

is the function of the density of quantum states, i.e. the number of quantum states accounting for a unit range of energy taken close to E, and $G(E)$ is the total number of all the quantum states with energy less than E.

Even without knowing the explicit form of the distribution function $w(E)$, it can be asserted that a subsystem in thermodynamic equilibrium ought to be found most of the time in states close to the mean value of energy $\overline{E}$.

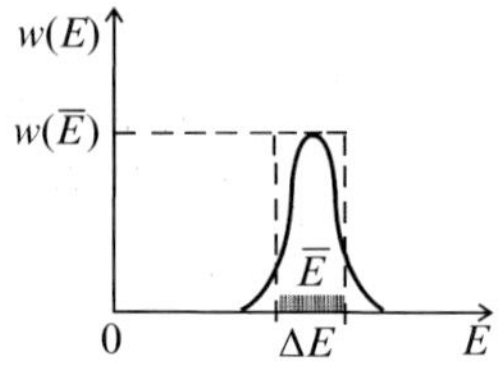

Fig. 1.6. The distribution function over energy

Therefore, the distribution function $w(E)$ over energy ought to have a sharp maximum at $E = \overline{E}$ (Fig. 1.6).

According to the normalization condition

$$\int w(E)\mathrm{d}E = 1. \tag{1.104}$$

This geometrically means that the area under the curve $w(E)$ ought to be equal unity.

If the curve depicted in Fig. 1.6 is approximately replaced by a rectangle with the height $w(\overline{E})$ and width ΔE, condition (1.104) can be presented in the form

$$w(\overline{E})\Delta E = 1, \tag{1.105}$$

where ΔE is called *the width of the distribution curve over energy.*

Taking into account the distribution function (1.102) in (1.105), we get

$$W(\overline{E})\Delta G = 1. \tag{1.106}$$

Here,

$$\Delta G = g(\overline{E})\Delta E \tag{1.107}$$

is the number of microstates accounting for the range of energies ΔE of the subsystem and is called *the statistical weight* of a macrostate of the system with energy $E = \overline{E}$.

The statistical weight ΔG shows the number of microstates corresponding to one pre-assigned macrostate of the system. Therefore, ΔG characterizes the degree of the internal chaoticity of the system.

The statistical weight of a closed system, in conformity with (1.97), can be presented as the product of statistical weights of the individual subsystems:

$$\Delta G = \prod_{\alpha} \Delta G_{\alpha}, \tag{1.108}$$

where $\Delta G_{\alpha} = \Delta G(\overline{E}_{\alpha})$ is the statistical weight of the α-subsystem.

In statistical physics, apart from the statistical weight, a more convenient function, also characterizing the degree of chaoticity of a system, is accepted. This function is defined as a logarithm of the statistical weight:

$$S = k_0 \ln \Delta G \tag{1.109}$$

and is called the *entropy of a system*. Here k_0 is the Boltzmann constant. As is seen, entropy cannot be negative, i.e. $S \geq 0$, since the statistical weight $\Delta G \geq 1$. Note that entropy of a system, as also the energy, in conformity with (1.108) and (1.109), possesses the property of additivity, i.e.

$$S = \sum_\alpha S_\alpha. \tag{1.110}$$

Here, $S_\alpha = k_0 \ln \Delta G_\alpha$ is entropy of the α-subsystem.

Thus, it can be asserted that *entropy is a function of the state of a macroscopic system and characterizes the degree of its internal chaoticity; entropy has only a statistical sense; entropy cannot be spoken of separately for a given particle.*

If we take into account (1.106) in (1.109), entropy can be expressed by the distribution function

$$S = -k_0 \ln W(\overline{E}). \tag{1.111}$$

Inasmuch as, according to (1.96), the logarithm of the distribution function is a linear function of energy, $\ln W(\overline{E})$ can be replaced with the mean $\overline{\ln W(E)}$, i.e.

$$\ln W(\overline{E}) = \overline{\ln W(E)}. \tag{1.112}$$

Then, the expression of entropy (1.111) takes the form

$$S = -k_0 \sum_n W_n \ln W_n. \tag{1.113}$$

We now consider classical systems in the quasi-classical approximation. In this case, using the normalization condition for classical systems consisting of N particles,

$$\int \rho(q,p)\mathrm{d}\Gamma = 1, \tag{1.114}$$

where $\mathrm{d}\Gamma = \mathrm{d}q\ \mathrm{d}p = \prod_i^{3N} \mathrm{d}q_i\ \mathrm{d}p_i$ is an element of volume of the phase space, and therefore we can pass from the distribution over microstates $\rho(q,p)$ to the distribution over energies $\rho(E)$. To do this, rewrite the condition (1.114) in the form

$$\int \rho(E(q,p))\frac{\mathrm{d}\Gamma}{\mathrm{d}E}\mathrm{d}E = 1, \tag{1.115}$$

or

$$\int \rho(E(q,p))(2\pi\hbar)^{3N} g_0(E)\mathrm{d}E = 1. \tag{1.116}$$

Here,

$$g_0(E) = (2\pi\hbar)^{-3N}\frac{\mathrm{d}\Gamma}{\mathrm{d}E} \tag{1.117}$$

is the function of the density of states of quasi-classical systems.

As is seen from (1.116),

$$\rho(E) = (2\pi\hbar)^{3N} \rho(E(q,p)) g_0(E) \tag{1.118}$$

is the distribution function over energy. Then condition (1.116) takes the form

$$\int \rho(E)\mathrm{d}E = 1. \tag{1.119}$$

Taking into account that $\rho(E)$ takes on a maximum value at $E = \overline{E}$, (1.119) can be roughly presented as

$$\rho(\overline{E})\Delta E = 1. \tag{1.120}$$

Here ΔE is the width of the distribution curve (Fig. 1.6). If substitute (1.118) into (1.120), we get

$$(2\pi\hbar)^{3N} \overline{\rho(E(q,p))} \Delta G = 1, \tag{1.121}$$

where

$$\Delta G = g_0(\overline{E})\Delta E \tag{1.122}$$

is the number of microstates accounting for the range of energy ΔE taken close to energy $E = \overline{E}$ in the quasi-classical case, i.e. the statistical weight of the macrostate with energy $E = \overline{E}$. Then entropy can be presented in the form

$$S = k_0 \ln \Delta G = k_0 \ln g_0(\overline{E})\Delta E. \tag{1.123}$$

If the value of $g_0(E)$ from (1.117) is taken into account, for the quasi-classical case entropy can be presented as

$$S = k_0 \ln \frac{\Delta q \Delta p}{(2\pi\hbar)^{3N}}. \tag{1.124}$$

In the quasi-classical case, entropy can be also expressed by the distribution function over microstates. To do this, take into account (1.121) and (1.123). Then we get

$$S = -k_0 \ln[(2\pi\hbar)^{3N} \overline{\rho(E(q,p))}]. \tag{1.125}$$

On the basis of property (1.58) of the distribution function, we can replace

$$\ln \overline{\rho(E(q,p))} = \overline{\ln \rho(E(q,p))}. \tag{1.126}$$

As a result, for entropy we get

$$S = -k_0 \int \rho(q,p) \ln \left[(2\pi\hbar)^{3N} \rho(q,p)\right] \mathrm{d}q\, \mathrm{d}p. \tag{1.127}$$

From additivity of entropy (1.110), one more its property stems. If we divide the width of the distribution function ΔE (Fig. 1.6) by the number of energy levels ΔG in this range, we get the distance between adjacent energy levels:

$$D(E) = \frac{\Delta E}{\Delta G} = \Delta E \, \mathrm{e}^{-S(E)/k_0}. \tag{1.128}$$

From the property of additivity, it follows that the more the amount of substance (the number of particles N) in a system, the more is the entropy of the system $S(E)$ and the denser the energy levels. Thus, with increasing amount of substance in a system, the distance between adjacent energy levels exponentially decreases.

In conclusion, once more recall the basic properties of entropy:

1. Entropy is a function of state and characterizes the degree of its internal chaoticity. The more the entropy, the more is the chaoticity, and vice versa; the less the entropy, the more is the ordering of the system.
2. Entropy has only a statistical sense; it cannot be spoken of for a separate particle.
3. Entropy cannot be negative, i.e. $S \geq 0$.
4. Entropy is an additive quantity, i.e. $S = \sum_{\alpha} S_{\alpha}$.
5. Entropy characterizes the density of levels in the energy spectrum of a system. The more the entropy, the denser are the energy levels.
6. Entropy of a system in the thermodynamic equilibrium state takes on a maximum value. This property follows from the law of increasing entropy, which is further discussed in Sect. 1.7 below.

1.7 Law of Increasing Entropy: Reversible and Irreversible Processes

In Sect. 1.6, we introduced the concept of entropy of an isolated system found in thermodynamic equilibrium. The question arises whether entropy can be spoken of for a system that is not in thermodynamic equilibrium. A positive answer to this question can be given, i.e. the concept of entropy for thermodynamic non-equilibrium systems can also be introduced. To substantiate this, imagine that the considered closed system is not found in the thermodynamic equilibrium state, and its relaxation time is τ. If we study the system in the time range $\Delta t < \tau$, it is evident that the system is in the non-equilibrium state (Fig. 1.7a) and, naturally, we cannot speak of a specified value of entropy of the system. However, if we subdivide the considered system into small subsystems, the relaxation time τ_{α}[7] of each of them (assume with number α) will be less than the time of observation Δt, i.e. $\tau_{\alpha} < \Delta t$ (Fig. 1.7b). As can be seen from the figure, for a very small time τ_{α} the subsystem passes onto its local equilibrium state (the quantity L_{α} tends to its local equilibrium value $L_{0\alpha}$). And it can be said that for all the time of observation Δt, the subsystem

[7] With a decrease in dimensions of the subsystem its relaxation time decreases.

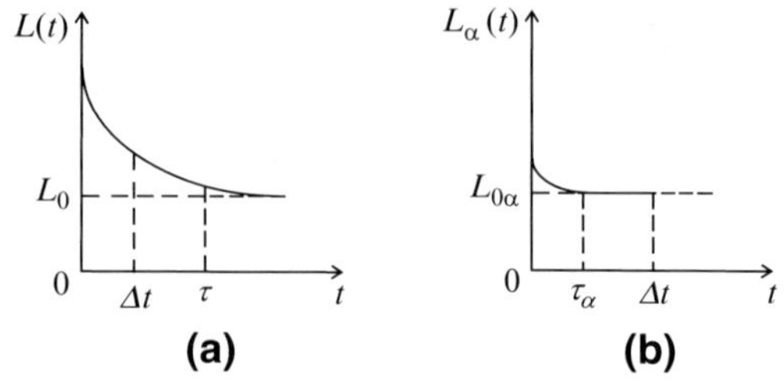

Fig. 1.7. The relaxation of the thermodynamical parameter: (a) - for system; (b) - for subsystem

is found in this local equilibrium state. Consequently, the concept of entropy $S_\alpha = S(E_\alpha)$ can be introduced for each subsystem found in the local equilibrium state. Inasmuch as entropy is an additive quantity, we can speak of the instantaneous value of entropy of a large non-equilibrium system, having defined it as $S(t) = \sum_\alpha S_\alpha(E_\alpha)$.

According to the first postulate of thermodynamics, in the course of time an isolated non-equilibrium system ought to pass into the equilibrium state. The question arises as to how entropy of the system $S(t)$ changes as a result of this process.

In order to answer this question, use the microcanonical distribution function (1.99) for isolated systems with energy E_0 and pass from the distribution over microstates to the distribution over energy. Then, (1.99) takes the form

$$\mathrm{d}W = \mathrm{const}\,\delta(E - E_0) \prod_\alpha \frac{\mathrm{d}G_\alpha}{\mathrm{d}E_\alpha} \mathrm{d}E_\alpha. \tag{1.129}$$

If we replace the derivative $\mathrm{d}G_\alpha/\mathrm{d}E_\alpha$ in this distribution by the ratio $\Delta G_\alpha/\Delta E_\alpha$ and make use of the expression of entropy of the α-subsystem, stemming from definition (1.109)

$$\Delta G_\alpha = \exp\left(\frac{S_\alpha(E_\alpha)}{k_0}\right), \tag{1.130}$$

we get

$$\mathrm{d}W = \mathrm{const}\,\delta(E - E_0)\,\mathrm{e}^{S/k_0} \prod_\alpha \frac{\mathrm{d}E_\alpha}{\Delta E_\alpha}, \tag{1.131}$$

where $S = \sum_\alpha S_\alpha(E_\alpha)$ is entropy of the isolated system. Inasmuch as the range of energy ΔE_α very weakly depends on energy compared with the factor e^{S/k_0}, it can be regarded as constant and the distribution function over energy (1.131) takes the form

$$\mathrm{d}W = \mathrm{const}\,\delta(E - E_0)\mathrm{e}^{S/k_0} \prod_\alpha \mathrm{d}E_\alpha. \tag{1.132}$$

The obtained distribution function (1.132) is the probability of the subsystems found in states with energies $E_1, E_2, \ldots, E_\alpha, \ldots$ in the ranges of energy $E_1 + \mathrm{d}E_1, E_2 + \mathrm{d}E_2, \ldots, E_\alpha + \mathrm{d}E_\alpha, \ldots$, respectively. Here the $\delta(E - E_0)$ function provides the isolatedness of the system, i.e. $E = \sum_\alpha E_\alpha = E_0 = \text{const}$.

As seen from (1.132), the distribution function over energies of an isolated system having the above-indicated physical sense very strongly and exponentially depends on entropy of the system, i.e. $S = S(E_1, E_2, \ldots, E_\alpha, \ldots)$. Therefore, the greater the probability of the considered macrostate of the system, the higher is the entropy of this state. It is known that the probability of an isolated system found in thermodynamic equilibrium state is maximum. In this state, energy of subsystems ought to be equal to its mean value, $E_\alpha = \overline{E}_\alpha$. Thus, entropy of an isolated system in thermodynamic equilibrium state has the maximum value:

$$S(\overline{E}_1, \overline{E}_2, \ldots, \overline{E}_\alpha, \ldots) = S_{\max}. \tag{1.133}$$

Conclusion: *If an isolated system at a certain instant of time is not found in thermodynamic equilibrium state, in the course of time internal processes proceed in such a direction, as a result of which the system comes to its equilibrium state and at that entropy of the system increases, reaching its maximum value. This assertion is called the Law of Increasing Entropy or the Second Law of Thermodynamics.* The law in such a form was formulated by Clausius in 1865, and statistically substantiated by Boltzmann in 1870.

Note that on a system such a process can be performed at which each macroscopic state will be in thermodynamic equilibrium and entropy does not change. Taking into account this case also, the law of increasing entropy in the general form can be formulated as follows: *Entropy of an isolated system never decreases; it either increases or, in the particular case, remains constant.*

In conformity with this, processes proceeding in all macroscopic systems can be subdivided into two groups (Fig. 1.8):

$$\begin{aligned} \mathrm{d}S/\mathrm{d}t &> 0 \text{ – irreversible process} \\ \mathrm{d}S/\mathrm{d}t &= 0 \text{ – reversible process.} \end{aligned} \tag{1.134}$$

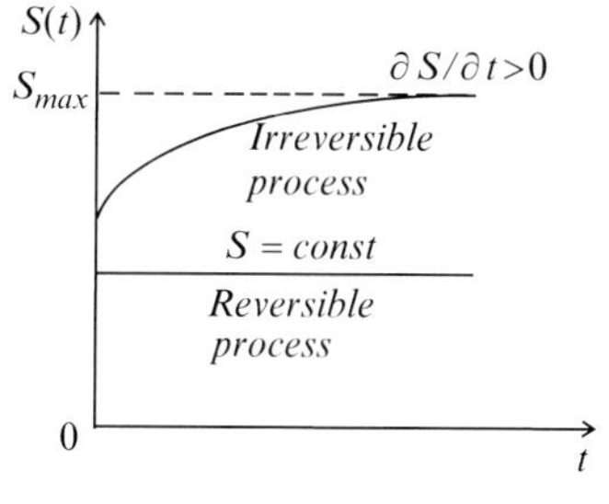

Fig. 1.8. The time dependence of entropy for irreversible and reversible processes

Irreversible processes cannot proceed in the reverse direction, inasmuch as in that case the entropy decreases, but this contradicts the law of increasing entropy. In nature, we frequently come across irreversible processes. Diffusion, thermal conductivity, expansion of a gas, etc. can serve as examples of irreversible processes.

A reversible process can proceed in direct and reverse directions. In this case, a system passes through identical equilibrium macroscopic states, since in this case entropy in both the directions does not change. In nature, it can be said that reversible processes do not exist, and they can be created only approximately by artificial methods. Compression or expansion of a gas found under a piston in a cylinder is one example of the simple adiabatic process.

An adiabatic process is a reversible process. Processes proceeding *sufficiently slowly* in adiabatically isolated ($\Delta Q = 0$) systems are called adiabatic. We can show that in such processes entropy does not change: $\mathrm{d}S/\mathrm{d}t = 0$, i.e. the process is reversible. To do this, consider the simplest case: a gas under a piston in an adiabatically isolated cylinder (Fig. 1.9).

As the external parameter, take the volume of the gas under the piston, which for the given cylinder is determined by the height l. By changing this height, we can increase or decrease the volume of the gas. Inasmuch as the change in entropy with time is related to the change in volume, we can write

$$\frac{\mathrm{d}S}{\mathrm{d}t} = \frac{\mathrm{d}S}{\mathrm{d}l} \cdot \frac{\mathrm{d}l}{\mathrm{d}t}. \tag{1.135}$$

Suppose that the piston moves sufficiently slowly; then the change in entropy with time can be expanded into a series in powers of $\dot{l} = \mathrm{d}l/\mathrm{d}t$:

$$\frac{\mathrm{d}S}{\mathrm{d}t} = A_0 + A_1 \frac{\mathrm{d}l}{\mathrm{d}t} + A_2 \left(\frac{\mathrm{d}l}{\mathrm{d}t}\right)^2 + \cdots \tag{1.136}$$

The constants A_0 and A_1 in this series ought to be equal to zero. The constant A_0 equals zero because at $\dot{l} = 0$ the state of the system does not change and entropy remains constant, i.e. $\mathrm{d}S/\mathrm{d}t = 0$; A_1 equals zero because when changing the sign of $\dot{l}$ (the motion of the piston down and up), $\mathrm{d}S/\mathrm{d}t$ changes its sign, which contradicts the law of increasing entropy ($\mathrm{d}S/\mathrm{d}t \geq 0$). Thus,

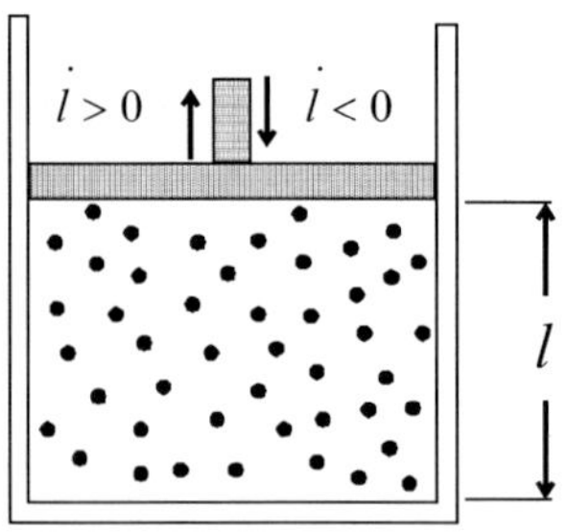

Fig. 1.9. An adiabatic process

$$\frac{\mathrm{d}S}{\mathrm{d}t} = A_2 \left(\frac{\mathrm{d}l}{\mathrm{d}t}\right)^2. \tag{1.137}$$

If we take into account (1.135), we get

$$\frac{\mathrm{d}S}{\mathrm{d}l} = A_2 \frac{\mathrm{d}l}{\mathrm{d}t}. \tag{1.138}$$

Hence it is seen that as the velocity of the piston tends to zero ($\dot{l} \to 0$), the change in entropy of the system with respect to the external parameter l tends to zero: $\mathrm{d}S/\mathrm{d}l = 0$, i.e. when the change in volume of the adiabatically isolated gas is slow, the processes of expansion or compression are adiabatically reversible.

Thus, we have seen that the processes of expansion or compression are reversible ($\mathrm{d}S/\mathrm{d}l = 0$) if the piston moves with a very small velocity. Then the following question arises: In reference to which velocity ought the velocity of the piston be small? In order to answer this question, take a look at the process of compression or expansion of the gas. If the piston moves down ($\dot{l} < 0$), immediately under the piston the instantaneous density of the gas increases. The velocity of motion of the piston ought to be such that at any instance of time, i.e. at any position of the piston, the density of the gas everywhere would be identical, i.e. the gas would be found in thermodynamic equilibrium. An analogous situation arises also during the upward motion of the piston ($\dot{l} > 0$): as the piston moves up, the gas ought to fill up the vacuum forming under the piston, in order that the density would be identical everywhere and equilibrium would be attained. It is apparent that during the motion of the piston the process of regaining uniform distribution of particles of the gas occurs with the speed of sound. Therefore, in order that the process would be adiabatic, the velocity of the motion of the piston ought to be less than the speed of propagation of sound v_0 in the gas:

$$\left|\frac{\mathrm{d}l}{\mathrm{d}t}\right| \ll v_0. \tag{1.139}$$

If we take into account that the speed of sound in most gases is on the order of 350 m/s, the piston can be moved with a sufficiently large velocity, satisfying condition (1.139), and the adiabaticity of the process is not violated.

1.8 Absolute Temperature and Pressure: Basic Thermodynamic Relationship

In Sect. 1.7 we became acquainted with three thermodynamic quantities: volume V, internal energy $E(q, p)$ and entropy $S(E)$. Of these, volume is an external parameter, and energy and entropy are internal ones. Energy has both a mechanical and a statistical sense, and entropy has only a statistical

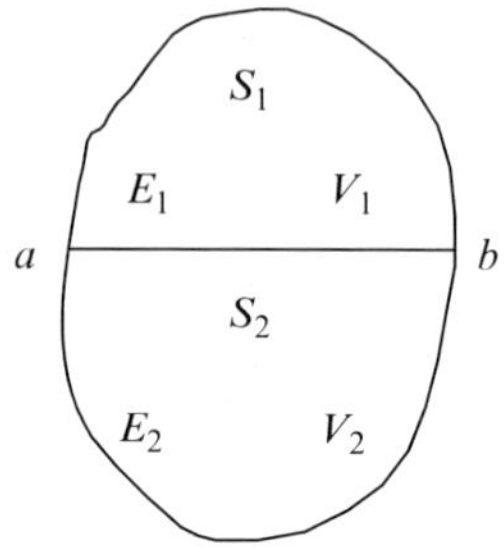

Fig. 1.10. The thermal equilibrium of two subsystems

sense. The change in one of these quantities induces a change in the others. Derivatives of these quantities are also thermodynamic parameters: for instance, absolute temperature and pressure. We consider them separately.

Absolute temperature. Assume that an isolated system with energy $E_0 = \text{const}$ in thermodynamic equilibrium consists of two subsystems with energy, entropy and volume E_1, S_1, V_1 and E_2, S_2, V_2, respectively (Fig. 1.10). As a consequence of the additivity of energy and entropy, we have

$$E = E_1 + E_2, \tag{1.140}$$

and

$$S = S_1(V_1, E_1) + S_2(V_2, E_2). \tag{1.141}$$

Assume that the boundary of division of the two subsystems ab (Fig. 1.10) is immovable; therefore, volumes V_1 and V_2 do not change but energies of the systems E_1 and E_2 and also entropies S_1 and S_2 can change.

Taking into account $E_2 = E - E_1$, notice that entropy of an isolated system depends only on one independent variable E_1, i.e. $S = S(E_1)$. According to the law of increasing entropy, in thermodynamic equilibrium entropy of an isolated system ought to be maximum. For this, it is necessary to fulfill the condition

$$\frac{\partial S}{\partial E_1} = \left(\frac{\partial S_1}{\partial E_1}\right)_{V_1} + \left(\frac{\partial S_2}{\partial E_2}\right)_{V_2} \frac{\partial E_2}{\partial E_1} = 0. \tag{1.142}$$

Since $\partial E_2/\partial E_1 = -1$, from (1.142) we get[8]

$$\left(\frac{\partial S_1}{\partial E_1}\right)_{V_1} = \left(\frac{\partial S_2}{\partial E_2}\right)_{V_2}. \tag{1.143}$$

If an isolated system in thermodynamic equilibrium is subdivided into n arbitrary subsystems, the condition of thermodynamic equilibrium (1.143) in the general form can be presented as

[8] Here and everywhere a quantity, indicated at the foot of the bracket when taking the derivate, is regarded as constant.

$$\left(\frac{\partial S_1}{\partial E_1}\right)_{V_1} = \left(\frac{\partial S_2}{\partial E_2}\right)_{V_2} = \cdots = \left(\frac{\partial S_n}{\partial E_n}\right)_{V_n}. \tag{1.144}$$

Hence it is seen that if an isolated system is found in thermodynamic equilibrium, there exists the quantity $(\partial S/\partial E)_V$, which is identical in any part of the system. The reciprocal value of this quantity is denoted by T:

$$\left(\frac{\partial E}{\partial S}\right)_V = T \tag{1.145}$$

and is called *the absolute temperature* or, in brief, *temperature*. Then condition of thermodynamic equilibrium or maximum entropy (1.144) takes the form

$$T_1 = T_2 = \cdots = T_n = T. \tag{1.146}$$

If we take into account the definition of entropy (1.109), according to (1.145) temperature is measured in *degrees*.

If entropy is defined not by the expression (1.109) but as $S = \ln \Delta G$, temperature (1.145) ought to be measured in energetic units. Hence, it is seen that the Boltzmann constant k_0 relates energy to temperature. For instance, $1\,\text{erg} = \kappa_0 \cdot 1\,\text{deg}$. From experiment it has been determined that $\kappa_0 = 1.38 \times 10^{-16}\,\text{erg/deg}$. Thus, *the Boltzmann constant* k_0 *is the number of ergs corresponding to one degree.*

Note some properties of the absolute temperature:

1. In a system which is in thermodynamic equilibrium, the temperature at all points is identical (1.146).
2. As with entropy, temperature has only a statistical sense, and we cannot speak of the temperature of one isolated particle but we can speak of its energy.
3. Temperature is an internal intensive parameter. This property stems from the fact that we can subdivide the system into n arbitrary subsystems, but the definition of the absolute temperature (1.146) remains invariable since this definition does not depend upon the number n.
4. *The absolute temperature* of a steady macroscopic system is positive.[9] If, on the contrary, $T < 0$, the state of the system is non-steady and, breaking into individual parts, it is destroyed. In order to prove this, subdivide the system into a multitude of subsystems. Let the α-subsystem have the total energy $E_{0\alpha}$, mass M_α, impulse $\boldsymbol{P}_\alpha$ and entropy S_α. If it is taken into account that entropy of the subsystem depends on the internal energy, entropy of the large system takes the form

$$S = \sum_\alpha S_\alpha \left(E_{0\alpha} - \frac{P_\alpha^2}{2M_\alpha}\right) \equiv \sum_\alpha S_\alpha(E_\alpha), \tag{1.147}$$

[9] Artificially, a system with negative absolute temperature can be created, but such a metastable state is non-steady and as a result of the minor external intervention the system, having come out of this state, passes into the normal state.

where E_α is the internal energy of the subsystem. If it is supposed that $T < 0$, then $(\partial S/\partial E)_V < 0$. According to the law of increasing entropy, for an increase in S it is necessary to have a decrease in the internal energy E_α. It is possible in that case that, within the limits of the law of conservation of an impulse $\sum_\alpha \boldsymbol{P}_\alpha = 0$, impulses of the subsystems grow, and thereby the subsystems macroscopically move in different directions, i.e. the state of the system is non-steady. Thus, temperature cannot be negative.

5. The attainment of identical temperature in a system determines the direction of the flow of energy in it: energy from the high-temperature region passes into the low-temperature one. In order to show this, assume that the considered isolated system consists of two parts with temperatures T_1 and T_2 (Fig. 1.10). The system, being in the non-equilibrium thermodynamic state, in the course of time ought to pass into the thermodynamic equilibrium and in both the parts the temperature ought to equalize. During this passage, as consequence of increasing entropy, we have

$$\frac{\mathrm{d}S}{\mathrm{d}t} = \left(\frac{\partial S_1}{\partial E_1}\right)_{V_1} \frac{\mathrm{d}E_1}{\mathrm{d}t} + \left(\frac{\partial S_2}{\partial E_2}\right)_{V_2} \frac{\mathrm{d}E_2}{\mathrm{d}t} > 0. \tag{1.148}$$

Inasmuch as in isolated systems $E = \text{const}$, on the strength of (1.140) we can write

$$\frac{\mathrm{d}E_2}{\mathrm{d}t} = -\frac{\mathrm{d}E_1}{\mathrm{d}t}. \tag{1.149}$$

Then, taking into account the definition of the absolute temperature (1.145), condition of irreversibility of the passage process (1.148) takes the form

$$\left(\frac{1}{T_1} - \frac{1}{T_2}\right) \frac{\mathrm{d}E_1}{\mathrm{d}t} > 0. \tag{1.150}$$

Hence it is seen that if $T_1 > T_2$, then $\mathrm{d}E_1/\mathrm{d}t < 0$, and if $T_1 < T_2$, then $\mathrm{d}E_1/\mathrm{d}t > 0$.

Thus, in the process of the passage of the system from the non-equilibrium state into the equilibrium one, energy is transferred from the high-temperature part into the low-temperature one.

6. The value of the absolute temperature of a system determines the degree of the change in entropy (chaoticity) with the change in its energy ΔE. Indeed, according to (1.145), the change in entropy can be presented as $\Delta S = \Delta E/T$. Hence it is seen that at the pre-assigned change in energy, entropy at low temperatures changes more than that at high temperatures.[10]

Pressure. To define this thermodynamic parameter, we make use of the known formula

$$\boldsymbol{F} = -\frac{\partial E(p, q, \boldsymbol{r})}{\partial \boldsymbol{r}} \tag{1.151}$$

[10] With an increase in energy, the volume of the system ought to remain constant.

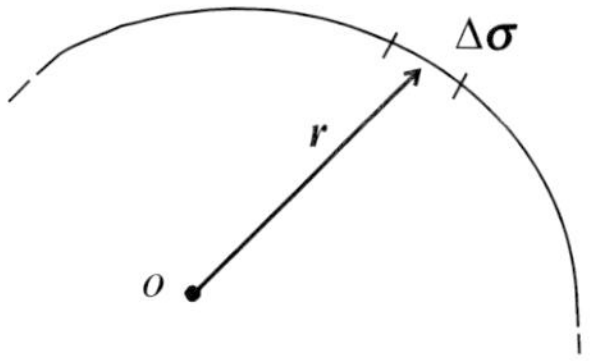

Fig. 1.11. A fragment of the surface of the system

from classical mechanics. Here $E(p, q, \boldsymbol{r})$ is the internal energy (the Hamilton function) of the system; p and q are generalized impulses and coordinates of particles of the system; $\boldsymbol{r}$ is the radius vector of an element of the surface $\Delta\sigma$ of the system; and $\boldsymbol{F}$ is the force acting on this element of the surface.

In Fig. 1.11, a fragment of the surface of the system is shown. In the course of time, $\boldsymbol{r}(t)$ changes, and thereby $\Delta\boldsymbol{\sigma}$ and volume of the system change. Assume that this process is adiabatic. Then $\boldsymbol{r}(t)$ plays the role of an external parameter. It is clear that the change in $\boldsymbol{r}(t)$ induces the change in energy $E(q, p, \boldsymbol{r})$. From classical mechanics it is known that the total derivation of energy (the Hamilton function) of a system can be replaced with the partial derivative

$$\frac{\mathrm{d}E(q,p,\boldsymbol{r})}{\mathrm{d}t} = \frac{\partial E(q,p,\boldsymbol{r})}{\partial t}. \tag{1.152}$$

Owing to the fact that the change in energy with respect to time is due to the change in $\boldsymbol{r}(t)$, we can write

$$\frac{\partial E(q,p,\boldsymbol{r})}{\partial t} = \frac{\partial E(q,p,\boldsymbol{r})}{\partial \boldsymbol{r}} \frac{\mathrm{d}\boldsymbol{r}}{\mathrm{d}t}. \tag{1.153}$$

Denote the mean value of the energy of the system by $E = \overline{E(q, p, \boldsymbol{r})}$ and take its time derivative. By interchanging the position of the operation of averaging and the time derivative, we get

$$\frac{\mathrm{d}E}{\mathrm{d}t} = \frac{\mathrm{d}\overline{E(q,p,\boldsymbol{r})}}{\mathrm{d}t} = \overline{\frac{\partial E(q,p,\boldsymbol{r})}{\partial t}} = \overline{\frac{\partial E(q,p,\boldsymbol{r})}{\partial \boldsymbol{r}}} \frac{\mathrm{d}\boldsymbol{r}}{\mathrm{d}t}. \tag{1.154}$$

On the other hand, the mean energy E is a function of volume V and entropy S. If the process of the change in volume is regarded as adiabatic ($S = \text{const}$), (1.154) can be presented in the form

$$\frac{\mathrm{d}E}{\mathrm{d}t} = \left(\frac{\partial E}{\partial \boldsymbol{r}}\right)_S \frac{\mathrm{d}\boldsymbol{r}}{\mathrm{d}t}. \tag{1.155}$$

Comparing (1.154) and (1.155), we get

$$\overline{\frac{\partial E(q,p,\boldsymbol{r})}{\partial \boldsymbol{r}}} = \left(\frac{\partial E}{\partial \boldsymbol{r}}\right)_S. \tag{1.156}$$

By substituting this expression into (1.151), for the mean value of the force acting on the element of the surface $\Delta\boldsymbol{\sigma}$ we get

$$\overline{\boldsymbol{F}} = -\left(\frac{\partial E}{\partial \boldsymbol{r}}\right)_S = -\left(\frac{\partial E}{\partial V}\right)_S \frac{\mathrm{d}V}{\mathrm{d}\boldsymbol{r}}. \tag{1.157}$$

Taking into account that $\mathrm{d}V = \Delta\boldsymbol{\sigma}\mathrm{d}\boldsymbol{r}$, for pressure $P = \overline{\boldsymbol{F}}/\Delta\boldsymbol{\sigma}$ we have

$$P = -\left(\frac{\partial E}{\partial V}\right)_S. \tag{1.158}$$

Definitions of the absolute temperature (1.145) and pressure (1.158) can be combined as follows:

$$\mathrm{d}E = T\,\mathrm{d}S - P\,\mathrm{d}V. \tag{1.159}$$

This relationship is called *the basic thermodynamic relationship*, since thermodynamics as a whole is based on this.

Note some properties of pressure:

1. Pressure has both a mechanical and statistical sense. The statistical sense of pressure resides in the fact that it is defined as the derivative of the mean value of energy with respect to volume (1.158).
2. Pressure is an internal intensive parameter.
3. In a system in thermodynamic equilibrium, pressure at each point is identical. In order to show this, imagine that an isolated system is divided into two parts (Fig. 1.10), the boundary of division *ab* being movable; i.e. volumes V_1 and V_2 can change but the total volume of the system $V = V_1 + V_2$ does not change ($V = \text{const}$). In order that entropy $S = S_1(V_1) + S_2(V_2)$ of a system in thermodynamic equilibrium be maximum, it is necessary to fulfill the condition

$$\left(\frac{\partial S}{\partial V_1}\right)_{E_0} = \left(\frac{\partial S_1}{\partial V_1}\right)_{E_1} + \left(\frac{\partial S_2}{\partial V_2}\right)_{E_2} \frac{\partial V_2}{\partial V_1} = 0. \tag{1.160}$$

Taking into account that $\partial V_2/\partial V_1 = -1$, we get

$$\left(\frac{\partial S_1}{\partial V_1}\right)_{E_1} = \left(\frac{\partial S_2}{\partial V_2}\right)_{E_2}. \tag{1.161}$$

According to the basic thermodynamic relationship (1.159), $(\partial S/\partial V)_E = P/T$. Then the condition for thermodynamic equilibrium (1.161) takes the form $P_1/T_1 = P_2/T_2$. Since in equilibrium $T_1 = T_2$, necessarily $P_1 = P_2$. If we divide the system into n subsystems, the condition of the equilibrium takes the form

$$P_1 = P_2 = \cdots = P_n = P. \tag{1.162}$$

Recall that expression (1.146) is the condition of thermal equilibrium, and (1.162) is that of mechanical equilibrium. But also note that the mechanical equilibrium in the system is attained faster than the thermal one, inasmuch as the creation of the thermal equilibrium is related to the very slow process of thermal conductivity.

4. In the equilibrium state pressure is positive. From the relationship $(\partial S/\partial V)_E = P/T$, it follows that if $P > 0$, then $(\partial S/\partial V)_E > 0$. In this case, at an increase in entropy the system ought to expand, but other bodies surrounding it hinder this. On the contrary, if $P < 0$, then $(\partial S/\partial V)_E < 0$. Hence it follows that at an increase in entropy the system ought to be spontaneously compressed, converting it into a point, which is impossible.

Notice that in nature metastable states of the system with negative pressure are possible. The breaking away from the walls of the vessels of some liquids can serve as an example of this.

2

Law of Thermodynamics: Thermodynamic Functions

Summary. Usually, when we speak of the laws of thermodynamics, we have in mind its three laws. In fact, there are four laws of thermodynamics. One of them is *the zeroth law.* This law is expressed in the form of the second postulate (see Sect. 1.1) and is a law about temperature.

In this chapter, the three basic laws of thermodynamics are expounded. The first law is about internal energy and its conservation; the second and third laws are about entropy and its change. The method of thermodynamic functions and the finding of general thermodynamic relationships based on them comprise the basic content of the chapter.

2.1 First Law of Thermodynamics: Work and Amount of Heat: Heat Capacity

Once more note that the first law of thermodynamics is about internal energy, which is a function of the state of a macroscopic system and its conservation. It is known that any macroscopic system can interact with its surrounding systems, i.e. with the surrounding medium, by three channels: mechanical ($\Delta A \neq 0$), thermal ($\Delta Q \neq 0$) and material ($\Delta N \neq 0$) (see Sect. 1.1).

To begin with, consider closed systems ($N = \text{const}, \Delta N = 0$).[1] In this case, only two types of interactions, i.e. mechanical and thermal, are possible.

Mechanical interaction. Work. Consider an adiabatically isolated ($\Delta Q = 0$) closed system. For such systems, only mechanical interaction is possible. In such an interaction, two cases are possible: a system at the expense of internal energy performs work and increases its volume ($\Delta V > 0$); and a system with external forces exerting on it decrease its volume ($\Delta V < 0$).

When a system performs work, its internal energy decreases, therefore the elementary work being performed is regarded as negative ($\mathrm{d}A < 0$), and similarly, the elementary work being performed by external forces on a system is regarded as positive ($\mathrm{d}A > 0$).

[1] Thermodynamics of open systems ($\Delta N \neq 0$) will be considered in Sect. 3.1.

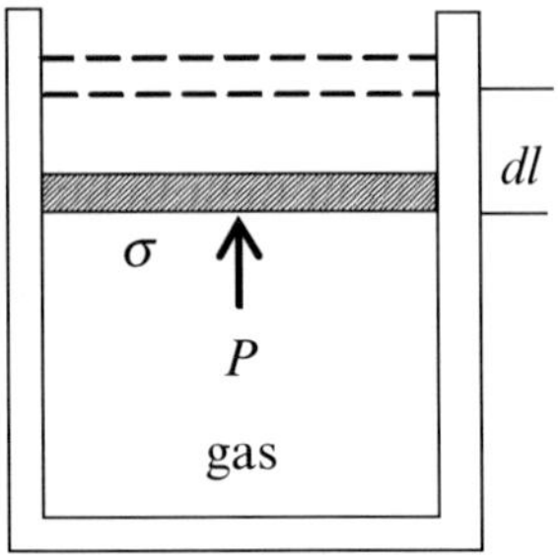

Fig. 2.1. A gas in a cylindrical vessel under a piston

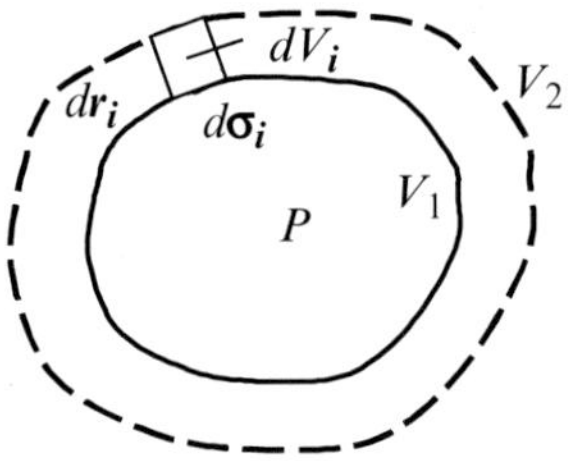

Fig. 2.2. An adiabatically isolated system with boundaries of an arbitrary shape

To compute the elementary work, for simplicity assume that the considered system represents a gas in a cylindrical vessel under a piston (Fig. 2.1).

If we pressurise the gas to P, and assume the cross-sectional area of the piston is σ, the gas acts on the piston with a force $P\sigma$, which is directed upwards. This force moves the piston up by a magnitude $\mathrm{d}l$ and performs the elementary work $P\sigma\mathrm{d}l = P\mathrm{d}V$. According to our condition, the elementary work being performed by the system ought to be negative. Therefore, the elementary work can be written down in the form[2]

$$\mathrm{d}A = -P\,\mathrm{d}V, \tag{2.1}$$

Where, in the given case, $\mathrm{d}V > 0$.

It can be shown that the expression for work (2.1) is just for macroscopic systems of any form. To do this, consider an adiabatically isolated system with boundaries of an arbitrary shape (Fig. 2.2). Assume that under the influence of pressure P the system increases its volume from V_1 to V_2. To calculate the work being performed in this process, subdivide boundaries of the system into small portions and denote the area of an arbitrarily taken portion by $\mathrm{d}\boldsymbol{\sigma}_i$. Then, on the element of the surface $\mathrm{d}\boldsymbol{\sigma}_i$, the force $P\mathrm{d}\boldsymbol{\sigma}_i$ acts. If under the action of this

[2] If under action of external forces the volume of the gas decreases ($\mathrm{d}V < 0$), then $\mathrm{d}A > 0$. Thus, (1.1) is the general expression for the elementary work.

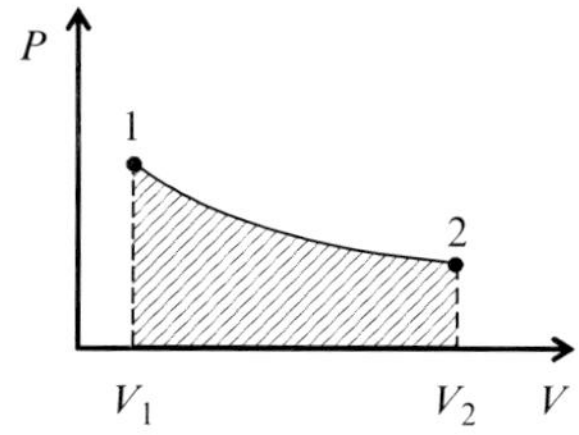

Fig. 2.3. The work being performed by the gas in the process of expansion

force the element of the surface $\mathrm{d}\boldsymbol{\sigma}_i$ is moved by the magnitude $\mathrm{d}\boldsymbol{r}_i$, the most elementary work being performed equals $\mathrm{d}A_i = -P\,\mathrm{d}\boldsymbol{\sigma}_i\,\mathrm{d}\boldsymbol{r}_i = -P\,\mathrm{d}V_i$, where $\mathrm{d}V_i = \mathrm{d}\boldsymbol{\sigma}_i\,\mathrm{d}\boldsymbol{r}_i$. If we add up over all the surface of the system, the elementary work being performed by the gas in the process of expansion from V_1 to V_2 can be written down in the form

$$\mathrm{d}A = \sum_i \mathrm{d}A_i = -P\sum_i \mathrm{d}V_i = -P\,\mathrm{d}V, \tag{2.2}$$

which coincides with (2.1). The same expression for the elementary work is obtained in the case of action of external forces, leading to a decrease in volume from V_2 to $V_1\,(V_1 < V_2)$, too.

Inasmuch as the internal energy of an adiabatically isolated closed system changes only at the expense of the work being performed, the law of conservation of energy in this case has the appearance

$$\mathrm{d}E = \mathrm{d}A = -P\,\mathrm{d}V. \tag{2.3}$$

If the system, expanding from the state with energy E_1 and volume V_1, passes into the state with energy E_2 and volume V_2, the law of conservation (2.3) can be written down in the integral form

$$E_2 - E_1 = -\int_{V_1}^{V_2} P(V)\mathrm{d}V. \tag{2.4}$$

It is evident that the work being performed and, consequently, the change in energy (2.4) of the system in the process of expansion numerically equal the cross-hatched area under the curve $P = P(V)$ on the $P - V$ plane (Fig. 2.3).

Thermal interaction. Amount of heat. Consider a system in a thermostat with immovable boundaries. If temperatures of the thermostat and system are different, the exchange of a certain amount of heat $\mathrm{d}Q$ occurs between them. We arrange to regard the amount of heat that is transferred from the system into the thermostat as negative ($\mathrm{d}Q < 0$), and the amount of heat that is transferred from the thermostat into the system as positive ($\mathrm{d}Q > 0$).

Recall that at the mechanical interaction the external parameter of the system (volume V) changes, and at the thermal interaction its internal parameter (temperature T) changes. The process of the heat exchange between the system and thermostat occurs as a result of thermal conductivity. This is a very complex molecular-kinetic process: particles of the system and thermostat that are found close to the boundary of the system with the thermostat interact (collide) and exchange energy, and this process of collision (the exchange of energy) propagates inside the system (the thermostat). As a result, a certain amount of heat $\mathrm{d}Q$ transfers from the system to the thermostat or, vice versa, depending on where the temperature is higher. Inasmuch as the amount of heat $\mathrm{d}Q$ received as a result of the exchange changes the internal energy by $\mathrm{d}E$, the law of conservation can be presented in the form

$$\mathrm{d}E = \mathrm{d}Q. \tag{2.5}$$

However, equality (2.5) can be written only in the case where the amount of heat is equivalent to the mechanical energy (work) and is measured in identical units. The equivalence of heat and mechanical energy was determined for the first time by the German physiologist Julius Robert von Mayer in 1842, and in 1843 the English physicist James Prescott Joule determined the coefficient of equivalence, which is

$$1\,\mathrm{cal} = 4.184\,\mathrm{J} = 4.184 \times 10^7\,\mathrm{erg}. \tag{2.6}$$

By the discovery of equivalence of the amount of heat and work, von Mayer and Joule laid down the basis of the First Law of Thermodynamics. The mathematical expression of this law was proposed by the German physicist and physiologist Hermann Ludwig Ferdinand von Helmholtz in 1847. Thus, three outstanding scientists: i.e. von Mayer, Joule and von Helmholtz, are regarded as the discoverers of the First Law of Thermodynamics.

Assume that a closed system, thanks to mechanical and thermal interactions, passes from the initial state 1 into the final state 2. Comprehensively having studied processes of such a type, and as a result of the generalisation of numerous experimental data for closed systems ($N = \mathrm{const}$), the First Law of Thermodynamics can be formulated as follows: *As a closed system passes from the initial state 1 with energy E_1 into the final state 2 with energy E_2, the change in energy $(E_2 - E_1)$ equals the sum of the amount of heat received as a result of the exchange and the work being carried out; this sum depends only on these states and does not depend on the mode of passage of the system from the initial state into the final state* (Fig. 2.4).

Hence it follows that the internal energy is a function of state. Thus, it can be written as

$$\int_1^2 \mathrm{d}Q + \int_1^2 \mathrm{d}A = E_2 - E_1. \tag{2.7}$$

From the first law, the following conclusion is drawn: for closed systems there exists such a function of state (the internal energy E), single-valuedly

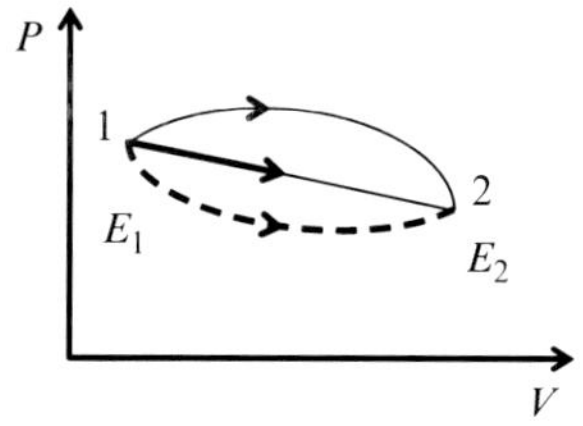

Fig. 2.4. The illustration of the First Law of Thermodynamics

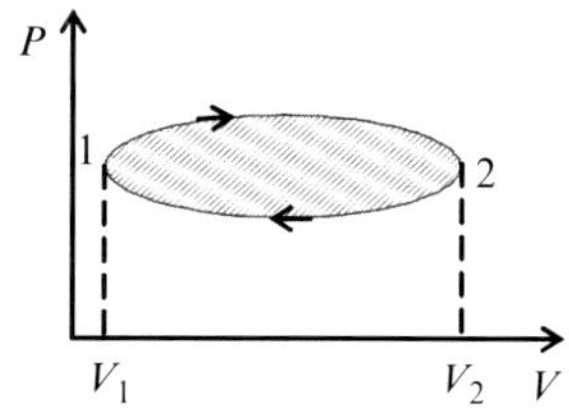

Fig. 2.5. The work being performed at the circular process

characterizing the state of the system, that its change is a total differential:

$$\int_1^2 \mathrm{d}E = E_2 - E_1. \tag{2.8}$$

Note that (2.7) is the integral form of the First Law of Thermodynamics, i.e. the law of conservation of energy for closed systems. By combining (2.7) and (2.8), the differential form of the First Law of Thermodynamics can be presented in the form

$$\mathrm{d}A + \mathrm{d}Q = \mathrm{d}E, \tag{2.9}$$

i.e. *the change in energy of a closed system at the elementary process equals the sum of the elementary work being carried out and the amount of heat received as a result of the exchange.*

Note a certain subtlety. Though in (2.9) $\mathrm{d}A$ and $\mathrm{d}Q$ separately are not total differentials, their sum $\mathrm{d}A + \mathrm{d}Q$, i.e. the change in energy $\mathrm{d}E$, is a total differential. This means that E is a function of state, but A and Q are not. We can say what the internal energy E in the given state of the system is, but we do not know which part of this energy is mechanical and which part is thermal. We can only say which part in the change in internal energy in passing from one state to another accounts for the share of the mechanical work, and which one for the share of the change in thermal energy (see Sect. 1.9).

There exists one more definition of the First Law of Thermodynamics. If a system participating in the circular process returns to the starting state, its internal energy does not change (Fig. 2.5).

Then, according to (2.7), for the work being carried out we get

$$\oint \mathrm{d}A = -\oint \mathrm{d}Q \tag{2.10}$$

or

$$\oint p(V)\mathrm{d}V = \oint \mathrm{d}Q. \tag{2.11}$$

The work being performed numerically equals the cross-hatched area of the circular process (Fig. 2.5), and the sign of work depends on the direction of the process.

From First Law of Thermodynamics for circular process (2.10), it follows that during the process the system receiving a certain amount of heat and converting it into equivalent work returns to its starting state.

Thus, in the circular process the system performs work only at the expense of the thermal energy. If $\mathrm{d}Q=0$, from (2.10) it follows that

$$\oint \mathrm{d}A = 0, \tag{2.12}$$

i.e. without spending thermal energy, work cannot be performed.

For many years, attempts were made to create a machine to carry out work without expenditure of thermal energy. Such a machine would be called *the perpetual engine (perpetuum mobile) of the first kind.* From the above, the First Law of Thermodynamics can also be formulated as follows: *It is impossible to create a perpetual engine (perpeiuum mobile) of the first kind.*

Heat capacity. Among thermodynamic coefficients, the heat capacity occupies a special place. Here we will give only its definition and find the general expression on the basis of the First Law of Thermodynamics.

The heat capacity numerically equals the amount of heat necessary to increase the temperature of the substance (the system) by one degree. The specific (molar) heat capacity numerically equals the amount of heat necessary to increase the temperature of 1 g (1 mole) of substance by 1°.

In practice, the heat capacity is measured under two conditions: at constant volume ($V=\text{const}$), which is the isochoric heat capacity C_V; and at constant pressure ($P=\text{const}$) which is the isobaric heat capacity C_P.

Using the differential form of the First Law of Thermodynamics (2.9), the amount of heat received by the system can be written as

$$\mathrm{d}Q = \mathrm{d}E - \mathrm{d}A. \tag{2.13}$$

If V, T are taken as independent parameters determining the state of the system, then $E=E(V,T)$. Then the change in energy is

$$\mathrm{d}E = \left(\frac{\partial E}{\partial V}\right)_T \mathrm{d}V + \left(\frac{\partial E}{\partial T}\right)_V \mathrm{d}T. \tag{2.14}$$

If we substitute (2.14) and expression for work (2.1) into (2.13), we get

$$\mathrm{d}Q = \left(\frac{\partial E}{\partial T}\right)_V \mathrm{d}T + \left[\left(\frac{\partial E}{\partial V}\right)_T + P\right] \mathrm{d}V. \tag{2.15}$$

Hence, for the isochoric heat capacity C_V we have

$$C_V = \left(\frac{\partial Q}{\partial T}\right)_V = \left(\frac{\partial E}{\partial T}\right)_V . \tag{2.16}$$

At constant pressure P, according to the equation of state $V = V(P, T)$, volume V depends only on temperature T and therefore the change in volume is $\mathrm{d}V = (\partial V/\partial T)_P \,\mathrm{d}T$. Then, according to (2.15), in the condition $P = \text{const}$, the expression for the amount of heat takes the form

$$\mathrm{d}Q = \left(\frac{\partial E}{\partial T}\right)_V \mathrm{d}T + \left[\left(\frac{\partial E}{\partial V}\right)_T + P\right] \left(\frac{\partial V}{\partial T}\right)_P \mathrm{d}T. \tag{2.17}$$

Hence, for the isobaric heat capacity $C_P = (\partial Q/\partial T)_P$ we can write

$$C_P = C_V + \left[\left(\frac{\partial E}{\partial V}\right)_T + P\right] \left(\frac{\partial V}{\partial T}\right)_P . \tag{2.18}$$

Here, the term $(\partial E/\partial V)_T (\partial V/\partial T)_P$ determines the change in energy of the system for a change in temperature by one degree at constant pressure ($P = \text{const}$), and the term $P\,(\partial V/\partial T)_P$ corresponds to the amount of heat spent on performing work to increase the volume for a change in temperature by one degree at $P = \text{const}$. If energy of the system does not depend on volume, i.e. $(\partial E/\partial V)_T = 0$ (an ideal gas), the difference $C_P - C_V = = P\,(\partial V/\partial T)_P$ is determined by the amount of heat spent on performing work.

From (2.16) and (2.18), it follows that to calculate the isochoric heat capacity it is sufficient to know the caloric equation of the state of the system, i.e. the dependence of the energy $E = E(T, V)$, and to calculate the isobaric heat capacity it is necessary to know also the thermal equation of the state of the system $P = P(T, V)$.

We now show that the difference of heat capacities $C_P - C_V$ is determined only by the thermal equation of the state of the system $P = P(V, T)$. The basic thermodynamic relationship $\mathrm{d}E = T\,\mathrm{d}S - P\,\mathrm{d}V$ (see Sect. 1.8) can be presented in the form

$$\left(\frac{\partial E}{\partial V}\right)_T = T\left(\frac{\partial S}{\partial V}\right)_T - P. \tag{2.19}$$

In Sect. 2.3 [see (2.62)], we will show that $(\partial S/\partial V)_T = (\partial P/\partial T)_V$. By substituting this expression into (2.19), we get

$$\left(\frac{\partial E}{\partial V}\right)_T + P = T\left(\frac{\partial P}{\partial T}\right)_V . \tag{2.20}$$

With regard to (2.18) and (2.20), $C_P - C_V$ finally takes the form

$$C_P - C_V = T\left(\frac{\partial P}{\partial T}\right)_V \left(\frac{\partial V}{\partial T}\right)_P . \tag{2.21}$$

As seen from (2.21), in order to calculate the difference $C_P - C_V$ it is sufficient to know the explicit form of the thermal equation of the state of the system $P = P(V, T)$. In particular, for an ideal gas, using the equation $PV = RT$, from (2.21) the von Mayer equation follows:[3]

$$C_P - C_V = R, \tag{2.22}$$

where $R = N_A k_0 = 8.31 \times 10^7\,\mathrm{erg/(K\,mol)} = 8.31\,\mathrm{J/(K\,mol)} = 1.92\,\mathrm{cal/(K\,mol)}$ is the universal gas constant, $k_0 = 1.38 \times 10^{-16}\,\mathrm{erg/K}$ is the Boltzmann constant and $N_\mathrm{A} = 6.026 \times 10^{23}\,\mathrm{mol}^{-1}$ is the Avogadro number.

As can be seen, for ideal gases $C_P > C_V$. This is explained by the fact that at $P = \mathrm{const}$, a certain part of the thermal energy being supplied is expended on performing work in the expansion of the gas. And in the case of $V = \mathrm{const}$, owing to the absence of expansion, the thermal energy is spent only on increasing the internal energy and, consequently, the temperature of the system. Note that inequality $C_P > C_V$ holds true not only for ideal gases but also for any macroscopic systems (see Sect. 2.4).

2.2 Second Law of Thermodynamics: Carnot Cycle

In the general form, the formulation of the Second Law of Thermodynamics – the law of increasing entropy – was given in Sect. 1.7. Still earlier, in Sect. 1.6 the definition of the statistical sense of entropy was given. Inasmuch as entropy of a system is determined by the number of microstates corresponding to the pre-assigned macrostate, or in other words, is determined as the logarithm of the statistical weight of the pre-assigned macrostate, it characterises the state of a system and is a function of state.

We introduce the definition of the Second Law of Thermodynamics as proposed by Clausius in 1865: *Possible internal processes of an isolated system not in thermodynamic equilibrium ought to proceed in the direction of the increase in entropy; when reaching the thermodynamic equilibrium, these processes cease and entropy takes on a maximum value* (Fig. 1.8).

This is the statistically substantiated general form of the Second Law of Thermodynamics. However, it should be noted that the second law, as also the first one, is an experimental law based on generalisation of test data. At the beginning of the nineteenth century, with the aim of increasing the efficiency of heat machines, processes of heat transfer into the mechanical work and

[3] Just on the basis of this equation, in 1842 von Mayer determined the mechanical equivalent of heat.

vice versa were intensively studied. It was established that the processes of transfer of *the amount of heat* $\rightarrow$ *work* (*the mechanical energy*) and *work* $\rightarrow$ *the amount of heat* are not symmetric. It was experimentally proved that all mechanical energy (work) can be converted into an equivalent amount of heat, but it is impossible to completely convert all heat into an equivalent effective work:

$$\underrightarrow{\Delta A = \Delta Q\,;} \quad \underrightarrow{\Delta Q > \Delta A\,.} \tag{2.23}$$

Hence, it follows that it is impossible to create a heat machine that converts all heat into mechanical energy (work) with the efficiency equal to unity, i.e. it is impossible to create a perpetual engine of the second kind.

At the same time, it became known that heat spontaneously (without performing work) cannot transfer from a cold body to a hot one. As a result of a generalisation of such experimental conclusions, two different but equivalent formulations of the Second Law of Thermodynamics arose:

Clausius postulate (1850). It is impossible to have a process the only final result of which is the transfer of heat from a low-temperature body to a high-temperature one.

Kelvin postulate (1852). It is impossible to have a circular process the only final result of which is the conversion into work of all heat received from the system with a pre-assigned temperature.

Indeed, from the Kelvin postulate it follows that it is impossible to create a machine, working on the basis of a circular cycle, in which heat received from the only source completely converts into work.

Hence, it is seen that it is impossible to create a machine (the perpetual engine of the second kind) with the efficiency

$$\eta = \frac{\Delta A}{\Delta Q}, \tag{2.24}$$

equal to unity; here, ΔQ is the amount of the expended heat, and ΔA is the performed work. The question arises whether it is impossible to create a machine with $\eta = 1$, and whether it is only possible to create a machine with the maximum efficiency $\eta = \eta_{\text{max}} < 1$. It was established that the machine with the maximum efficiency ought to work in a reversible circular cycle. Such a reversible circular cycle was proposed for the first time in 1824 by the French engineer Sadi Carnot.

The Carnot cycle represents a reversible circular process consisting of two isothermal and two adiabatic processes. At first, a working body is brought into contact with a heater with temperature T_2 and allowed to isothermally expand (Fig. 2.6a). Thus, the system receives an amount of heat ΔQ_2 from the heater. Thereupon, the system parts from the heater and adiabatically expands (Fig. 2.6b). Thereafter, the working body is brought into contact with the cooler with temperature T_1 and, being isothermally compressed, gives up the amount of heat ΔQ_1 to the cooler (Fig. 2.6c). Then, the system parts from the cooler (Fig. 2.6d) and, being adiabatically compressed, returns

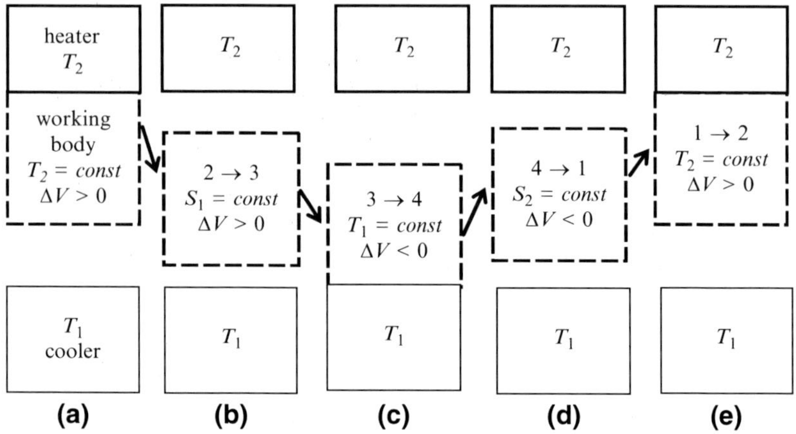

Fig. 2.6. The scheme of the Carnot Cycle

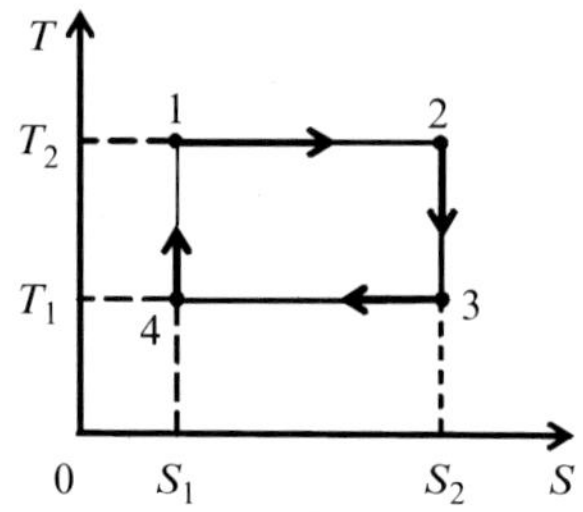

Fig. 2.7. The Carnot Cycle on (T, S) plane

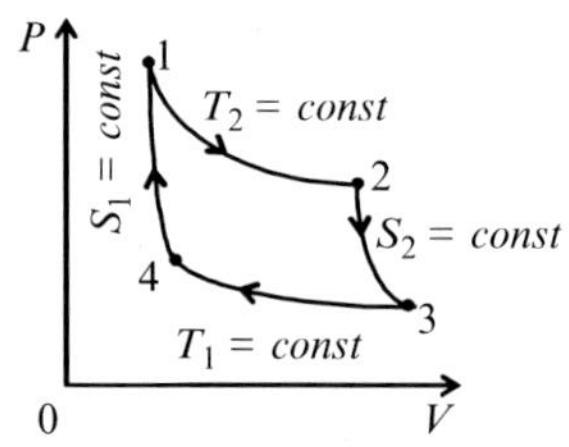

Fig. 2.8. The Carnot Cycle on (P, V) plane

to the starting position (Fig. 2.6e). As a result, the working body performs a reversible circular process.

Processes comprising the whole Carnot cycle are graphically shown on (T, S) and (P, V) planes in Figs. 2.7 and 2.8, respectively. In these figures, processes $1 \to 2$ and $3 \to 4$ are isothermal, and $2 \to 3$ and $4 \to 1$ are adiabatic.

Inasmuch as in the course of one Carnot cycle work $\Delta A = |\Delta Q_2| - |\Delta Q_1|$ is performed, the efficiency of the heat machine, working on the basis of this cycle, equals

$$\eta = \frac{|\Delta Q_2| - |\Delta Q_1|}{|\Delta Q_2|} \tag{2.25}$$

And we now prove the known theorem, characteristic for the Carnot cycle. To do this, using the formula for work (2.1), write down the First Law of Thermodynamics (2.9) in the form

$$\mathrm{d}E = \mathrm{d}Q - P\,\mathrm{d}V. \tag{2.26}$$

Once more, introduce the basic thermodynamic relationship (1.159), combining the First and Second Laws of Thermodynamics:

$$\mathrm{d}E = T\,\mathrm{d}S - P\,\mathrm{d}V. \tag{2.27}$$

From a comparison of the two latter expressions, the relation between the amount of heat and the change in entropy, i.e. the Clausius equality, can be found:

$$dQ = T\,\mathrm{d}S. \tag{2.28}$$

This equality is true only for reversible processes and relates the first two laws of thermodynamics. Hence, it is seen that when transferring the amount of heat $\mathrm{d}Q$ to the system with temperature T, its entropy increases by the magnitude

$$\mathrm{d}S = \frac{\mathrm{d}Q}{T}. \tag{2.29}$$

Inasmuch as entropy is a function of state, for a reversible circular process

$$\oint \mathrm{d}S = \oint \frac{\mathrm{d}Q}{T} = 0 \tag{2.30}$$

Hence, it follows that though $\mathrm{d}Q$ is not a total differential, $\mathrm{d}Q/T$ is a total differential.

If in the system an irreversible process is possible, the growth of entropy is associated not only with the delivered amount of heat $\mathrm{d}Q$, but also at the expense of irreversibility of possible processes. In this case[4]

$$\mathrm{d}S > \frac{\mathrm{d}Q}{T}. \tag{2.31}$$

This inequality is called the *Clausius inequality.*

We return once more to the Carnot cycle. Applying expression (2.28) to isothermal processes $1 \to 2$ and $3 \to 4$, for the amount of heat received by the system from the heater, we can write down $|\Delta Q_2| = T_2\,|\Delta S_2|$, and for the amount of heat given up to the cooler, we have $|\Delta Q_1| = T_1\,|\Delta S_1|$. Since the Carnot cycle is a circular reversible cycle, i.e. the working body returns to

[4] For instance, in the system, along with receiving heat, there occur irreversible processes of diffusion or heat transfer.

the starting state, the change in its entropy $\Delta S = \Delta S_2 + \Delta S_1 = 0$. Therefore $|\Delta S_2| = |\Delta S_1|$. Taking this into account in definition (2.25), the efficiency of the Carnot cycle takes the form

$$\eta = \frac{T_2 - T_1}{T_2}. \tag{2.32}$$

Hence stems the known Carnot theorem: *The efficiency of the Carnot machine does not depend on the nature of a working body and is determined only by temperature of a heater and a cooler.* It is seen that $\eta = 1$ is possible only in the case when the temperature of a cooler $T_1 = 0$. But in Sect. 2.6, we see from the Third Law of Thermodynamics that production of the absolute zero of temperature is impossible. Consequently, it is impossible to construct a machine with $\eta = 1$, in principle.

While for the First Law of Thermodynamics – the law about energy – there exists only one formulation, for the Second Law of Thermodynamics, i.e. the law about entropy, there exist several equivalent definitions. Further we will show that each of these definitions stems from the other, or, if one of them is wrong, the others are also wrong.

1. Note that the Clausius postulate follows from the law of increasing entropy. Indeed, in Sect. 1.8 we showed that energy (the amount of heat) can spontaneously transfer only from a high-temperature body to a low-temperature body; otherwise, entropy would decrease.
2. We prove that the Clausius and Kelvin postulates are equivalent. To do this, it is sufficient to show that if one of the postulates is wrong, the other is violated, too.

At first assume that the Clausius postulate is wrong. This means that a specified amount of heat Q_1 can spontaneously transfer from the system B_1 with temperature T_1 to the system B_2 with temperature $T_2 > T_1$ (Fig. 2.9).

Use the system B_2 in the capacity of a heater in the Carnot cycle. The working body takes away from B_2 the same amount of heat – Q_1, a part of which the Carnot machine converts into work $Q_1 - Q_2 = A$, transferring at the same time the amount of heat $Q_2 < Q_1$ to the system B_1. As a result,

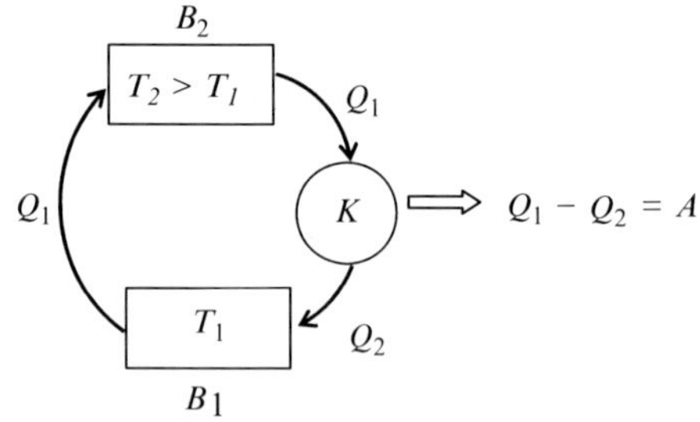

Fig. 2.9. The illustration of the equivalency of the Clausius and Kelvin hypotheses

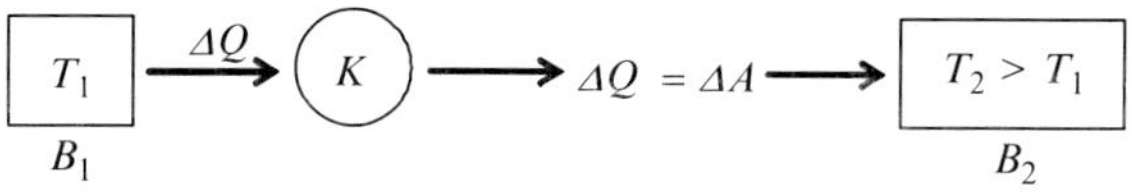

Fig. 2.10. The illustration of the equivalency of the Kelvin and Clausius hypotheses

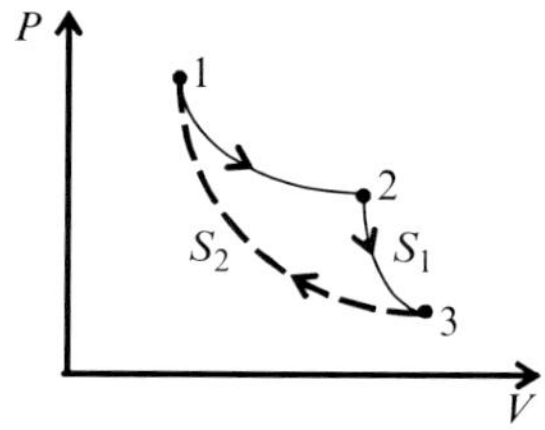

Fig. 2.11. A hypothetical circular process

the state of system B_2 does not change, and the system B_1 loses the amount of heat $Q_1 - Q_2 = A$, i.e. at the expense of heat received from the system B_2, work $Q_1 - Q_2 = A$ is performed. Thus, a process takes place, the only result of which is the performance of work A at the expense of heat solely of the system B_1. It is evident that this conclusion contradicts the Kelvin postulate.

We now show that if the Kelvin postulate is wrong, the Clausius postulate is wrong, too. Assume that the Kelvin postulate is wrong, i.e. the amount of heat ΔQ received from the system with temperature T_1 can be completely converted into work: $\Delta Q = \Delta A$ (Fig. 2.10). All the work received ΔA by some means (for instance, by friction) can be completely converted into heat and transferred to a system, including the system B_2, with temperature $T_2 > T_1$. Hence it follows that it is possible to have a process, the only result of which is the transfer of heat from the low-temperature body B_1 to the high-temperature body B_2, which contradicts the Clausius postulate.

From the Second Law of Thermodynamics, there follows one more important result: *entropy is a single-valued function of the state of the system, and, consequently, different adiabats cannot intersect.* To prove this conclusion, assume the reverse, i.e entropy is not a single-valued function of state and different adiabats can intersect. If this is right, we can create a circular process, shown in Fig. 2.11. This circular process consists of three processes. In the process of the isothermal expansion 1–2 of the system, a certain amount of heat ΔQ received from the thermostat is completely converted into equivalent work.

Thereupon, as a result of two adiabatic processes 2–3 and 3–1, the system returns to the starting state. Such a circular process is possible if adiabats S_1 and S_2 intersect in state 3. Hence, it follows that two different values of entropy S_1 and S_2 correspond to the same state 3. Thus, as a result of such a hypothetical process, the amount of heat received from the thermostat completely gets converted into work, and this means there is a possibility of creating the perpetual engine of the second kind, which contradicts the Second Law of Thermodynamics. Consequently, our supposition is wrong, i.e. different adiabats cannot intersect, and entropy is a single-valued function of state.

So, the Second Law of Thermodynamics – a law about entropy – in the general form can be formulated as follows: *Entropy is a single-valued function of the state of a system; as a result of irreversible processes proceeding in an isolated non-equilibrium system, entropy of the system increases and reaches a maximum in the equilibrium state.*

Note that this definition is the most general formulation of the Second Law of Thermodynamics, since it follows the Clausius postulate. On the other hand, the Kelvin postulate is equivalent to the Clausius postulate, from which the impossibility of creating the perpetual engine of the second kind stems.

2.3 Thermodynamic Functions of Closed Systems: Method of Thermodynamic Potentials

The basic task of thermodynamics and statistical physics is the finding of physical quantities characterizing a macroscopic state of a system and determination of the relation between them (e.g. the equation of state). In thermodynamics this task is solved through experimental methods and in statistical physics through analytical methods. However, the determination of quantities possessing a physical sense is not always possible directly from experiments or theoretical computations. Therefore, auxiliary, so-called thermodynamic functions are introduced. These functions are also called thermodynamic potentials. Knowing their explicit form, thermodynamic parameters and general relationships between them can be determined. Such a method is called *the method of thermodynamic potentials.*

Here we will consider closed systems ($N = \text{const}$). Chapter 3 is devoted to open systems, i.e. systems with a variable number of particles ($N \neq \text{const}$).

From the preceding chapter (Sects. 1.3 and 1.5), it is known that to determine the mean value of a physical quantity L for classical systems it is necessary to calculate the integral

$$\bar{L} = \int L(q,p)\rho(q,p)\mathrm{d}q\,\mathrm{d}p, \tag{2.33}$$

and for quantum systems to do the sum

$$\bar{L} = \sum_n W_n L_{nn}. \tag{2.34}$$

To solve this problem from classical mechanics, it is necessary to know the explicit form of the function $L(q,p)$ as well as the distribution function $\rho(q,p)$, and from quantum mechanics it is essential to know the diagonal matrix elements L_{nn} and diagonal elements of the statistical matrix $W_n \equiv W_{nn}$. Inasmuch as for any physical parameter it is not possible to find the explicit form of $L(q, p)$ and L_{nn}, it is not possible to immediately calculate $\bar{L}$. Therefore, we are enforced to use the method of thermodynamic potentials.

For closed systems, there exist four thermodynamic functions – potentials. We introduce them below.

1. *Internal energy.* Recall that the expression for the change in the internal energy of a system according to the First Law of Thermodynamics has the appearance

$$\mathrm{d}E = \mathrm{d}Q + \mathrm{d}A. \tag{2.35}$$

With regard to the Second Law

$$\mathrm{d}Q = T\,\mathrm{d}S, \tag{2.36}$$

formula (2.35) acquires the form

$$\mathrm{d}E = T\,\mathrm{d}S - P\,\mathrm{d}V, \tag{2.37}$$

where $\mathrm{d}A = -P\,\mathrm{d}V$.

Hence it follows that

$$\mathrm{d}E = \mathrm{d}Q - P\,\mathrm{d}V. \tag{2.38}$$

As can be seen, for isochoric ($V = \text{const}$) processes

$$\mathrm{d}E = \mathrm{d}Q \tag{2.39}$$

i.e. in isochoric processes the change in the internal energy of a system equals the amount of heat received by the system as a result of the exchange. Since the internal energy is a function of state, i.e. $\mathrm{d}E$ is a total differential, from (2.36) it follows that

$$T = \left(\frac{\partial E}{\partial S}\right)_V; \quad P = -\left(\frac{\partial E}{\partial V}\right)_S. \tag{2.40}$$

From (2.36), it is seen that V and S are independent variables for the function of the internal energy:

$$E = E(V, S). \tag{2.41}$$

Knowing the explicit form of this function, on the basis of (2.40) we can calculate the temperature and pressure. If from definition (2.40) we take the derivative of T with respect to V and of P with respect to S, we get

$$\left(\frac{\partial T}{\partial V}\right)_S = \frac{\partial^2 E}{\partial S \partial V}; \left(\frac{\partial P}{\partial S}\right)_V = -\frac{\partial^2 E}{\partial V \partial S}. \tag{2.42}$$

Hence follows the first thermodynamic relationship

$$\left(\frac{\partial T}{\partial V}\right)_S = -\left(\frac{\partial P}{\partial S}\right)_V . \tag{2.43}$$

Inasmuch as relationship (2.43) is obtained independently of the explicit form of the function $E(V, S)$, it bears the general character and relates two properties of the system – the change in temperature at the adiabatic expansion to the change in pressure at the isochoric receiving of heat.

From (2.39) follows the definition of the isochoric heat capacity:

$$C_V = \left(\frac{\partial E}{\partial T}\right)_V . \tag{2.44}$$

The internal energy as a function of variables S and V can be called *the thermodynamic potential.*

2. *Enthalpy or heat function.* As is seen from (2.39), in isochoric ($V = \text{const}$) processes the amount of heat supplied to the system equals the increase in the internal energy. It asks, if the process occurs isobarically ($P = \text{const}$), to the change in which function does the amount of heat delivered to the system be equal. To elucidate this question, write down relationship (2.38) at $P = \text{const}$ in the form

$$\mathrm{d}(E + PV) = \mathrm{d}Q \tag{2.45}$$

and introduce the notation

$$W = E + PV. \tag{2.46}$$

From relationship (2.45), it follows that for isobaric processes

$$\mathrm{d}W = \mathrm{d}Q. \tag{2.47}$$

The function W is called *enthalpy* or *the heat function*[5]. of the system. Thus, in isobaric processes the heat delivered equals the increase in enthalpy. If on the right-hand side of thermodynamic relationship (2.37) we take into account the identity $P\,\mathrm{d}V + V\,\mathrm{d}P = \mathrm{d}(PV)$, for the change in enthalpy we get

$$\mathrm{d}W = T\,\mathrm{d}S + V\,\mathrm{d}P. \tag{2.48}$$

Hence it follows that entropy S and pressure P are independent variables for enthalpy:

$$W = W(S, P). \tag{2.49}$$

Knowing the explicit form of the function $W(S, P)$, on the basis of (2.48) the temperature and volume of the system can be determined:

[5] This function is also called the heat content.

$$T = \left(\frac{\partial W}{\partial S}\right)_P; \quad V = \left(\frac{\partial W}{\partial P}\right)_S. \tag{2.50}$$

In definitions (2.50) of T, we take the derivative with respect to pressure and for V with respect to entropy. Then we get

$$\left(\frac{\partial T}{\partial P}\right)_S = \frac{\partial^2 W}{\partial P \partial S}; \quad \left(\frac{\partial V}{\partial S}\right)_P = \frac{\partial^2 W}{\partial S \partial P}. \tag{2.51}$$

Hence follows the second thermodynamic relationship:

$$\left(\frac{\partial T}{\partial P}\right)_S = \left(\frac{\partial V}{\partial S}\right)_P, \tag{2.52}$$

which relates two properties of the system – the change in temperature T to the change in pressure in adiabatic processes, and the change in volume V at the delivery of a certain amount of heat in isobaric processes.

From (2.47) stems the definition of the isobaric heat capacity:

$$C_P = \left(\frac{\partial W}{\partial T}\right)_P. \tag{2.53}$$

Consequently, in isobaric processes the heat capacity is determined by the change in enthalpy of the system. In this case ($P = \text{const}$), W plays the role of the internal energy.

3. *Free energy or von Helmholtz potential.* The above-indicated thermodynamic functions – the internal energy and enthalpy – are inconvenient from the practical point of view, since one of their arguments – entropy – is not measurable by experiment. Therefore, it is necessary to introduce a characteristic function of the system, independent parameters of which are P, T, V. One of such functions is the free energy.

On the basis of (2.35) and (2.36), the basic thermodynamic relationship can be presented in the form

$$\mathrm{d}E = T\,\mathrm{d}S + \mathrm{d}A. \tag{2.54}$$

For isothermal processes ($T = \text{const}$), (2.54) takes the form

$$\mathrm{d}(E - TS) = \mathrm{d}A. \tag{2.55}$$

Denote

$$F = E - TS, \tag{2.56}$$

then we get

$$\mathrm{d}F = \mathrm{d}A. \tag{2.57}$$

The function F is called *the free energy* or *the von Helmholtz potential.* From (2.57) it follows that in isothermal processes ($T = \text{const}$) the work being performed equals the change in the free energy, i.e. the free energy is a part of the internal energy, which can be converted into work. Then TS can be called *the bound energy.* In order to find a differential of the free energy on the right-hand side of (2.37), we take into account the identity $\mathrm{d}(TS) = T\,\mathrm{d}S + S\,\mathrm{d}T$. Then we get

$$\mathrm{d}F = -P\,\mathrm{d}V - S\,\mathrm{d}T. \tag{2.58}$$

It is seen that V and T are independent variables of the free energy F:

$$F = F(V,T). \tag{2.59}$$

Knowing the explicit form of this function with the aid of the expression, stemming from (2.58) the pressure and entropy of the system can be computed:

$$P = -\left(\frac{\partial F}{\partial V}\right)_T; \quad S = -\left(\frac{\partial F}{\partial T}\right)_V. \tag{2.60}$$

In (2.60), the first expression represents the thermal equation of state in the general form $P = P(V,T)$.

In definitions (2.60), taking the derivative of P with respect to T and of S with respect to V, we get

$$\left(\frac{\partial P}{\partial T}\right)_V = -\frac{\partial^2 F}{\partial T \partial V}; \quad \left(\frac{\partial S}{\partial V}\right)_T = -\frac{\partial^2 F}{\partial V \partial T}. \tag{2.61}$$

Hence stems the third thermodynamic relationship

$$\left(\frac{\partial P}{\partial T}\right)_V = \left(\frac{\partial S}{\partial V}\right)_T, \tag{2.62}$$

relating two properties of the system. In (2.62), the left-hand side represents the change in pressure at an isochoric change in temperature, and the right-hand side the change in entropy at an isothermal change in volume.

4. *Gibbs thermodynamic potential.* Among the above-considered thermodynamic functions, the function in which the pressure P and temperature T are independents is absent. In order to find such a function, in (2.58) take into account the identity $P\,\mathrm{d}V + V\,\mathrm{d}P = \mathrm{d}(PV)$. Then we get

$$\mathrm{d}(F + PV) = -S\,\mathrm{d}T + V\,\mathrm{d}P. \tag{2.63}$$

Denoting

$$\Phi = F + PV = E - TS + PV = W - TS, \tag{2.64}$$

we get

$$\mathrm{d}\Phi = -S\,\mathrm{d}T + V\,\mathrm{d}P. \tag{2.65}$$

The function of which T and P are independent variables,

$$\Phi = \Phi(T, P) \tag{2.66}$$

is called *the Gibbs thermodynamic potential.*

Knowing the explicit form of this function, entropy and volume can be found:

$$S = -\left(\frac{\partial \Phi}{\partial T}\right)_P; \quad V = \left(\frac{\partial \Phi}{\partial P}\right)_T. \tag{2.67}$$

In definitions (2.67), the second expression represents the thermal equation of state in the general form: $V = V(P, T)$. If we take in (2.67) the derivative of S with respect to P and of V with respect to T, we get:

$$\left(\frac{\partial S}{\partial P}\right)_T = -\frac{\partial^2 \Phi}{\partial P \partial T}; \quad \left(\frac{\partial V}{\partial T}\right)_P = \frac{\partial^2 \Phi}{\partial T \partial P}. \tag{2.68}$$

Hence stems the fourth thermodynamic relationship:

$$\left(\frac{\partial S}{\partial P}\right)_T = -\left(\frac{\partial V}{\partial T}\right)_P. \tag{2.69}$$

It should be noticed that the above-introduced thermodynamic relationships (2.43), (2.52), (2.62) and (2.69) are obtained independently of the explicit form of thermodynamic functions and therefore bear the general thermodynamic character. These relationships, and also definitions (2.40), (2.50), (2.60) and (2.67), constitute the basis of the Gibbs method of thermodynamic potentials.

In conclusion, we note that the free energy F and the Gibbs free potential Φ among thermodynamic functions occupy a special place for two reasons.

1. The internal energy and enthalpy are expressed by these potentials. From the relationship $E = F + TS$ and (2.60), it follows that the internal energy is expressed by the free energy F:

$$E = F - T\left(\frac{\partial F}{\partial T}\right)_V = -T^2\left(\frac{\partial}{\partial T}\frac{F}{T}\right)_V. \tag{2.70}$$

This equation bears the name *the Gibbs–von Helmholtz equation.*

From the relationship $W = \Phi + TS$ and definition (2.67), it follows that enthalpy is expressed by the Gibbs potential Φ:

$$W = \Phi - T\left(\frac{\partial \Phi}{\partial T}\right)_P = -T^2\left(\frac{\partial}{\partial T}\frac{\Phi}{T}\right)_P. \tag{2.71}$$

2. To elucidate the second reason for the importance of the free energy F and the Gibbs free potential Φ, rewrite relationship (2.37) for irreversible processes:

$$\frac{\mathrm{d}E}{\mathrm{d}t} + P\frac{\mathrm{d}V}{\mathrm{d}t} < T\frac{\mathrm{d}S}{\mathrm{d}t}. \tag{2.72}$$

If a process proceeds at constant volume ($V = \text{const}$) and constant temperature ($T = \text{const}$), from (2.72) it follows that

$$\frac{\mathrm{d}}{\mathrm{d}t}(E - TS) < 0; \quad \left(\frac{\mathrm{d}F}{\mathrm{d}t}\right)_{V,T} < 0. \tag{2.73}$$

Consequently, irreversible processes at constant T and V ought to proceed in the direction of the decrease in the free energy of the system. Thus, in the equilibrium state the free energy ought to take on the minimum value:

$$(F)_{T,V} \Rightarrow F_{\min}. \tag{2.74}$$

If a process proceeds at constant pressure ($P = \text{const}$) and temperature ($T = \text{const}$), from (2.72) we get

$$\frac{\mathrm{d}}{\mathrm{d}t}(E + PV - TS) < 0; \quad \left(\frac{\mathrm{d}\Phi}{\mathrm{d}t}\right)_{P,T} < 0. \tag{2.75}$$

Consequently, irreversible processes at constant P and T ought to proceed in the direction of the decrease in the Gibbs potential. In the equilibrium state, the Gibbs potential ought to take on the minimum value:

$$(\Phi)_{T,P} \Rightarrow \Phi_{\min}. \tag{2.76}$$

At the end of this section, we once more recall in the compact form the thermodynamic functions and thermodynamic relationships:

Internal energy

$$\mathrm{d}E = T\,\mathrm{d}S - P\,\mathrm{d}V.$$

Enthalpy

$$\mathrm{d}W = T\,\mathrm{d}S + V\,\mathrm{d}P.$$

Free energy

$$F = E - TS; \quad \mathrm{d}F = -P\,\mathrm{d}V - S\,\mathrm{d}T. \tag{2.77}$$

Gibbs thermodynamic potential

$$\Phi = W - TS; \quad \mathrm{d}\Phi = -S\,\mathrm{d}T + V\,\mathrm{d}P.$$

Hence, four thermodynamic relationships follow:

$$\begin{aligned} &\left(\frac{\partial T}{\partial V}\right)_S = -\left(\frac{\partial P}{\partial S}\right)_V; \quad \left(\frac{\partial T}{\partial P}\right)_S = \left(\frac{\partial V}{\partial S}\right)_P; \\ &\left(\frac{\partial P}{\partial T}\right)_V = \left(\frac{\partial S}{\partial V}\right)_T; \quad \left(\frac{\partial V}{\partial T}\right)_P = -\left(\frac{\partial S}{\partial P}\right)_T. \end{aligned} \tag{2.78}$$

2.4 Thermodynamic Coefficients and General Relationships Between Them

Thermodynamic coefficients are quantities that determine how changes in some parameters effect changes in other parameters that characterize the macroscopic state of a system and that can be measured on test.

The most important thermodynamic coefficient is the heat capacity of a system.

The heat capacity numerically equals the amount of heat (energy) necessary to raise the temperature of a system by one degree. From the definition of the heat capacity (2.16) and from the definition of heat (2.28), it follows that the heat capacity is a measure of the change in entropy of a system when its temperature changes by one degree.

The heat capacity can be determined at two conditions: at constant volume (isochoric) and at constant pressure (isobaric). Then, according to (2.28), we get for the isochoric heat capacity

$$C_V = T\,(\partial S/\partial T)_V\,, \tag{2.79}$$

and for the isobaric heat capacity

$$C_P = T\,(\partial S/\partial T)_P\,. \tag{2.80}$$

We also introduce definitions of the other thermodynamic coefficients:

Isobaric coefficient of thermal expansion is

$$\alpha_P = 1/V\,(\partial V/\partial T)_P\,; \tag{2.81}$$

The coefficient characterizing the relative change in temperature at an adiabatic expansion or compression is

$$\alpha_S = -1/T\,(\partial T/\partial V)_S\,; \tag{2.82}$$

The isochoric thermal coefficient of pressure is

$$\beta_V = 1/P\,(\partial P/\partial T)_V\,; \tag{2.83}$$

The coefficient characterizing the relative change in temperature of a system at an adiabatic change in pressure is

$$\beta_S = 1/T\,(\partial T/\partial P)_S\,; \tag{2.84}$$

The coefficient of isothermal compression is

$$\gamma_T = -1/V\,(\partial V/\partial P)_T\,; \tag{2.85}$$

The coefficient of adiabatic compression is

$$\gamma_S = -1/V\,(\partial V/\partial P)_S\,. \tag{2.86}$$

The reverse coefficients of compression

$$\gamma_T^{-1} = B_T = -V\,(\partial P/\partial V)_T \text{ and } \gamma_S^{-1} = B_S = -V\,(\partial P/\partial V)_S\,,$$

are respectively called the isothermal and the adiabatic bulk modulus or the statistical modulus.

The physical meaning of the above-indicated coefficients is clear from their definitions. Each of them can be measured by experiments and characterises a specified property of the system.

Thermodynamic coefficients, as well as macroscopic parameters, can be calculated on the basis of Gibbs thermodynamic potentials. According to relationships (2.60) and (2.67), heat capacities C_V and C_P are expressed by the free energy and Gibbs potential as follows:

$$C_V = -T\left(\partial^2 F/\partial T^2\right)_V; \quad C_P = -T\left(\partial^2 \Phi/\partial T^2\right)_P. \tag{2.87}$$

The isobaric coefficient of thermal expansion is, according to (2.68) and (2.81)

$$\alpha_P = 1/V\left(\partial^2 \Phi/\partial T \partial P\right); \tag{2.88}$$

The isochoric thermal coefficient of pressure is, according to (2.61) and (2.83)

$$\beta_V = 1/P\left(\partial^2 F/\partial T \partial V\right). \tag{2.89}$$

The coefficient of isothermal compression is, according (2.67) and (2.85)

$$\gamma_T = -1/V\left(\partial^2 \Phi/\partial P^2\right)_T. \tag{2.90}$$

If the explicit forms of functions $F = F(V, T)$ and $\Phi = \Phi(P, T)$ are known, we can calculate the above-introduced thermodynamic coefficients and compare the obtained results with experimental facts. The finding of the explicit form of functions $F = F(V, T)$ and $\Phi = \Phi(P, T)$ is the task of statistical physics. Later (see Chaps. 5–7) we will find the explicit forms of these functions for different systems. Here, we will determine general relationships between different thermodynamic coefficients independently of the explicit forms of functions F and Φ.

To do this, we introduce every possible derivative of thermodynamic quantities that determine different thermodynamic coefficients. It is known that the state of any macroscopic system is determined by the four thermodynamic parameters S, V, P, T. We isolate, in turn, one of them and, having the three others constant, write every possible derivative of thermodynamic parameters.

$$\textcircled{S},\quad V,\ P\ T,\quad \left(\frac{\partial V}{\partial P}\right)_T,\quad \left(\frac{\partial P}{\partial T}\right)_V,\quad \left(\frac{\partial T}{\partial V}\right)_P, \tag{2.91}$$

$$S,\quad \textcircled{V},\ P,\ T\quad \boxed{\left(\frac{\partial S}{\partial P}\right)_T,}\quad \left(\frac{\partial P}{\partial T}\right)_S,\quad \left(\frac{\partial T}{\partial S}\right)_P, \tag{2.92}$$

$$S,\quad V,\ \textcircled{P},\ T\quad \boxed{\left(\frac{\partial S}{\partial V}\right)_T,}\quad \left(\frac{\partial V}{\partial T}\right)_S,\quad \left(\frac{\partial T}{\partial S}\right)_V, \tag{2.93}$$

$$S,\quad V,\ P,\ \textcircled{T}\quad \boxed{\left(\frac{\partial S}{\partial V}\right)_P,}\quad \left(\frac{\partial V}{\partial P}\right)_S,\quad \boxed{\left(\frac{\partial P}{\partial S}\right)_V.} \tag{2.94}$$

Note that a derivative in each line is obtained from the preceding one by clockwise rotation.

The above-presented tableau contains 12 partial derivatives of thermodynamic parameters. Only eight of them, according to relationships (2.79)–(2.86), determine thermodynamic coefficients that possess a physical sense and can be measured by experiment. The remaining four partial derivatives (enclosed in the dashed frame), according to thermodynamic relationship (2.78), are related to one of the eight above-indicated thermodynamic coefficients.

In order to find the general relationships between thermodynamic coefficients (2.79)–(2.86), i.e. between partial derivatives (2.91)–(2.94), we make use of the method of the Jacobian.

Having multiplied derivatives of each line (2.91)–(2.94) and using properties of the Jacobian (see Appendix D), we get the following four thermodynamic relationships:

$$\left(\frac{\partial V}{\partial P}\right)_T \cdot \left(\frac{\partial P}{\partial T}\right)_V \cdot \left(\frac{\partial T}{\partial V}\right)_P = -1, \tag{2.95}$$

$$\left(\frac{\partial S}{\partial P}\right)_T \cdot \left(\frac{\partial P}{\partial T}\right)_S \cdot \left(\frac{\partial T}{\partial S}\right)_P = -1, \tag{2.96}$$

$$\left(\frac{\partial S}{\partial V}\right)_T \cdot \left(\frac{\partial V}{\partial T}\right)_S \cdot \left(\frac{\partial T}{\partial S}\right)_V = -1, \tag{2.97}$$

$$\left(\frac{\partial S}{\partial V}\right)_P \cdot \left(\frac{\partial V}{\partial P}\right)_S \cdot \left(\frac{\partial P}{\partial S}\right)_V = -1. \tag{2.98}$$

Taking into account properties of the Jacobian (see Appendix D), the thermodynamic relationships (2.43), (2.52), (2.62) and (2.69) can be presented in the form

$$\left(\frac{\partial T}{\partial V}\right)_S = -\left(\frac{\partial P}{\partial S}\right)_V \text{ or } \frac{\partial(T,S)}{\partial(V,S)} \cdot \frac{\partial(V,S)}{\partial(P,V)} = 1, \quad (2.99)$$

$$\left(\frac{\partial T}{\partial P}\right)_S = \left(\frac{\partial V}{\partial S}\right)_P \text{ or } \frac{\partial(T,S)}{\partial(P,S)} \cdot \frac{\partial(P,S)}{\partial(P,V)} = 1, \quad (2.100)$$

$$\left(\frac{\partial P}{\partial T}\right)_V = \left(\frac{\partial S}{\partial V}\right)_T \text{ or } \frac{\partial(P,V)}{\partial(T,V)} \cdot \frac{\partial(T,V)}{\partial(T,S)} = 1, \quad (2.101)$$

$$\left(\frac{\partial S}{\partial P}\right)_T = -\left(\frac{\partial V}{\partial T}\right)_P \text{ or } \frac{\partial(T,S)}{\partial(P,T)} \cdot \frac{\partial(P,T)}{\partial(P,V)} = 1. \quad (2.102)$$

The above-indicated equalities derive from the fact that each thermodynamic function is a function of state, i.e. the change in each of them is a total differential.

Having made use of the property of the Jacobian (see Appendix D), each of the conditions (2.99)–(2.102) can be presented in the general form

$$\frac{\partial(T,S)}{\partial(P,V)} = 1. \quad (2.103)$$

Thus, (2.103) is the general condition of the fact that all thermodynamic functions are functions of state.

Taking into account relationships (2.99)–(2.102) in (2.95)–(2.98), we get

$$\left(\frac{\partial V}{\partial P}\right)_T \cdot \left(\frac{\partial P}{\partial T}\right)_V \cdot \left(\frac{\partial T}{\partial V}\right)_P = -1, \quad (2.104)$$

$$\left(\frac{\partial V}{\partial T}\right)_P \cdot \left(\frac{\partial P}{\partial T}\right)_S \cdot \left(\frac{\partial T}{\partial S}\right)_P = 1, \quad (2.105)$$

$$\left(\frac{\partial P}{\partial T}\right)_V \cdot \left(\frac{\partial V}{\partial T}\right)_S \cdot \left(\frac{\partial T}{\partial S}\right)_V = -1, \quad (2.106)$$

$$\left(\frac{\partial P}{\partial T}\right)_S \cdot \left(\frac{\partial V}{\partial P}\right)_S \cdot \left(\frac{\partial T}{\partial V}\right)_S = 1. \quad (2.107)$$

Each of the derivates entering into these equations determines one of the eight thermodynamic coefficients being measured. Therefore, taking into account definitions (2.79)–(2.86), from (2.104) to (2.107) the general relationships between thermodynamic coefficients can be found:

$$\frac{\alpha_P}{\beta_V \gamma_T} = P, \quad (2.108)$$

$$\frac{C_P \beta_S}{\alpha_P} = V, \quad (2.109)$$

$$\frac{C_V \alpha_S}{\beta_V} = P, \quad (2.110)$$

$$\frac{\beta_S}{\alpha_s \gamma_s} = V. \tag{2.111}$$

From (2.108) and (2.110) and also from (2.109) and (2.111) we get general relationships containing only thermodynamic coefficients:

$$\frac{\alpha_P}{C_V \gamma_T \alpha_S} = 1; \quad \frac{\alpha_P}{C_P \gamma_S \alpha_S} = 1. \tag{2.112}$$

Using the relationships (2.108)–(2.112), one isobaric α_P and three adiabatic $\alpha_S, \beta_S, \gamma_S$ coefficients can be expressed by the remaining β_V, γ_T, C_V and C_P. As a result, we get

$$\alpha_P = P\beta_V \gamma_T; \quad \beta_S = V\frac{\alpha_P}{C_P} = VP\frac{\beta_V \gamma_T}{C_P}; \tag{2.113}$$

$$\alpha_S = P\frac{\beta_V}{C_V}; \quad \gamma_S = \frac{C_V}{C_P}\gamma_T. \tag{2.114}$$

In order to find the relation between isobaric and isochoric heat capacities, we make use of the property of the Jacobian (Appendix D):

$$C_P = T\left(\frac{\partial S}{\partial T}\right)_P = T\frac{\partial(S,P)}{\partial(T,P)} = T\frac{\partial(S,P)}{\partial(T,V)}\frac{\partial(T,V)}{\partial(T,P)} = T\left(\frac{\partial V}{\partial P}\right)_T \frac{\partial(S,P)}{\partial(T,V)}. \tag{2.115}$$

If in (2.115) we take into account the definition of the Jacobian (see Appendix D)

$$\frac{\partial(S,P)}{\partial(T,V)} = \left(\frac{\partial S}{\partial T}\right)_V \left(\frac{\partial P}{\partial V}\right)_T - \left(\frac{\partial S}{\partial V}\right)_T \left(\frac{\partial P}{\partial T}\right)_V \tag{2.116}$$

and use (2.101), we get

$$C_P - C_V = -T\left(\frac{\partial V}{\partial P}\right)_T \left(\frac{\partial P}{\partial T}\right)_V^2 = TVP^2\gamma_T\beta_V^2. \tag{2.117}$$

Having rewritten relationship (2.104) in the form

$$\left(\frac{\partial V}{\partial P}\right)_T \left(\frac{\partial P}{\partial T}\right)_V = -\left(\frac{\partial V}{\partial T}\right)_P, \tag{2.118}$$

and substituted it into (2.117), we get

$$C_P - C_V = T\left(\frac{\partial V}{\partial T}\right)_P \left(\frac{\partial P}{\partial T}\right)_V = TVP\alpha_P\beta_V. \tag{2.119}$$

Rewrite relationship (2.104) in the form

$$\left(\frac{\partial P}{\partial T}\right)_V = -\left(\frac{\partial V}{\partial T}\right)_P \left(\frac{\partial P}{\partial V}\right)_T. \tag{2.120}$$

Take into account (2.117), and for the difference $C_P - C_V$ we get one more expression

$$C_P - C_V = -T\left(\frac{\partial P}{\partial V}\right)_T \left(\frac{\partial V}{\partial T}\right)_P^2 = \frac{TV\alpha_P^2}{\gamma_T}. \tag{2.121}$$

Note that expressions (2.117), (2.119) and (2.121) for the difference $C_P - C_V$, naturally, are equivalent.

In conclusion, we briefly analyse the obtained general results.

1. According to the thermodynamic inequality $(\partial V/\partial P)_T < 0$ (see Sect. 2.5), from (2.117) it follows that $C_P > C_V$. This result is explained by the fact that at isobaric conditions ($P=\text{const}$) with the raise of temperature by one degree volume ought to increase in order that the pressure remains constant. The amount of heat being received is expended on the raising of the temperature and also on performing work to increase volume.
2. From the inequality $C_V < C_P$, according to (2.114), it follows that $\gamma_S < \gamma_T$, i.e. at adiabatic conditions ($S=\text{const}$) a system is compressed with more difficulty than at isothermal ($T=\text{const}$) conditions, since in adiabatic compression the system is heated and, naturally, is compressed with more difficulty.
   Then the inequality for the adiabatic and the isothermal bulk modulus can be written as
   $$\mathrm{B}_S = \frac{C_P}{C_V}\mathrm{B}_T; \quad \mathrm{B}_S > \mathrm{B}_T. \tag{2.122}$$
3. The relationships (2.108)–(2.112) between thermodynamic coefficients do not depend on the explicit form of thermodynamic functions and therefore bear the general character. Knowing one or several of them, the rest can be found.
4. From (2.113) and (2.114), it is seen that coefficients α_S and β_S, characterizing the change in temperature at adiabatic conditions, are inversely proportional to heat capacities C_V and C_P, which is physically clear.
5. Finally, note the most important obtained result. From relationships (2.113), (2.114) and (2.117), it follows that *knowing C_V and the thermal equation of state $P=P(V,T)$, all thermodynamic coefficients of a closed system can be calculated*, because in order to find coefficients β_V, γ_T and the difference $C_P - C_V$, it is sufficient to know the thermal equation of state $P=P(V,T)$.

In order to calculate the isochoric heat capacity $C_V = (\partial E/\partial T)_V$, it is necessary to know the caloric equation of state $E=E(V,T)$. And the dependence of the internal energy on temperature $E=E(T,V)$ is determined by the structure as well as by the nature of the motion and interaction of particles forming the system. *For this reason, the heat capacity C_V occupies a special place among all thermodynamic coefficients.*

Thus, we come to the most important conclusion: *In order to calculate all thermodynamic coefficients measured on test, it is sufficient to know only the explicit form of thermal and caloric equations of states* $P = P(T, V)$ *and* $E = E(T, V)$, *respectively.*

In order to find these equations, as it follows from (2.54) and (2.68), it is necessary to know the explicit form of only the function of the free energy $F = F(V, T)$

$$P = -\left(\frac{\partial F}{\partial V}\right)_T; \quad E = F - T\left(\frac{\partial F}{\partial T}\right)_V. \tag{2.123}$$

General conclusion: *In order to construct the theoretical thermodynamics of a closed system, i.e. to find thermal and caloric equations of state, and also to calculate all thermodynamic coefficients, it is sufficient to know only one function – the free energy* $F = F(V, T)$.

Finding the explicit form of this function is the basic task of statistical physics. This question in statistical physics is solved on the basis of the Gibbs method (see Sect. 4.3).

2.5 Thermodynamic Inequalities: Stability of Equilibrium State of Homogeneous Systems

The condition of homogeneous closed macroscopic systems found in thermodynamic equilibrium was considered in Sect. 1.8 on the basis of the law of increasing entropy. From the necessary condition of maximum entropy $S = S_{\max}$, it follows that temperature and pressure ought to be identical at all points of the system: $T = \text{const}$, $P = \text{const}$.

The question arises as to what conditions thermodynamic parameters and their derivatives ought to satisfy in order that the equilibrium state be steady, i.e. so that continually arising small fluctuations cannot bring the system out of the equilibrium state. In order to answer the posed question, we make use of the condition of the minimum of the Gibbs potential $\Phi(T, P) \Rightarrow \Phi_{\min}$ (see Sect. 2.3) in the equilibrium state.

According to the property of the thermodynamic potential, in the steady state at pre-assigned values of independent variables T and P, the Gibbs potential $\Phi(T = \text{const}, P = \text{const})$ in reference to other variables ought to be minimum.

Consider a system in a thermostat. Let temperature and pressure of the thermostat be T_0 and P_0. Inasmuch as the system is found in equilibrium with the thermostat, its temperature and pressure ought to be the same: $T = T_0, P = P_0$. As the other variables of the system we take entropy S and volume V, and, naturally, its energy depends on these variables: $E = E(V, S)$. Then, the Gibbs thermodynamic potential can be written down in the following form:

$$\Phi = E(V, S) + P_0V - T_0S. \tag{2.124}$$

As can be seen, Φ depends on variables V and S at constant temperature and pressure: $\Phi_{P_0,T_0} = \Phi(V,S)$. Consider the condition of the minimum of a function $\Phi = \Phi(V,S)$ of two variables.

The condition of the extremum of the function Φ is obtained by equating the first partial derivative with respect to variables V and S to zero:

$$\left(\frac{\partial \Phi}{\partial S}\right)_V = \left(\frac{\partial E}{\partial S}\right)_V - T_0 = 0, \tag{2.125}$$

$$\left(\frac{\partial \Phi}{\partial V}\right)_S = \left(\frac{\partial E}{\partial V}\right)_S + P_0 = 0. \tag{2.126}$$

If we take into account that $(\partial E/\partial S)_V = T$ and $(\partial E/\partial V)_S = -P$, from (2.125) and (2.126) we get the known condition of the equilibrium: $T = T_0$ and $P = P_0$.

In order that an extremum of a function of two variables would be the minimum, it is necessary that its second partial derivative with respect to any of the variables (for instance, S)

$$\left(\frac{\partial^2 \Phi}{\partial S^2}\right)_V = \left(\frac{\partial^2 E}{\partial S^2}\right)_V > 0, \tag{2.127}$$

and also the determinant, composed of second derivatives

$$\begin{vmatrix} \left(\frac{\partial^2 E}{\partial S^2}\right)_V & \frac{\partial^2 E}{\partial V \partial S} \\ \frac{\partial^2 E}{\partial S \partial V} & \left(\frac{\partial^2 E}{\partial V^2}\right)_S \end{vmatrix} > 0 \tag{2.128}$$

would be positive.

Inequalities (2.127) and (2.128) are sufficient conditions of a minimum of the Gibbs thermodynamic potential Φ.

We rewrite condition (2.127) in the form

$$\left(\frac{\partial^2 E}{\partial S^2}\right)_V = \frac{\partial}{\partial S}\left(\frac{\partial E}{\partial S}\right)_V = \left(\frac{\partial T}{\partial S}\right)_V = \frac{T}{C_V} > 0. \tag{2.129}$$

Inasmuch as $T > 0$, we get the first thermodynamic inequality

$$C_V > 0, \tag{2.130}$$

determining stability of the equilibrium state of a system. Since $C_P > C_V$ from (2.130), it follows that $C_P > 0$.

Now the second condition of the minimum of the function Φ (2.128) can be presented in the following form:

$$\begin{vmatrix} \frac{\partial}{\partial S}\left(\frac{\partial E}{\partial S}\right)_V & \frac{\partial}{\partial V}\left(\frac{\partial E}{\partial S}\right)_V \\ \frac{\partial}{\partial S}\left(\frac{\partial E}{\partial V}\right)_S & \frac{\partial}{\partial V}\left(\frac{\partial E}{\partial V}\right)_S \end{vmatrix} = -\begin{vmatrix} \left(\frac{\partial T}{\partial S}\right)_V & \left(\frac{\partial T}{\partial V}\right)_S \\ \left(\frac{\partial P}{\partial S}\right)_V & \left(\frac{\partial P}{\partial V}\right)_S \end{vmatrix} > 0. \tag{2.131}$$

Having rewritten determinant (2.131) in the form of the Jacobian, for condition (2.128) we get

$$\frac{\partial(T,P)}{\partial(S,V)} < 0. \tag{2.132}$$

With regard to properties of the Jacobian (see Appendix D), inequality (2.132) takes the form

$$\frac{\partial(T,P)}{\partial(S,V)} = \frac{\partial(T,P)}{\partial(T,V)}\frac{\partial(T,V)}{\partial(S,V)} = \left(\frac{\partial P}{\partial V}\right)_T \frac{T}{C_V} < 0. \tag{2.133}$$

Since $C_V > 0$, from (2.133) follows the second thermodynamic inequality

$$\left(\frac{\partial P}{\partial V}\right)_T < 0. \tag{2.134}$$

Conditions (2.130) and (2.134) determining the stability of the equilibrium state of homogeneous closed systems are called *thermodynamic inequalities*.

We show that as conditions (2.130) and (2.134) are fulfilled, the equilibrium state of a homogeneous system is steady, i.e. the natural fluctuations arising in the course of time fade, and as conditions inverse to (2.130) and (2.134) are fulfilled, natural fluctuations that arise are strengthened, bringing the system out of the equilibrium state.

Consider each of the inequalities separately.

1. If the condition of stability (2.130) is fulfilled, the fluctuations arising fade and the system remains in the state of equilibrium. Separate a small region in a homogeneous system. If as a result of the fluctuation, energy grows by the magnitude ΔE, inasmuch as $C_V > 0$, the local temperature of this region also grows by the magnitude ΔT. As a result, energy from the "hot" region transfers into the cold surroundings, the fluctuation fades and the system remains in the equilibrium state. On the contrary, if as a result of the fluctuation in the small region, energy decreases by the magnitude ΔE, inasmuch as $C_V > 0$, the local temperature also decreases by the magnitude ΔT, and energy from the surroundings transfers into the "cold" small region and the temperature over the entire system equalises. As a result, the equilibrium is regained.

On the contrary, if $C_V < 0$, at a fluctuation in energy $\Delta E > 0$, the local temperature in the small region would decrease by the magnitude $\Delta T < 0$. As a result, energy from the surroundings transfers into this "cold" region and the fluctuation in energy strengthens. Thus, small fluctuations bring the system out of equilibrium. In the case of a decrease in energy, at the expense of the fluctuation $\Delta E < 0$, temperature grows by $\Delta T > 0$, since it is supposed that $C_V < 0$. Then energy from the considered high-temperature region transfers into the surrounding region and, thereby, fluctuations in the side of the decrease in energy strengthen, and the temperature rises. Thus, an unsteady state arises.

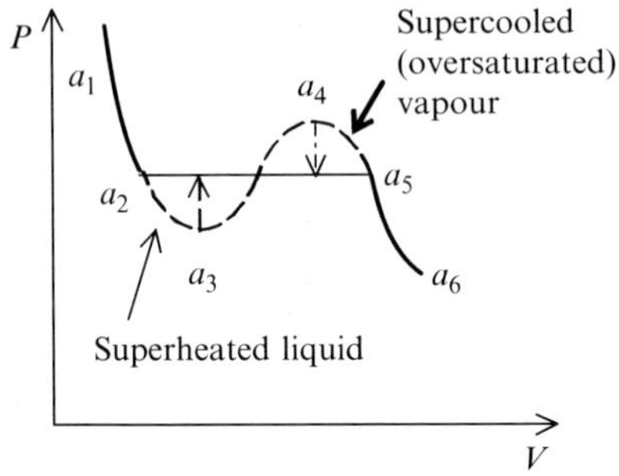

Fig. 2.12. The van der Waals isotherm for a temperature below the critical temperature

2. From the second condition of stability (2.134), it follows that at the isothermal and the adiabatic compression of a homogeneous system [see (2.114)] found in the steady state, its pressure ought to increase and, conversely, at expansion of the system, the pressure ought to decrease.

With the example of the van der Waals gas, we show that this condition actually provides stability of the thermodynamic state in reference to small fluctuations. In Fig. 2.12 the van der Waals isotherm for a temperature below the critical temperature is shown. Here, the solid curve $a_1a_2a_5a_6$ is the real isotherm, the portion a_6a_5 corresponds to the gas, a_2a_1 is the liquid phase, and the horizontal straight line a_2a_5 is the two-phase state (liquid + gas) of the system.

The portions of the isotherm a_2a_1 and a_5a_6 describe the homogeneous state, and condition (2.134) is fulfilled. If in a small portion of these homogeneous states (liquid or gas), as a result of weak fluctuations, the density increases (volume decreases), the local pressure grows, inasmuch as $(\partial P/\partial V)_T < 0$, and the volume increases, and thereby the fluctuation in the density fades. On the contrary, if the local density decreases (volume increase), pressure in a small region drops, and therefore the surroundings compress the considered small region (volume decreases). Thereby, the fluctuation in the density fades, and the system conserves its stability.

In the portions a_2a_3 and a_5a_4 of the theoretical van der Waals isotherm, condition (2.134) is fulfilled, but states corresponding to these portions are metastable and as a result of a minor external coercion the homogeneity is violated, and the points a_3 and a_4 turn out to be in the horizontal portion of the isotherm–isobar a_2a_5 and the system becomes two-phased (liquid + gas).

In the portion a_3a_4 of the isotherm (Fig. 2.12), condition (2.134) is violated and the inverse inequality $(\partial P/\partial V)_T > 0$ takes place. Therefore, states corresponding to this portion are unsteady. If such states even arise, as a result of strengthening of fluctuations these states instantaneously disappear and the system passes into the two-phase state (liquid + gas), and the homogeneity is violated. Indeed, if in states of the system corresponding to the portion a_3a_4, the substance is compressed as a result of fluctuations in the local region (the local specific volume decreases), pressure ought to also decrease inasmuch as $(\partial P/\partial V)_T > 0$. Then, the external pressure compresses this region till the appearance of a drop of liquid, i.e. till the system does not pass into the two-phase state (liquid + gas). On the contrary, if in states of the system corresponding to the same portion a_3a_4 the density decreases as a result of the fluctuations (the local specific volume increases), pressure in the local region also increases inasmuch as $(\partial P/\partial V)_T > 0$ and the expansion continues as long as the system does not pass into the normal state, in order that the condition $(\partial P/\partial V)_T < 0$ be fulfilled.

Thus, states of a homogeneous system satisfying the condition $(\partial P/\partial V)_T > 0$ are always absolutely unsteady in reference to small fluctuations in the density or decay into two phases.

Note some general conclusions following from thermodynamic inequalities.

From the thermodynamic inequality $C_P > C_V > 0$, and also different definitions of the heat capacity

$$\left[C_V = \left(\frac{\partial E}{\partial T}\right)_V, \ C_P = \left(\frac{\partial W}{\partial T}\right)_P, \ C_{V,P} = T\left(\frac{\partial S}{\partial T}\right)_{V,P}\right],$$

it follows that *the internal energy E at $V =$* const, *enthalpy W at $P =$* const, *and also entropy at constant V or P are monotonic functions of temperature.*

From the second thermodynamic inequality (2.134) [with regard to relationships (2.85), (2.86) and (2.114)], it follows that *in homogeneous systems at both isothermal and adiabatic processes volume $V = V(P)$ is a monotonically decreasing function of pressure, and, conversely, pressure P is a monotonically decreasing function of volume.* The portions a_1a_2 (liquid) and a_5a_6 (vapour) of the van der Waals isotherm, shown in Fig. 2.12, can serve as examples.

In conclusion, consider the relation of the first thermodynamic inequality $C_V > 0$ to the known Le Chatelier–Brown principle.

The Le Chatelier–Brown principle holds: If external factors coerce a system that is in the equilibrium state, tending to bring it out of this state, there arise processes in the system tending to decrease this coercion, i.e. the system, using all its possibilities, endeavors to conserve its equilibrium state. This principle was formulated in 1884 by Le Chatelier and substantiated in 1887 by Brown.

Using a simple example, we show that the inequality $C_V > 0$ corresponds to the Le Chatelier–Brown principle. Suppose that the relatively small system 1 is found in equilibrium with the thermostat 2 with temperature T (Fig. 2.13). Assume that the amount of heat ΔQ (the coercion) communicates to the system 1 externally. Inasmuch as $C_V > 0$, the temperature of the system 1 increases by the magnitude ΔT and the equilibrium with the thermostat is violated. In order to decrease this coercion, a certain amount of heat $\Delta Q'$ transfers from the system 1 into the thermostat 2. Therefore, the increase in ΔT weakens.

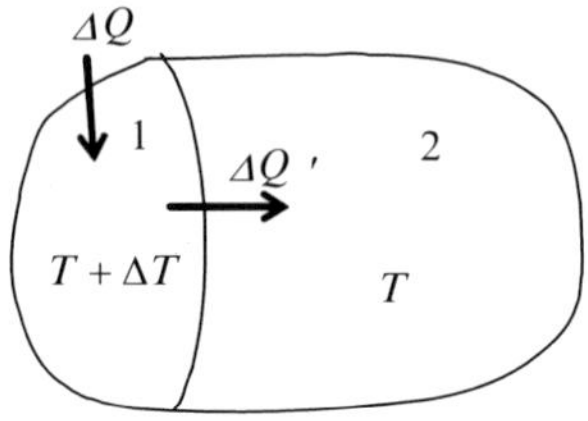

Fig. 2.13. The illustration of the Le Chatelier-Brown principle in the case of the immediate external coercion

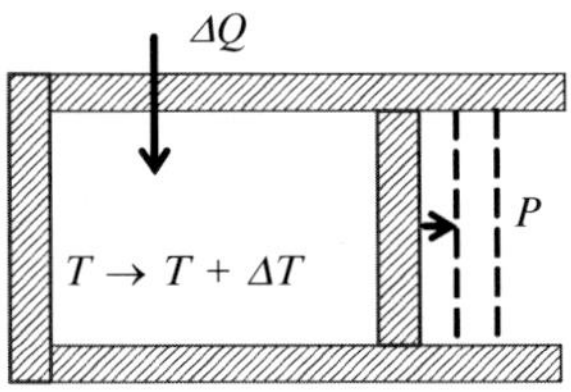

Fig. 2.14. The illustration of the Le Chatelier-Brown principle in the case of the indirect external coercion

In some cases, a system weakens an external coercion not immediately, but indirectly. We show that the inequality $C_P > C_V$ corresponds to this case. Imagine that in a cylinder under a piston there is a gas (Fig. 2.14). It is assumed that the piston is movable and its equilibrium position is determined by the external pressure P and pressure of the gas. We supply to the system externally an amount of heat ΔQ. As a result of this coercion, entropy and temperature of the system increase by magnitudes $\Delta S = \Delta Q/T$ and ΔT, respectively.

The inequality $C_P > C_V$ can be rewritten in the form $(\Delta S/\Delta T)_P > (\Delta S/\Delta T)_V$, and we take into account that the increase in entropy in both the cases is identical, i.e. $\Delta S = \Delta Q/T$. Then we get

$$(\Delta T)_V > (\Delta T)_P. \tag{2.135}$$

Thus, when supplying the amount of heat ΔQ, the increase in temperature in the case of $P = const$ is small, i.e. the gas expanding (the position of the piston is shown by the dashed line) is heated a little, and thereby weakens the external coercion (the increase in temperature as a result of ΔQ), because a certain part of the heat being supplied is expended on performing work in the expansion of the gas.

2.6 Third Law of Thermodynamics: Nernst Principle

Thermodynamics as a phenomenological theory is based on four laws, stemming from generalisation of experimental facts.

The zeroth law is a law about temperature and the equilibrium;
The first law is a law about the internal energy and its conservation;
The second law is a law about entropy and its increase;
The third law does not introduce a new concept in the theory but is a law about the behaviour of entropy of a system at low temperatures.

In thermodynamics, in solving some questions a large meaning is given to the behaviour of entropy at low temperatures, especially at the temperature of absolute zero. For instance, to determine some chemical constants it is necessary to know the value of entropy at absolute zero $S(0)$.

From the thermodynamic inequality $C_P > C_V > 0$, it follows that $(\partial S/\partial T)_{V,P} = C_{V,P}/T > 0$. Thus, in all cases entropy is a monotonic function of temperature: as temperature tends to absolute zero, entropy decreases. The question arises as to what is the value of entropy at temperature absolute zero: $S(0) = ?$ The answer to this question is given by *the third law of thermodynamics – the Nernst principle.*

Having generalised numerous experimental data, the German chemist–physicist Nernst in 1906 formulated the following principle: *As temperature tends to absolute zero, entropy of a system, independently of values of external parameters determining its state, tends to a certain constant value. For definiteness, this constant is regarded as equal to zero.*

In order to write the mathematical expression of this principle, denote by x_i a multitude of external parameters characterizing the ith state of a system (volume, pressure, a magnetic field, and so on), and by $S(x_i, T)$ entropy of the ith state of the system. According to the Nernst principle

$$S(x_1, 0) = S(x_2, 0) = \ldots = S(x_n, 0) = \text{const} = 0, \tag{2.136}$$

where $x_1, x_2, \ldots, x_n$ are values of external parameters in different states $1, 2, \ldots, n$.

From (2.136), it follows that the differences of entropies of two arbitrary states as $T \to 0$ ought also to tend to zero:

$$\lim_{T\to 0} \Delta S = \lim_{T\to 0} [S(x_i, T) - S(x_k, T)] = 0, \tag{2.137}$$

i.e. from the mathematical form of the Nernst principle it follows that at absolute zero at different values of external parameters, entropy of states of the system and also the difference of entropies ought to equal zero. This result is schematically depicted in Fig. 2.15.

Note that in classical physics it is not possible to substantiate the Nernst principle. This principle in some way can be substantiated only in quantum mechanics, taking into account the discreteness of energy levels. To do this, subdivide the considered system into small but macroscopic parts (subsystems). If we take into account the discreteness of the energy spectrum, each subsystem at $T = 0$ is found at its lowest energy level (quantum state). Thus, the statistical weight of each subsystem is $\Delta G_\alpha = 1$. This means that only one microstate corresponds to the macroscopic state of the subsystem at $T = 0$.

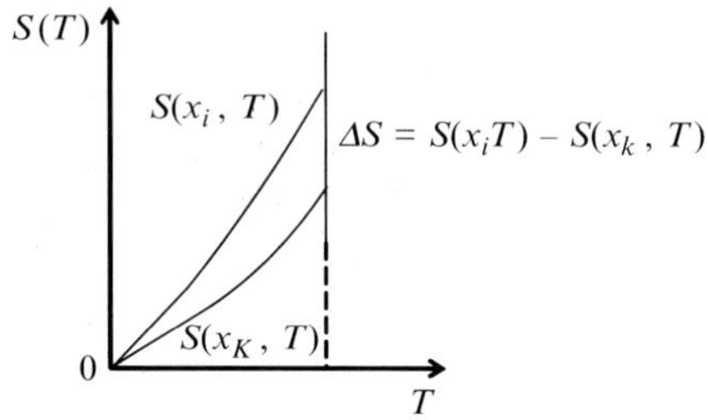

Fig. 2.15. The temperature dependence of the entropy at different values of external parameters

Then the statistical weight of the system as a whole with regard to (1.108) ought to be equal to unity:

$$\Delta G = \prod_{\alpha} \Delta G_{\alpha} = 1. \tag{2.138}$$

Taking into account the definition of entropy (1.109), we get

$$\lim_{T\to 0} S = k_0 \lim_{T\to 0} \left[\ln \prod_{\alpha} \Delta G_{\alpha} \right] = k_0 \lim_{T\to 0} [\ln 1] = 0, \tag{2.139}$$

i.e. *the Nernst principle.*

Note some conclusions following from the Nernst principle.

1. Heat capacities (C_V and C_P) of all systems at the absolute zero are equal to zero:

$$\lim_{T\to 0} C_V(T) = 0; \quad \lim_{T\to 0} C_P(T) = 0. \tag{2.140}$$

This result can be obtained in two ways:

– Firstly, from the definition of the heat capacity in the general form:

$$S_{V,P}(T) = \int_0^T \frac{C_{V,P}(T')}{T'} \mathrm{d}T'. \tag{2.141}$$

According to the Nernst principle, the numerical value of the lower boundary of integral (2.141) ought to be equal to zero, and to do this it is necessary to fulfil the equality $C_{V,P}(0) = 0$;

– Secondly, in a particular case assume that entropy depends on temperature T as follows:

$$S(P, V, T) = A(P, V) T^n, \tag{2.142}$$

where $A(P, V)$ is any function of volume and pressure, and $n > 0$ is a positive real number. Then according to the definition of the heat capacity

$$C_{V,P} = T\left(\frac{\partial S}{\partial T}\right)_{V,P} = nA(P,V)T^n \sim T^n, \tag{2.143}$$

i.e. as $T \to 0$ heat capacities tend to zero by the same law as entropy.

2. As $T \to 0$, the difference $C_P - C_V$ tends to zero faster than heat capacities themselves. Indeed, taking into account relationships (2.101) and (2.102) in (2.119) and using (2.142), we get

$$C_P - C_V = -T\left(\frac{\partial S}{\partial V}\right)_T\left(\frac{\partial S}{\partial P}\right)_T = -\left(\frac{\partial A}{\partial V}\right)\left(\frac{\partial A}{\partial P}\right)T^{2n+1}. \tag{2.144}$$

From (2.143) and (2.144) it follows that

$$\frac{C_P - C_V}{C_{V,P}} = -\frac{1}{nA}\left(\frac{\partial A}{\partial V}\right)\left(\frac{\partial A}{\partial P}\right)T^{n+1} \sim T^{n+1}. \tag{2.145}$$

As can be seen, the difference $(C_P - C_V)$ tends to zero faster than $C_{V,P}$.
3. As $T \to 0$, coefficients of thermal expansion α_P and the thermal coefficient of pressure β_V also tend to zero. Indeed, using definitions (2.81) and (2.83), relationship (2.78), and mathematical form of the Nernst principle (2.137), we get

$$\lim_{T\to 0}\alpha_P = \frac{1}{V}\lim_{T\to 0}\left(\frac{\partial V}{\partial T}\right)_P = -\frac{1}{V}\lim_{T\to 0}\left(\frac{\partial S}{\partial P}\right)_T = 0, \tag{2.146}$$

$$\lim_{T\to 0}\beta_V = \frac{1}{P}\lim_{T\to 0}\left(\frac{\partial P}{\partial T}\right)_V = \frac{1}{P}\lim_{T\to 0}\left(\frac{\partial S}{\partial V}\right)_T = 0. \tag{2.147}$$

Here when using principle (2.137), it was taken into account that in our case $x = V$ and $x = P$.

4. The isotherm and adiabat corresponding to the temperature of absolute zero coincide. Indeed, inasmuch as at the absolute zero $\gamma = C_P/C_V = 1$, the equation of the adiabat $PV^\gamma = const$ coincides with the equation of the isotherm $PV = \text{const}$.
5. *It is not possible to produce the absolute zero of temperature; we can only asymptotically approach it.*

Note that this conclusion is one of the definitions of the Third Law of Thermodynamics, i.e. it is equivalent to it.

The impossibility of producing the absolute zero of temperature can be substantiated in different ways:

(a) It is known that cooling of the system occurs as a result of repetition of two consecutive processes: adiabatic expansion (temperature drops) and isothermal compression (entropy decreases).According to the Third Law of Thermodynamics, as $T \to 0$, entropy does not depend on compression (on the external pressure) and the isotherm coincides with the adiabat; therefore it is impossible to produce the state with $S=0$ as a result of the finite number of the above-indicated processes. Then it turns out that the state with $T=0$ cannot be produced either, since according to the third law, the state with $T=0$ and the state with $S=0$ are the same.

(b) According to the Nernst principle, as $T \to 0$, heat capacities C_P and C_V also tend to zero. Therefore, close to temperature $T=0$, any fluctuation in energy raises temperature, and thus it is impossible to produce the state with $T=0$.

(c) Imagine the reverse, i.e. assume that it is possible to produce a system with $T=0$. Use this system as a cooler in the Carnot cycle. Let the temperature of a heater be $T_1 \neq 0$, and that of the cooler be $T_2=0$. In this imaginary reversible circular cycle, the working body in the first isothermal process $1 \to 2$ (see Fig. 2.7) receives from the heater the amount of heat ΔQ_1 and entropy increases by the magnitude

$$\Delta S_{12} = \frac{\Delta Q_1}{T_1}. \tag{2.148}$$

Inasmuch as processes $2 \to 3$ and $4 \to 1$ are adiabatic, $\Delta S_{23} = \Delta S_{41} = 0$. The second isothermal compression $3 \to 4$ proceeds at $T_2 = 0$ and, according to the third law, ought to be isoenthropic, i.e. adiabatic, such that $\Delta S_{34} = 0$. Then the total change in entropy in the circular Carnot cycle equals

$$\Delta S = \Delta S_{12} + \Delta S_{23} + \Delta S_{34} + \Delta S_{41} = \frac{\Delta Q_1}{T_1}. \tag{2.149}$$

It is known that in the circular process entropy ought not to change, i.e.

$$\Delta S = \oint \mathrm{d}S = 0. \tag{2.150}$$

Inconformity of (2.149) and (2.150) (since $\Delta Q_1 \neq 0$) shows that our supposition is wrong, i.e. it is impossible to produce a system with the temperature of absolute zero.

In conclusion, note that if the dependence of the heat capacity over a wide temperature range is known, then, taking into account the Nernst principle ($S(0)=0$), the absolute value of entropy and other thermodynamic functions can be calculated.

For instance, the dependence of entropy on temperature at pre-assigned volume and pressure has the appearance

$$S_{V,P}(T) = \int_0^T \frac{C_{V,P}(T')}{T'} \mathrm{d}T'. \tag{2.151}$$

For other thermodynamic functions we have:
Internal energy

$$E(T) = E(0) + \int_0^T C_V(T')\mathrm{d}T', \tag{2.152}$$

Enthalpy

$$W(T) = W(0) + \int_0^T C_P(T')\mathrm{d}T', \tag{2.153}$$

Free energy

$$F(T) = E(0) + \int_0^T C_V(T')\mathrm{d}T - T\int_0^T \frac{C_V(T')}{T'}\mathrm{d}T', \tag{2.154}$$

Thermodynamic potential

$$\Phi(T) = W(0) + \int_0^T C_P(T')\mathrm{d}T' - T\int_0^T \frac{C_P(T')}{T'}\mathrm{d}T', \tag{2.155}$$

where $E(0)$ and $W(0)$ are energy and enthalpy of the system at $T=0$.

From the above formulae, once more the importance of the heat capacity of a system is seen.

2.7 Thermodynamic Relationships for Dielectrics and Magnetics

Up to now we considered closed systems, macroscopic states of which are determined by such parameters as $P,\ V,\ T,\ S$, and we have found their thermodynamic relationships and thermodynamic functions (see Sect. 2.3). However, there exist such systems in which a macroscopic state is also determined by other specific parameters besides the indicated parameters. Dielectrics and magnetics can serve as examples of such systems. At first we consider dielectrics.

A dielectric is a material medium consisting of bound charges but not conducting an electrical current. Dielectrics can be non-polar, polar and ferroelectric. Their macroscopic state apart from S, T, V and P is determined by two other parameters: an electric field $\boldsymbol{\mathcal{E}}$ as an external parameter, and the polarisation vector $\mathscr{P}$ as an internal parameter. We elucidate the way in which these new parameters enter into thermodynamic relationships.

According to the First Law of Thermodynamics, the total change in the internal energy has the appearance

$$dE = dQ + dA, \tag{2.156}$$

where dA is work being performed. In usual cases, $dA = -P\,dV$. For dielectrics, it is necessary also to take into account work being performed by an external electric field. Here we will not consider the ferroelectric state of dielectrics.

Polar and non-polar dielectrics can be considered as neutral systems consisting of bound electric charges. In such systems, electric charges, grouping themselves, form a neutral atom or molecule. Depending on the symmetry of distribution of the charges, atoms and molecules may or may not possess an intrinsic electric dipole moment. Dielectrics consisting of atoms and molecules and possessing an intrinsic electric dipole moment are called *polar dielectrics*, and dielectrics consisting of atoms and molecules and not possessing an intrinsic electric dipole moment are called *non-polar dielectrics.*

An external electric field displaces bound charges and, thereby, performs a certain work. We calculate this work. On each bound charge e_i, the electric field $\boldsymbol{\mathcal{E}}$ acts with the force $e_i\boldsymbol{\mathcal{E}}$. Inasmuch as under action of this force the charge is displaced by the displacement vector $d\boldsymbol{r}_i$, work being performed at that is $e_i d\boldsymbol{r}_i \mathcal{E}$. If it is added up over all charges of the system in unit volume, work being performed by the electric field is

$$dA' = \sum_i e_i \boldsymbol{\mathcal{E}}\, d\boldsymbol{r}_i = \boldsymbol{\mathcal{E}} \sum_i e_i d\boldsymbol{r}_i = \boldsymbol{\mathcal{E}}\, d\mathscr{P}, \tag{2.157}$$

where $d\mathscr{P} = \sum\limits_i e_i dr_i$ is the electric dipole moment arising as a result of displacement of electric charges, i.e. the polarisation vector of the medium. Comparing work performed by the electric field with work in the usual case $dA = -P\,dV$, notice the correspondence $-P \to \boldsymbol{\mathcal{E}}$ and $V \to \mathscr{P}$.

1. *Internal energy.* If in (2.156) we take into account that $dQ = T\,dS$, and also the expression for work being performed by the electric field (2.157), for the change in the internal energy we get

$$dE = T\,dS - P\,dV + \boldsymbol{\mathcal{E}}\,d\mathscr{P}. \tag{2.158}$$

The latter item presents as $\boldsymbol{\mathcal{E}}\,d\mathscr{P} = d(\boldsymbol{\mathcal{E}}\mathscr{P}) - \mathscr{P}\,d\boldsymbol{\mathcal{E}}$. Then (2.177) takes the form

$$dE' = T\,dS - P\,dV - \mathscr{P}\,d\boldsymbol{\mathcal{E}}. \tag{2.159}$$

Here

$$E' = E - \boldsymbol{\mathcal{E}}\mathscr{P} \tag{2.160}$$

is the internal energy of a dielectric in an external electric field.

If $V = \text{const}$ and $\boldsymbol{\mathcal{E}} = \text{const}$, from (2.159) it follows that $dE' = T\,dS = dQ$.

2. *Enthalpy.* In the energy expression (2.159), we carry out the replacement $P\,\mathrm{d}V = \mathrm{d}(PV) - V\,\mathrm{d}P$. Then we get

$$\mathrm{d}W' = T\,\mathrm{d}S + V\,\mathrm{d}P - \mathscr{P}\,\mathrm{d}\boldsymbol{\mathcal{E}}. \tag{2.161}$$

Here

$$W' = E' + PV = E + PV - \mathscr{P}\boldsymbol{\mathcal{E}} = W - \mathscr{P}\boldsymbol{\mathcal{E}} \tag{2.162}$$

is enthalpy of a dielectric in an electric field.
If $P = \text{const}$ and $\boldsymbol{\mathcal{E}} = \text{const}$, from (2.161) it follows that $\mathrm{d}W' = \mathrm{d}Q = T\,\mathrm{d}S$.
3. *Free energy.* By replacing $T\,\mathrm{d}S = \mathrm{d}(TS) - S\,\mathrm{d}T$ in the expression for the internal energy (2.159), we get

$$\mathrm{d}F' = -S\,\mathrm{d}T - P\,\mathrm{d}V - \mathscr{P}\,\mathrm{d}\boldsymbol{\mathcal{E}} \tag{2.163}$$

Here

$$F' = E' - TS = E - \mathscr{P}\boldsymbol{\mathcal{E}} - TS = F - \mathscr{P}\boldsymbol{\mathcal{E}} \tag{2.164}$$

is the free energy of a dielectric in an electric field.
4. *Gibbs potential.* In the expression of the free energy (2.163), by replacing $P\mathrm{d}V = \mathrm{d}(PV) - V\mathrm{d}P$, we get

$$\mathrm{d}\Phi' = -S\,\mathrm{d}T + V\,\mathrm{d}P - \mathscr{P}\,\mathrm{d}\boldsymbol{\mathcal{E}}. \tag{2.165}$$

Here

$$\Phi' = F' + PV = E - TS + PV - \mathscr{P}\boldsymbol{\mathcal{E}} = \Phi - \mathscr{P}\boldsymbol{\mathcal{E}} \tag{2.166}$$

is the Gibbs potential of a dielectric in an external electric field.

From the above relationships, it is seen that, when placing a dielectric in an electric field, all thermodynamic functions change by the quantity of the potential energy $(-\mathcal{P}\mathcal{E})$:

$$E' - E = W' - W = F' - F = \Phi' - \Phi = -\mathscr{P}\boldsymbol{\mathcal{E}}. \tag{2.167}$$

It is also seen that the polarisation vector can be computed with the aid of any thermodynamic function

$$\mathscr{P} = -\left(\frac{\partial E'}{\partial \boldsymbol{\mathcal{E}}}\right)_{S,V} = -\left(\frac{\partial W'}{\partial \boldsymbol{\mathcal{E}}}\right)_{S,P} = -\left(\frac{\partial F'}{\partial \boldsymbol{\mathcal{E}}}\right)_{V,T} = -\left(\frac{\partial \Phi'}{\partial \boldsymbol{\mathcal{E}}}\right)_{P,T}. \tag{2.168}$$

Knowing the explicit form of the Gibbs thermodynamic potential $\Phi' = \Phi'(T, P, \boldsymbol{\mathcal{E}})$, S, V and $\mathscr{P}$ can be found. For instance, from (2.165) it follows that

$$V = \left(\frac{\partial \Phi'}{\partial P}\right)_{T,\boldsymbol{\mathcal{E}}}; \quad \mathscr{P} = -\left(\frac{\partial \Phi'}{\partial \boldsymbol{\mathcal{E}}}\right)_{T,P}. \tag{2.169}$$

Hence we get

$$\left(\frac{\partial V}{\partial \boldsymbol{\mathcal{E}}}\right)_{P,T} = -\left(\frac{\partial \mathscr{P}}{\partial P}\right)_{\boldsymbol{\mathcal{E}},T}. \tag{2.170}$$

The derivative $(\partial V/\partial \boldsymbol{\mathcal{E}})_{P,T}$ entering into this expression characterises the change in volume of a dielectric under the action of an electric field, i.e. *the electrostriction phenomenon*; and the derivative $(\partial \mathscr{P}/\partial P)_{\boldsymbol{\mathcal{E}},T}$ is a coefficient characterizing the change in the electric polarisation under the action of pressure, i.e. *the piezoelectric effect.* From (2.170), it is seen that both effects are associated with each other.

Magnetics come in diamagnetic, paramagnetic and ferromagnetic states. For dia- and paramagnetics, by analogy with dielectrics, thermodynamic relationships can be written. To do this, it is sufficient to replace in corresponding functions the electric field strength with the magnetic field strength, and the electric polarisation vector with the magnetic polarisation vector: $\boldsymbol{H} \to \boldsymbol{\mathcal{E}}; \mathscr{P} \to \boldsymbol{M}$. As a result, for magnetics the following relationships can be written:

$$\begin{aligned} \mathrm{d}E' &= T\,\mathrm{d}S - P\,\mathrm{d}V - \boldsymbol{M}\,\mathrm{d}\boldsymbol{H}, \\ \mathrm{d}W' &= T\,\mathrm{d}S + V\,\mathrm{d}P - \boldsymbol{M}\,\mathrm{d}\boldsymbol{H}, \\ \mathrm{d}F' &= -S\,\mathrm{d}T - P\,\mathrm{d}V - \boldsymbol{M}\,\mathrm{d}\boldsymbol{H}\,, \\ \mathrm{d}\Phi' &= -S\,\mathrm{d}T + V\,\mathrm{d}P - \boldsymbol{M}\,\mathrm{d}\boldsymbol{H}\,. \end{aligned} \tag{2.171}$$

Here $E' = E - \boldsymbol{MH}$, $W' = W - \boldsymbol{MH}$, $F' = F - \boldsymbol{MH}$, $\Phi' = \Phi - \boldsymbol{MH}$ are the thermodynamic functions of a magnetic material in an external magnetic field. Knowing the explicit form of the thermodynamic functions, and using the relationships (2.171), the different parameters can be determined. For instance, if the Gibbs potential $\Phi' = \Phi'(T, P, \boldsymbol{H})$ is known, V and $\boldsymbol{M}$ can be calculated:

$$V = \left(\frac{\partial \Phi'}{\partial P}\right)_{T,H}; \quad \boldsymbol{M} = -\left(\frac{\partial \Phi'}{\partial \boldsymbol{H}}\right)_{T,P}. \tag{2.172}$$

Hence we get

$$\left(\frac{\partial V}{\partial \boldsymbol{H}}\right)_{P,T} = -\left(\frac{\partial \boldsymbol{M}}{\partial P}\right)_{T,H}. \tag{2.173}$$

Here $(\partial V/\partial \boldsymbol{H})_{P,T}$ characterises the change in volume of a magnetic under the action of a magnetic field, i.e. *the magnetostriction effect*, and $(\partial \boldsymbol{M}/\partial P)_T$ characterises the change in the magnetisation vector under the action of pressure, i.e. *the piezomagnetic effect.*

In conclusion, note that the magnetisation vector can be found with the aid of any thermodynamic function:

$$\boldsymbol{M} = -\left(\frac{\partial E'}{\partial \boldsymbol{H}}\right)_{V,S} = -\left(\frac{\partial W'}{\partial \boldsymbol{H}}\right)_{S,P} = -\left(\frac{\partial F'}{\partial \boldsymbol{H}}\right)_{V,T} = -\left(\frac{\partial \Phi'}{\partial \boldsymbol{H}}\right)_{P,T}. \tag{2.174}$$

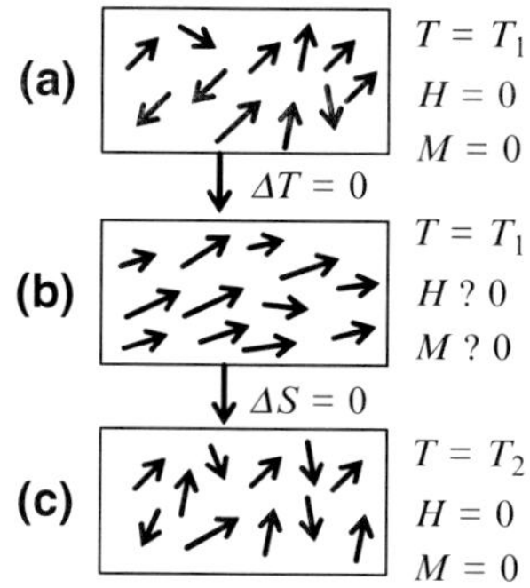

Fig. 2.16. The distribution of magnetic dipoles in a paramagnetic

2.8 Magnetocaloric Effect: Production of Ultra-Low Temperatures

To produce ultra-low temperatures (below 1 K) Debye in 1926 proposed to use the magnetocaloric effect. The essence of the magnetocaloric effect resides in the following: a solid paramagnetic salt with temperature $T = T_1$ isothermically is magnetised to the finite value H (Fig. 2.16), the transition $1 \to 2$). Thereupon, it is adiabatically ($\Delta S = 0$) demagnetised (the transition $2 \to 3$). During this transition, the temperature of the system (the paramagnetic salt) drops, i.e. $T_2 < T_1$. By repeating these processes many times, the temperature can be sufficiently lowered. Thus, the essence of the magnetocaloric effect resides in the fact that at the adiabatic demagnetisation of a paramagnetic its temperature drops.

The effect, i.e. the lowering of temperature, can be explained as follows: in the absence of an external magnetic field, in consequence of the chaotic distribution of magnetic dipoles in a paramagnetic, the mean magnetic moment of the system as a whole equals zero (Fig. 2.16, the state 1). As a result of the isothermal magnetisation, magnetic dipoles are ordered (the state 2). Thereupon, the system is adiabatically isolated and demagnetised, the order is violated and the chaoticity is retained (the state 3). At the transition $2 \to 3$, to destroy the ordering it is necessary to perform work. Since the process proceeds in adiabatic conditions, this work is performed at the expense of the internal energy of the system and, thus, the temperature drops.

The above-described processes on the S, T plane are schematically depicted in Fig. 2.17. Here temperature dependences of entropy in a magnetic field $S(H)$ and in its absence $S(0)$ are presented. When constructing the dependence, the inequality $C_V > 0$, the Nernst principle and the fact that in a magnetic field in a paramagnetic there exists the ordering of dipole moments and therefore $S(H) < S(0)$ were taken into account.

Assume that an external magnetic field is absent and the considered system (the paramagnetic salt) is found in the state with temperature T_1 and entropy

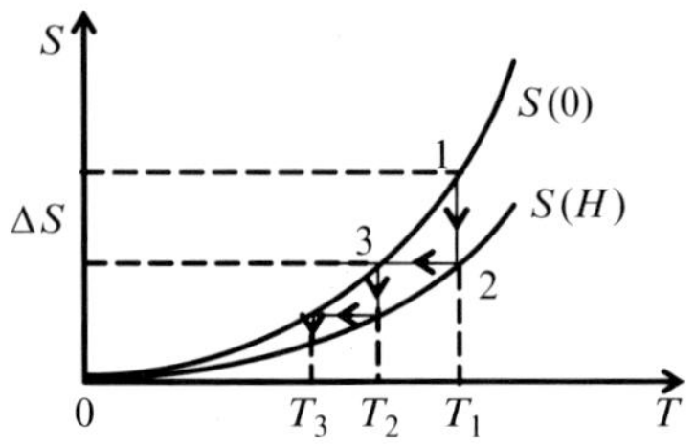

Fig. 2.17. The temperature dependence of the entropy of a paramagnetic at different values of magnetic field

$S(T_1, 0)$ (Fig. 2.17, state 1). Isothermally magnetise the system to the value $H \neq 0$ (the transition $1 \to 2$). During this time, its entropy decreases by the magnitude $\Delta S = S(T_1 H) - S(T_1 0)$. We calculate this change in entropy. Inasmuch as $T_1 = \text{const}$, S is a function only of H. Therefore

$$\int_1^2 \mathrm{d}S = \int_0^H \left(\frac{\partial S}{\partial H}\right)_T \mathrm{d}H, \tag{2.175}$$

or

$$\Delta S = S(T_1, H) - S(T_1, 0) = \int_0^H \left(\frac{\partial S}{\partial H}\right)_T \mathrm{d}H. \tag{2.176}$$

We use the thermodynamic relationship for magnetics from (2.171)

$$\mathrm{d}F' = -S\,\mathrm{d}T - P\,\mathrm{d}V - M\,\mathrm{d}H. \tag{2.177}$$

Hence we have

$$S = -\left(\frac{\partial F'}{\partial T}\right)_{V,H}; \quad M = -\left(\frac{\partial F}{\partial H}\right)_{V,T}. \tag{2.178}$$

or

$$\left(\frac{\partial S}{\partial H}\right)_T = \left(\frac{\partial M}{\partial T}\right)_H. \tag{2.179}$$

Taking into account (2.179) in (2.176), we get

$$S(T_1, H) = S(T_1, 0) + \int_0^H \left(\frac{\partial M}{\partial T}\right)_H \mathrm{d}H. \tag{2.180}$$

Using the Curie law for the magnetisation of paramagnetics

$$M = \frac{CH}{T}, \tag{2.181}$$

in (2.180), we get

$$S(T_1, H) = S(T_1, 0) - \frac{CH^2}{2T_1^2}, \tag{2.182}$$

where C is the Curie constant. As seen from (2.182), as a result of the rise of ordering in the magnetic field, the entropy of the system decreases by the magnitude $CH^2/2T_1$.

And we now calculate the change in temperature as a result of the adiabatic demagnetisation (the transition $2 \to 3$). This change is characterised by the quantity $(\partial T/\partial H)_S$. Use the thermodynamic relationship from (2.171),

$$\mathrm{d}E' = T\,\mathrm{d}S - P\,\mathrm{d}V - M\,\mathrm{d}H. \tag{2.183}$$

Hence we have

$$T = \left(\frac{\partial E'}{\partial S}\right)_{V,H}; \quad M = -\left(\frac{\partial E'}{\partial H}\right)_{S,V} \tag{2.184}$$

and, consequently,

$$\left(\frac{\partial T}{\partial H}\right)_S = -\left(\frac{\partial M}{\partial S}\right)_H. \tag{2.185}$$

Taking into account the property of the Jacobian (see Appendix B), transform the right-hand side of equality (2.185):

$$\begin{aligned}\left(\frac{\partial T}{\partial H}\right)_S &= -\left(\frac{\partial M}{\partial S}\right)_H = -\frac{\partial(M,H)}{\partial(S,H)} \\ &= -\frac{\partial(M,H)}{\partial(T,H)}\frac{\partial(T,H)}{\partial(S,H)} = -\left(\frac{\partial M}{\partial T}\right)_H\left(\frac{\partial T}{\partial S}\right)_H.\end{aligned} \tag{2.186}$$

If we take into account the definition $(\partial S/\partial T)_H = C_H/T$ and (2.181), we get

$$\left(\frac{\partial T}{\partial H}\right)_S = -\frac{T}{C_H}\left(\frac{\partial M}{\partial T}\right)_H = \frac{C}{C_H}\frac{H}{T}, \tag{2.187}$$

where C_H is the heat capacity of the paramagnetic in a constant magnetic field. It is known that at low temperatures $C_H = AT^3$ (A is the constant, not depending on temperature) and, therefore, applying (2.187) to the transition $2 \to 3$, for the change in temperature we get

$$(\Delta T)_S = \frac{C}{A}\frac{H}{T^4}(\Delta H)_S \tag{2.188}$$

or

$$T_2 = T_1 - \frac{C}{A}\frac{H}{T^4}\left|\Delta H\right|_S. \tag{2.189}$$

As is seen, at low temperatures ($T < 1\,\mathrm{K}$) the change in temperature is $\sim T^{-4}$; therefore it is sufficiently big. Note that using this method, i.e. many times repeating the indicated processes, temperature can be lowered to $10^{-3}\,\mathrm{K}$.

Note that at lower temperatures magnetic dipole moments associated with the spin of an electron strongly interact, and therefore magnetic dipoles are spontaneously ordered and an external magnetic field does not influence this ordering. In this case, the Curie law (2.181) does not hold, and the magnetisation M does not depend on temperature. Therefore, instead of (2.182) and (2.187), we obtain $\Delta S = S(T,H) - S(T,0) = 0$ and, consequently, $(\partial T/\partial H)_S = 0$.

To produce temperatures close to absolute zero, nuclear paramagnetics are used as a working body. Inasmuch as the interaction between nuclear magnetic moments is weaker than the interaction between electron magnetic moments, the spontaneous ordering of nuclear magnetic moments does not occur. The ordering of nuclear magnetic moments is governed by an external magnetic field. The magnetocaloric effect associated with the nuclear paramagnetism makes it possible to lower the temperature to 10^{-6} K.

In conclusion, note the following. From the above discussion it follows that by the multiple repetition of processes of the isothermal magnetisation and adiabatic demagnetisation it is possible to produce absolute zero of temperature. However, it is impossible because, according to the Nernst principle, at absolute zero entropy does not depend on external parameters (in the given case on H) and dependences $[S(H,T) = S(0,T)]$ and $S(H,T)$ close to absolute zero coincide. It is clear that the adiabatic process in this case coincides with the isothermal one and loses its sense; therefore the system cannot change its state to the side of low temperatures.

2.9 Thermodynamics of Systems with Variable Number of Particles: Chemical Potential

Up to now we were considering closed systems with an invariable number of particles ($N = \text{const}$). The state of such systems is determined by four parameters: entropy S, volume V, temperature T and pressure P. Of them, S and V are additive parameters, and P and T are intensive ones.

However, in nature, systems with a variable number of particles, i.e. open systems, exist, too. Thermodynamics and statistical physics thoroughly study such systems also. The following may be considered as examples of open systems: liquid with saturated vapour over it, solid and liquid phases that are in contact; components in chemical reactions; photon gas in equilibrium with a wall of the closed volume; systems of quasi-particles (when the temperature of a crystal changes, the number of phonons changes); electrons in the conduction band and holes in the valence band in semiconductors, i.e. an electron–hole gas (when the temperature of a semiconductor changes, the number of conduction electrons and free holes changes), etc.

The thermodynamic state of such systems, apart from the four above-indicated parameters, is determined also by the number of particles N. Thus, for open systems N is not a parameter, but a variable. We consider open systems and the dependence of thermodynamic functions on the number of particles N.

Note that the four thermodynamic functions (E, W, F and Φ) known to us possess one common property: they are all additive, i.e. as many times the number of particles in a system changes, so does the value of these quantities. Indeed, the internal energy E is an additive quantity according to the second postulate of thermodynamics, and other functions: enthalpy $W = E + PV$, the free energy $F = E - TS$ and Gibbs potential $\Phi = W - TS$, are additive inasmuch as they are expressed by additive quantities E, V and S.

The property of additivity of these functions gives the possibility to draw conclusions about their dependence on the number of particles N: *all thermodynamic functions are a uniform function of the first order of* N. Issuing from this, in the general form these dependences can be presented as follows:

The internal energy can be a function of entropy S, volume V and the number of particles N:

$$E = N\varphi_1\left(\frac{S}{N}, \frac{V}{N}\right); \tag{2.190}$$

Enthalpy can be a function of entropy S, pressure P and the number of particles N:

$$W = N\varphi_2\left(\frac{S}{N}, P\right); \tag{2.191}$$

The free energy can be a function of volume V, temperature T and the number of particles N:

$$F = N\varphi_3\left(\frac{V}{N}, T\right); \tag{2.192}$$

The Gibbs potential can be a function of temperature T, pressure P and the number of particles N:

$$\Phi = N\varphi_4(P, T). \tag{2.193}$$

It is clear that by virtue of additivity of S and V, functions φ_1, φ_2, φ_3 do not depend on N and all thermodynamic functions are proportional to the number of particles.

Inasmuch as thermodynamic functions themselves are proportional to the number of particles N, it is evident that their differentials change proportionally to $\mathrm{d}N$. Then to the differentials in Sect. 2.3 (2.77), it is necessary to add the item $\mu \mathrm{d}N$, proportional to $\mathrm{d}N$. Thus, the differentials of thermodynamic functions can be written as

$$\mathrm{d}E = T\,\mathrm{d}S - P\,\mathrm{d}V + \mu\,\mathrm{d}N, \tag{2.194}$$

$$\mathrm{d}W = T\,\mathrm{d}S + V\,\mathrm{d}P + \mu\,\mathrm{d}N, \tag{2.195}$$

$$\mathrm{d}F = -S\,\mathrm{d}T - P\,\mathrm{d}V + \mu\,\mathrm{d}N, \tag{2.196}$$

$$\mathrm{d}\Phi = -S\,\mathrm{d}T + V\,\mathrm{d}P + \mu\,\mathrm{d}N, \tag{2.197}$$

where the coefficient of proportionality μ is called *the chemical potential.* The chemical potential can be determined by differentiating any thermodynamic function with respect to N:

$$\mu = \left(\frac{\partial E}{\partial N}\right)_{S,V} = \left(\frac{\partial W}{\partial N}\right)_{S,P} = \left(\frac{\partial F}{\partial N}\right)_{T,V} = \left(\frac{\partial \Phi}{\partial N}\right)_{P,T}. \tag{2.198}$$

However, we need to take into account that in different cases μ will be a function of different parameters. For instance, from (2.193) it follows that

$$\mu = \left(\frac{\partial \Phi}{\partial N}\right)_{P,T} = \varphi_4(P,T), \tag{2.199}$$

i.e. in this case $\mu = \mu(P,T) = \varphi_4(P,T)$ is a function only of T and P and does not depend on the number of particles.

From (2.193) and (2.199), we get

$$\Phi = \mu N. \tag{2.200}$$

Hence follows the physical sense of the chemical potential: $\mu = \Phi/N$, i.e. *the chemical potential can be defined as the Gibbs potential, accounting for one particle.* From another definition of the chemical potential

$$\mu = \left(\frac{\partial F}{\partial N}\right)_{T,V}, \tag{2.201}$$

follows one more physical sense: *the chemical potential equals the free energy (the performed work) necessary to change the number of particles in a system by unity.*

Chemical potential is an intensive quantity, inasmuch as it does not depend on the number of particles. Having substituted the expression of the Gibbs potential (2.200) into (2.197), for a differential of the chemical potential we get

$$\mathrm{d}\mu = -\frac{S}{N}\mathrm{d}T + \frac{V}{N}\mathrm{d}P. \tag{2.202}$$

On denoting the intrinsic volume by $\upsilon = V/N$ and the intrinsic entropy by $s = S/N$, we get

$$\mathrm{d}\mu = -s\,\mathrm{d}T + \upsilon\,\mathrm{d}P. \tag{2.203}$$

We introduce one more thermodynamic function. To do this, in (2.196) carry out the replacement $\mu\,\mathrm{d}N = \mathrm{d}(\mu N) - N\,\mathrm{d}\mu$. Then we have

$$\mathrm{d}(F - \mu N) = -S\,\mathrm{d}T - P\,\mathrm{d}V - N\,\mathrm{d}\mu. \tag{2.204}$$

If we accept the notation

$$\boldsymbol{\Omega} = F - \mu N = F - \Phi = -PV, \tag{2.205}$$

from (2.204), for the new thermodynamic function we get

$$d\boldsymbol{\Omega} = -S\,dT - P\,dV - N\,d\mu. \tag{2.206}$$

As can be seen, T, V and μ are independent variables of the function, called *the grand thermodynamic potential or* $\boldsymbol{\Omega}$*-potential*: $\boldsymbol{\Omega} = \boldsymbol{\Omega}(T, V, \mu)$.

We have introduced two new functions μ and $\boldsymbol{\Omega}$. Thus, the total number of thermodynamic functions becomes six $(E, W, F, \Phi, \mu, \boldsymbol{\Omega})$.

If a system is found in an external magnetic field H, the change in its free energy, according to (2.171) and (2.196), has the appearance

$$dF' = -S\,dT - P\,dV - M\,dH + \mu\,dN. \tag{2.207}$$

Carrying out the replacement $\mu\,dN = d(\mu N) - N\,d\mu$, from (2.207) we have

$$d\boldsymbol{\Omega}' = -S\,dT - P\,dV - N\,d\mu - M\,dH. \tag{2.208}$$

Here

$$\boldsymbol{\Omega}' = F' - \mu N = F - MH - \Phi = \boldsymbol{\Omega} - MH \tag{2.209}$$

is the grand thermodynamic potential of an open system found in an external magnetic field, and $\boldsymbol{\Omega} = F - \mu N = F - \Phi$ is the thermodynamic potential at $H = 0$.

As is seen, T, V, μ and H are independent variables of $\boldsymbol{\Omega}' : \boldsymbol{\Omega}'(T, V, \mu, H)$.

Knowing the explicit form of this function, we can determine the number of particles, pressure, entropy and the magnetisation vector:

$$\begin{aligned} N &= -\left(\partial\boldsymbol{\Omega}'/\partial\mu\right)_{T,V,H};\, P = -\left(\partial\boldsymbol{\Omega}'/\partial V\right)_{T,\mu,H}; \\ S &= -\left(\partial\boldsymbol{\Omega}'/\partial T\right)_{V,\mu,H};\, M = -\left(\partial\boldsymbol{\Omega}'/\partial H\right)_{V,T,\mu}. \end{aligned} \tag{2.210}$$

Using a total differential of thermodynamic functions, i.e. expressions (2.194)–(2.197), the thermodynamic relationships for open systems can be found. For instance, from (2.194) we have

$$T = (\partial E/\partial S)_{V,N}; \quad P = -(\partial E/\partial V)_{S,N}; \quad \mu = (\partial E/\partial N)_{S,V}. \tag{2.211}$$

Hence

$$\begin{aligned} (\partial T/\partial V)_{S,N} &= -(\partial P/\partial S)_{V,N};\, (\partial T/\partial N)_{S,V} = (\partial\mu/\partial S)_{N,V}; \\ (\partial P/\partial N)_{S,V} &= -(\partial\mu/\partial V)_{S,N}. \end{aligned} \tag{2.212}$$

The above relationships can be presented also in the form of the Jacobian:

$$\frac{\partial(T,S)}{\partial(V,S)} = -\frac{\partial(P,V)}{\partial(S,V)} \text{ or } \frac{\partial(T,S)}{\partial(P,V)} = 1; \quad N = \text{const}, \tag{2.213}$$

$$\frac{\partial(T,S)}{\partial(N,S)} = \frac{\partial(\mu,N)}{\partial(S,N)} \text{ or } \frac{\partial(T,S)}{\partial(\mu,N)} = -1; \quad V = \text{const}, \tag{2.214}$$

$$\frac{\partial(P,V)}{\partial(N,V)} = -\frac{\partial(\mu,N)}{\partial(V,N)} \text{ or } \frac{\partial(P,V)}{\partial(\mu,N)} = 1; \quad S = \text{const}. \tag{2.215}$$

In a like manner, using differentials of other functions (2.195)–(2.197), other cases can also be considered. It can be shown that at $P=\text{const}$ (2.214) is obtained, and at $T=\text{const}$ (2.215) is obtained, i.e. the case of $P=\text{const}$ coincides with the case of $V=\text{const}$, and the case of $T=\text{const}$ coincides with the case of $S=\text{const}$.

2.10 Conditions of Equilibrium of Open Systems

It is known that the condition of equilibrium of systems in contact with each other is determined from the condition of maximum of entropy of the complete system. Inasmuch as for closed systems ($N=\text{const}$) entropy depends only on energy and volume $S=S(E,V)$, from the condition $S=S_{\max}$ (Sect. 1.8) it follows that two closed systems are found in equilibrium when their temperatures and pressures are identical: $T_1=T_2=T$ and $P_1=P_2=P$. These conditions provide thermal and mechanical equilibrium.

In open ($N \neq \text{const}$) systems, a new variable – the number of particles N – appears. Thus, in this case entropy depends on three variables $S=S(E,V,N)$. We find the condition of maximum of entropy relative to N.

Assume that two open systems with the number of particles N_1 and N_2 are in contact and form an isolated common system. The boundary *ab* dividing them (Fig. 2.18) transmits particles, i.e. N_1 and N_2 can change; however, their sum

$$N_1 + N_2 = N = \text{const} \tag{2.216}$$

remains constant.

In order that at pre-assigned volumes V_1, V_2 and energies E_1, E_2 of subsystems, entropy of the complete system

$$S = S_1(N_1) + S_2(N_2) \tag{2.217}$$

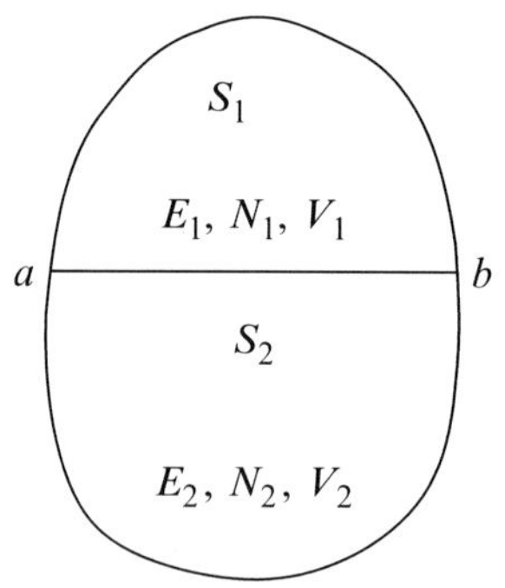

Fig. 2.18. The equilibrium of two open systems

would be maximum ($S = S_{\max}$), it is necessary to fulfil the condition

$$\frac{\partial S}{\partial N_1} = \frac{\partial S_1}{\partial N_1} + \frac{\partial S_2}{\partial N_2}\frac{\partial N_2}{\partial N_1} = 0. \tag{2.218}$$

If from (2.216), we take into account that $\partial N_2/\partial N_1 = -1$, we get

$$\left(\frac{\partial S_1}{\partial N_1}\right)_{V_1,E_1} = \left(\frac{\partial S_2}{\partial N_2}\right)_{V_2,E_2}. \tag{2.219}$$

From the basic thermodynamic relationship (2.194)

$$\mathrm{d}S = \frac{1}{T}\mathrm{d}E + \frac{P}{T}\mathrm{d}V - \frac{\mu}{T}\mathrm{d}N, \tag{2.220}$$

we have

$$\left(\frac{\partial S}{\partial N}\right)_{E,V} = -\frac{\mu}{T}. \tag{2.221}$$

On substituting this expression into (2.219), and taking into account that in equilibrium $T_1 = T_2 = T$, from the condition of maximality of entropy we get the condition of equilibrium of open systems

$$\mu_1 = \mu_2 = \mu \text{ or } \mu(P,T) = \text{const.} \tag{2.222}$$

Hence it follows that if open systems are found in equilibrium, their chemical potentials ought to be identical. Note that for open systems the chemical potential resembles the level of liquid in connected vessels. Thus, in the general case the following conditions

$$T_1 = T_2 = \ldots = T_n = T; \text{ thermal equilibrium}, \tag{2.223}$$

$$P_1 = P_2 = \ldots = P_n = P; \text{ mechanical equilibrium}, \tag{2.224}$$

$$\mu_1 = \mu_2 = \ldots = \mu_n = \mu; \text{ material equilibrium}, \tag{2.225}$$

are conditions of thermodynamic equilibrium of n number of contacting subsystems. If a system is found in an external field, the pressure depends on coordinates $P = P(x,y,z)$. In this case, condition (2.222) takes the form

$$\mu[P(x,y,z),T)] = \text{const.} \tag{2.226}$$

If we denote the potential energy of one particle in an external field by $u(x,y,z)$, the chemical potential equals

$$\mu = \mu_0(P,T) + u(x,y,z) = \text{const}, \tag{2.227}$$

where $\mu_0(P,T)$ is the chemical potential in the absence of an external field.

Consider the particular case of the atmospheric air in the gravitational field of the Earth, and assume that it is an ideal gas consisting of particles of the mass m. Then condition (2.227) can be presented in the following form:

$$\mu = \mu_0(P(z), T) + mgz = \text{const}, \tag{2.228}$$

where g is the acceleration due to gravity, and z is the distance from the surface of the Earth. Regarding temperature as constant and differentiating (2.228) with respect to z, we get

$$\left(\frac{\partial \mu_0}{\partial P}\right)_T \frac{\mathrm{d}P}{\mathrm{d}z} + mg = 0. \tag{2.229}$$

If we take into account that $(\partial \mu_0/\partial P)_T = \upsilon$, (2.229) takes the form

$$\upsilon \frac{\mathrm{d}P}{\mathrm{d}z} + mg = 0. \tag{2.230}$$

From the equation of the state of an ideal gas, we have

$$\upsilon = \frac{V}{N} = \frac{k_0 T}{P}. \tag{2.231}$$

Then (2.230) can be rewritten as

$$\frac{\mathrm{d}P}{P} = -\frac{mg}{k_0 T}\mathrm{d}z. \tag{2.232}$$

After integrating, we get the known barometric formula

$$P = P_0 \mathrm{e}^{-\frac{mgz}{k_0 T}}, \tag{2.233}$$

where P_0 is pressure of air on the surface of the Earth. Hence it follows that for an air layer to be found in equilibrium in the gravitational field, the pressure (the density) of air ought to exponentially depend on the height.

3

Canonical Distribution: Gibbs Method

Summary. It is known that the basic task of statistical physics is calculating the mean value of thermodynamic parameters, determining the macroscopic state of a system, the relation (the equation of state) between them, and also thermodynamic coefficients, measured on test. However, inasmuch as the immediate determination of these parameters is impossible, auxiliary functions – thermodynamic potentials – are introduced in theory. If these functions are known, we can find the equation of a state and compute thermodynamic coefficients (see Chap. 2). Thus, the problem is reduced to finding the explicit form of thermodynamic functions, especially free energy (the Helmholtz potential).

The present chapter is devoted to the Gibbs method, which is applied to find the function of free energy, on the basis of which the canonical distribution lies. On the basis of the microcanonical distribution, canonical distributions for closed and open systems in a thermostat are found and also the essence of the Gibbs method is expounded.

3.1 Gibbs Canonical Distribution for Closed Systems

It is known that to solve the basic task of statistical physics, i.e. to find the mean value of thermodynamic quantities for classical systems it is necessary to calculate the integral

$$\bar{L} = \int L(q,p)\rho(q,p)\mathrm{d}q\,\mathrm{d}p, \tag{3.1}$$

and for quantum systems to calculate the sum (see Sect. 1.3).

$$\bar{L} = \sum_n L_{nn} W_n. \tag{3.2}$$

Quantities $L(q,p)$ and L_{nn} are known from classical and quantum mechanics, respectively. The distribution function $\rho(q,p)$ and density matrix W_n are found from statistical physics. The explicit form of these functions depends on conditions in which the system is found. For instance, for completely isolated systems ρ and W_n are pre-assigned by expressions (1.66) and (1.99), respectively, i.e. the microcanonical distribution, obtained on the basis of the postulate of statistical physics.

It should be noted that a completely isolated system is an ideal case. In reality, however, there are conditions in which the system interacts with its surroundings, i.e. it is found in contact with the medium surrounding it (the thermostat). On the basis of properties, adduced in Sects. 1.4 and 1.5, the explicit form of the distribution function of a system in the thermostat can be presented in the form

$$\ln \rho(q,p) = A_0 + \beta E(q,p) \tag{3.3}$$

for classical systems and in the form

$$\ln W(E_n) = A_0 + \beta E_n \tag{3.4}$$

for quantum systems. These expressions can also be presented as

$$\rho(q,p) = \exp[A_0 + \beta E(q,p)] \tag{3.5}$$

and

$$W_n \equiv W(E_n) = \exp(A_0 + \beta E_n), \tag{3.6}$$

respectively.

To determine the explicit form and the physical sense of the constants A_0 and β, entering into the expressions (3.5) and (3.6), use the basic postulate of statistical physics, i.e. the microcanonical distribution. Assume that the considered system with the thermostat forms a completely isolated system (Fig. 3.1).

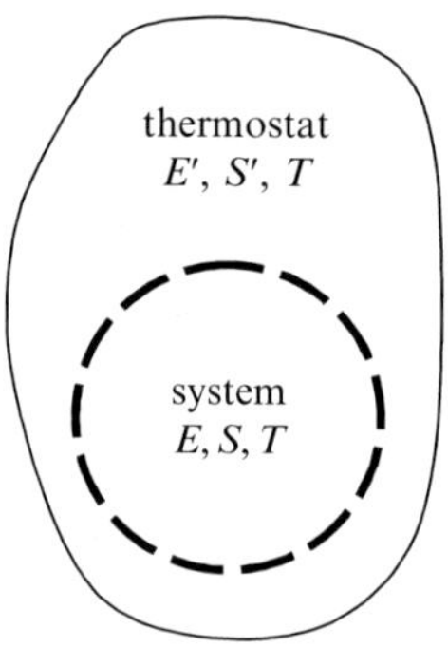

Fig. 3.1. The closed system in the thermostat

Separately consider classical and quantum systems.

Quantum systems. Assume that the considered system with the thermostat represents a quantum system. Let energy and entropy of the thermostat be E' and S', and energy and entropy of the system be E and S. At first consider closed (N = const) quantum systems. Inasmuch as the system is found in thermodynamic equilibrium with the thermostat, their temperatures T ought to be identical.

Inasmuch as it is supposed that the system is small compared with the thermostat, energy can be written as

$$E' + E = E_0 = \text{const};\, E \ll E', \tag{3.7}$$

where E_0 = const is the pre-assigned energy of the total isolated system. Having microcanonical distribution (1.99) applied to this isolated quantum system, we get

$$\mathrm{d}W = \text{const}\,\delta(E' + E - E_0)\mathrm{d}G'\,\mathrm{d}G, \tag{3.8}$$

where $\mathrm{d}G'$ and $\mathrm{d}G$ are the statistical weights of the macroscopic states of the thermostat and system with energies E' and E, respectively.

The following problem is stated: *What is the probability that the system is in a concrete microstate with energy $E = E_n$ at any microstate of the thermostat?* To find this probability it is necessary in (3.8), having accepted $E = E_n$ and $\mathrm{d}G = 1$, to integrate over microstates of the system, i.e. over $\mathrm{d}G'$. Then we get

$$W_n = \text{const}\int \delta(E' + E_n - E_0)\mathrm{d}G'. \tag{3.9}$$

In order to integrate with the aid of the δ-function, pass from the integration over $\mathrm{d}G'$ to the integration over energy $\mathrm{d}E'$. To do this present $\mathrm{d}G'$ in the form

$$\mathrm{d}G' = \frac{\mathrm{d}G'}{\mathrm{d}E'}\mathrm{d}E' \tag{3.10}$$

and replace the derivative $\mathrm{d}G'/\mathrm{d}E'$ by the ratio $\Delta G'/\Delta E'$

$$\mathrm{d}G' = (\Delta G'/\Delta E')\,\mathrm{d}E'. \tag{3.11}$$

If the expression $\Delta G' = \mathrm{e}^{S'(E')/k_0}$ is used, stemming from the definition of entropy (1.109), (3.9) takes the form

$$W_n = \text{const}\int \left(\frac{\mathrm{e}^{S'(E')/k_0}}{\Delta E'}\right)\delta(E' + E_n - E_0)\mathrm{d}E'. \tag{3.12}$$

Integrating with the aid of the δ-function over $\mathrm{d}E'$, we get

$$W_n = \text{const}\left(\frac{\mathrm{e}^{S'(E')/k_0}}{\Delta E'}\right)\Bigg|_{E'=E_0-E_n}. \tag{3.12a}$$

Inasmuch as $E_n \ll E_0$, the range of energies $\Delta E'$, entering into expression of the distribution function sought (3.12a), can be regarded as almost constant: $\Delta E'\,|_{E'=E_0-E_n} \approx \Delta E'\,|_{E'=E_0} = \text{const}$. Then the distribution function takes the form:

$$W_n = \text{const}\, \exp\left[S'(E_0 - E_n)/k_0\right]. \tag{3.13}$$

If, according to (3.7), one takes into account that $E_n \ll E' < E_0$, the function $S'(E_0 - E_n)$ can be expanded into a series in powers of E_n. Then entropy of the thermostat in the linear approximation takes the form

$$S'(E_0 - E_n) \cong S'(E_0) - E_n \left(\frac{\partial S'}{\partial E'}\right)\bigg|_{E'=E_0} + \cdots \tag{3.14}$$

According to the definition of temperature (1.145), we have

$$\left(\frac{\partial S'}{\partial E'}\right)\bigg|_{E'=E_0} = \frac{1}{T}, \tag{3.15}$$

where T is the absolute temperature of the thermostat (and also the temperature of the system). If you substitute the expression of entropy (3.14) into (3.13) and all constants, including also $e^{S'(E_0)/k_0} = \text{const}$, denoted by a new constant A, finally for the distribution function we get the following canonical expression

$$W_n = A \cdot e^{-E_n/k_0T}, \tag{3.16}$$

where A is the normalizing constant, not depending upon E_n, determined from the normalization condition

$$\sum_n W_n = 1. \tag{3.17}$$

Having substituted (3.16) into (3.17), for the normalizing constant we get

$$A^{-1} = \sum_n e^{-E_n/k_0T}. \tag{3.18}$$

Taking into account (3.18) in (3.16), for the distribution function we have

$$W_n = \frac{e^{-E_n/k_0T}}{\sum_n e^{-E_n/k_0T}}. \tag{3.19}$$

Having logarithmized both sides of expression (3.19), we get

$$\ln W_n = -\ln \sum_n e^{-E_n/k_0T} - \frac{E_n}{k_0T}. \tag{3.20}$$

If we compare expressions of the distribution function (3.19) and (3.20) with expressions (3.4) and (3.6), we find constants A_0 and β

$$A_0 = -\ln \sum_n \mathrm{e}^{-E_n/k_0 T} = \ln A, \beta = -\frac{1}{k_0 T}. \tag{3.21}$$

Expression (3.16) or in the explicit form (3.19) is called *the Gibbs canonical distribution function* for closed quantum systems found in the thermostat.

As is seen from (3.19), the distribution function over microstates of quantum systems is determined only by the energy spectrum of the system E_n.

Having substituted expression for the distribution function (3.19) into (3.2), we find the mean statistical value of the physical quantity

$$\bar{L} = \frac{\sum_n L_{nn} \mathrm{e}^{-E_n/k_0 T}}{\sum_n \mathrm{e}^{-E_n/k_0 T}}. \tag{3.22}$$

Hence it is seen that in order to calculate the mean statistical value of the quantity $\hat{L}$, characterizing quantum systems, it is sufficient to know only diagonal matrix elements L_{nn} of the operator $\hat{L}$ and the energy spectrum E_n. Both quantities L_{nn} and E_n ought to be known from quantum mechanics.

Classical systems. In order to find the expression of canonical distribution for such systems, it is sufficient in expression (3.16) for quantum systems to carry out the replacement

$$W_n \Rightarrow \rho(q,p); E_n \Rightarrow E(q,p); A \Rightarrow A_{\mathrm{cl}}. \tag{3.23}$$

As a result, we get the expression of the Gibbs distribution function over microstates for closed classical systems in the thermostat

$$\rho(q,p) = A_{\mathrm{cl}} \cdot \mathrm{e}^{-E(q,p)/k_0 T}. \tag{3.24}$$

However, it should be noted that the constant A_{cl}, entering here, is found from the normalization condition

$$\int \rho(q,p) dq dp = A_{cl} \int e^{-E(q,p)/k_0 T} dq dp = 1. \tag{3.25}$$

Hence we get the expression for the constant A_{cl}

$$A_{\mathrm{cl}}^{-1} = \int \mathrm{e}^{-E(q,p)/k_0 T} \mathrm{d}q\, \mathrm{d}p. \tag{3.25a}$$

and the explicit form of the distribution function

$$\rho(q,p) = \frac{\mathrm{e}^{-E(q,p)/k_0 T}}{\int \mathrm{e}^{-E(q,p)/k_0 T} \mathrm{d}q\, \mathrm{d}p}. \tag{3.26}$$

Canonical distribution over microstates for classical systems (3.24) was introduced in 1901 by the well-known American physicist Gibbs.

Having substituted (3.26) into (3.1), the mean statistical value for any physical quantities $L(q,p)$ of classical systems can be found:

$$\bar{L} = \frac{\int L(q,p)\mathrm{e}^{-E(q,p)/k_0T}\mathrm{d}q\,\mathrm{d}p}{\int \mathrm{e}^{-E(q,p)/k_0T}\mathrm{d}q\,\mathrm{d}p}. \tag{3.27}$$

As is seen from (3.27), in order to calculate $\bar{L}$, it is necessary to know $L(q,p)$ and energy of the microstate $E(q,p)$. Note that both functions $L(q,p)$ and $E(q,p)$ ought to be known from classical mechanics.

At the end of the section, note similarity and distinction between canonical and microcanonical distributions. The microcanonical distribution is the mathematical expression of the postulate of statistical physics about the identical probability of all microstates for completely isolated systems. The canonical distribution is obtained from the microcanonical one and is just for systems in the thermostat.

As is seen from Sects. 1.4 and 1.5, the microcanonical distribution function depends on energy in the form of the δ-function. Therefore the distribution function both over microstates and energies has a maximum at $E = E_0$ (Fig. 1.5).

In the case of the canonical distribution, the distribution function over microstates depends on energy exponentially [see (3.16) and (3.24)].

To determine the distribution over energies $w(E)$ for a system in the thermostat it is necessary to multiply the distribution function over microstates W_n by the function of the density of quantum states $g(E)$

$$w(E) = W_n g(E). \tag{3.28}$$

Inasmuch as the function of the density of states $g(E) \sim E^r$ (where $r > 0$) is an increasing function of energy E, and W_n is an exponentially decreasing function, their product $w(E)$ has a maximum (Fig. 3.2). It is evident that the function $w(E)$ takes on a maximum value at $E = \bar{E}$, where $\bar{E}$ is the mean value of energy of the system in the thermostat.

Thus, distribution functions over energies both for isolated systems (the microcanonical distribution) and systems in the thermostat (the canonical distribution) have a maximum (Figs. 1.5 and 3.2). However, in the case of the microcanonical distribution the width of the distribution curve is infinitely small (only one value $E = E_0$ corresponds to energy), and in the case of the canonical distribution the width of the distribution curve has a finite

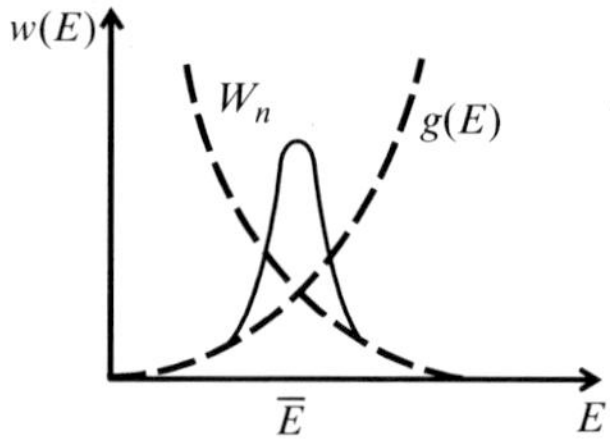

Fig. 3.2. The distribution over energies

value. The width of the distribution, presented in the form of a rectangle (see Fig. 1.6), is on the order of the fluctuation in energy of a system in the thermostat. Inasmuch as fluctuations in energy for macroscopic systems are insignificant, the canonical distribution can be applied also to isolated systems with a specified accuracy.

Note that known Maxwell and Boltzmann distributions as a particular case stem from Gibbs canonical distribution (3.24) for classical systems, i.e. the Gibbs canonical distribution is a generalized form of Maxwell and Boltzmann distributions.

3.2 Free Energy: Statistical Sum and Statistical Integral

The explicit form of normalizing factors A and A_{cl}, entering into the Gibbs distribution for quantum (3.16) and classical systems (3.24), is given by expressions (3.18) and (3.25a), respectively. In the given section, consider what physical sense these normalizing factors possess, i.e. through which known physical quantities they are expressed. To do this use the Boltzmann relationship relating entropy of the system to distribution function.

In the case of *quantum systems*, entropy S and the mean value of the distribution function are related by relationship (1.113), i.e.

$$S = -k_0\overline{\ln W_n}. \tag{3.29}$$

Having substituted the expression of the distribution function (3.16) into (3.29), we get

$$S = -k_0 \ln A + \frac{\overline{E_n}}{T}. \tag{3.30}$$

If the mean value of energy is denoted by $\overline{E_n} = E$, (3.30) takes the form

$$k_0 T \ln A = E - TS. \tag{3.31}$$

Knowing that $E - TS = F$, for the normalizing factor we have

$$A = \mathrm{e}^{F/k_0 T}. \tag{3.32}$$

If expression (3.32) is taken into account in (3.16), the Gibbs canonical distribution for quantum systems takes the entirely specified form

$$W_n = \mathrm{e}^{\frac{F-E_n}{k_0 T}}. \tag{3.33}$$

This form of the canonical distribution and the normalization condition (3.17) make it possible to calculate free energy:

$$F = -k_0 T \ln Z, \tag{3.34}$$

where

$$Z = \sum_n \mathrm{e}^{-E_n/k_0 T} \tag{3.35}$$

is called *the statistical sum.*

In the case of *classical systems*, the relation between entropy and distribution function is given by expression (1.125)

$$S = -k_0 \ln \left[(2\pi\hbar)^{3N} \overline{\rho(q,p)} \right]. \tag{3.36}$$

If we take into account the expression of the distribution function (3.24) in (3.36), we get

$$S = -k_0 \ln \left[(2\pi\hbar)^{3N} A_{\text{cl}} \right] + \frac{\overline{E(q,p)}}{T}. \tag{3.37}$$

Introducing the notation $\overline{E(q,p)} = E$ in expression (3.37), we can rewrite it in the form

$$k_0 T \ln \left[(2\pi\hbar)^{3N} A_{\text{cl}} \right] = E - TS = F. \tag{3.38}$$

Hence for the normalizing factor we get

$$A_{cl} = (2\pi\hbar)^{-3N} e^{F/k_0 T}. \tag{3.39}$$

Substitute this expression into (3.24). Then the distribution function for classical systems takes the form

$$\rho(q,p) = (2\pi\hbar)^{-3N} \mathrm{e}^{\frac{F-E(q,p)}{k_0 T}}. \tag{3.40}$$

Using normalization condition (3.25), for the function of free energy we get

$$F = -k_0 T \ln Z_{\text{cl}}, \tag{3.41}$$

where

$$Z_{\text{cl}} = \int{}' \mathrm{e}^{-E(q,p)/k_0 T} \frac{\mathrm{d}\Gamma}{(2\pi\hbar)^{3N}} \tag{3.42}$$

is called *the statistical integral;* $\mathrm{d}\Gamma = \prod_{i=1}^{3N} \mathrm{d}p_i \, \mathrm{d}q_i$ is an element of volume in the phase space.

The prime above the integral sign shows that the integration in the phase space is carried out only over phase points ЯHrresponding to nonequivalent physical microstates. In fact, when intercommutating two particles of a system with N number of identical particles, new phase points obtained are equivalent to preceding ones, i.e. both phase points correspond to the same physical microstate. It is evident that the number of such equivalent points equals $N!$. Therefore, in order to obtain the true expression of the statistical integral Z_{cl}, it is needed to take away the prime above the integral sign and conduct the integration over all phase points, and thereupon divide the obtained expression by $N!$.

Thus, the final expression for the statistical integral takes the form

$$Z_{\mathrm{cl}} = \frac{1}{N!}\int \mathrm{e}^{-E(q,p)/k_0 T}\frac{\mathrm{d}\Gamma}{(2\pi\hbar)^{3N}}. \tag{3.43}$$

Note that the consideration of $N!$ in formula (3.43) provides additivity of entropy and other thermodynamic quantities of the system. In particular, the true expression of statistical integral (3.43) circumvents the known *Gibbs paradox* (see Sect. 4.2).

In conclusion, consider one more substantiation of the Gibbs distribution and show that from it known thermodynamic relationships for the free energy can be obtained.

To do this use the normalization condition for the Gibbs distribution function (3.34):

$$\sum_n W_n = \sum_n \mathrm{e}^{\frac{F-E_n}{k_0 T}} = 1. \tag{3.44}$$

Naturally, energy of the system E_n ought to depend on external parameters. Assume that there is in all one external parameter and this parameter is volume: $E_n = E_n(V)$. Then the left-hand side of equality (3.44) is a function of temperature T and volume V. Knowing this, differentiate up both sides of (3.44). As a result, we get

$$\sum_n \frac{W_n}{k_0 T}\left[\mathrm{d}F - \frac{\partial E_n}{\partial V}\mathrm{d}V - \frac{F-E_n}{T}\mathrm{d}T\right] = 0. \tag{3.45}$$

Rewrite it in the following form

$$\mathrm{d}F\sum_n W_n = \mathrm{d}V\sum_n W_n\frac{\partial E_n}{\partial V} + \frac{\mathrm{d}T}{T}\left(F - \sum_n W_n E_n\right). \tag{3.46}$$

If we take into account that on the left-hand side of this inequality $\sum\limits_n W_n = 1$, and on the right-hand side

$$\sum_n W_n E_n = E, \sum_n W_n \frac{\partial E_n}{\partial V} = \frac{\partial E}{\partial V} = -P, F - E = -TS, \tag{3.47}$$

(3.46) takes the form of the known thermodynamic inequality

$$\mathrm{d}F = -P\,\mathrm{d}V - S\,\mathrm{d}T. \tag{3.48}$$

This conclusion shows that the Gibbs canonical distribution satisfies all basic principles and laws of thermodynamics and statistical physics.

3.3 Gibbs Method and Basic Objects of its Application

In Sect. 2.4 we came to the conclusion that if the explicit form of free energy F is known, we can find the equation of the state of the system, calculate its entropy and all thermodynamic coefficients:

thermal equation of state and entropy, respectively,

$$P = -\left(\frac{\partial F}{\partial V}\right)_T, S = -\left(\frac{\partial F}{\partial T}\right)_V, \tag{3.49}$$

caloric equation

$$E = F - T\left(\frac{\partial F}{\partial T}\right)_V; \tag{3.50}$$

heat capacity

$$C_V = T\left(\frac{\partial S}{\partial T}\right)_V, \quad C_P = T\left(\frac{\partial S}{\partial T}\right)_P; \tag{3.51}$$

coefficients of thermal expansion, thermal coefficients of pressure and isothermal coefficients of compression, respectively,

$$\alpha_P = \frac{1}{V}\left(\frac{\partial V}{\partial T}\right)_P, \quad \beta_V = \frac{1}{P}\left(\frac{\partial P}{\partial T}\right)_V, \quad \gamma_T = -\frac{1}{V}\left(\frac{\partial V}{\partial P}\right)_T. \tag{3.52}$$

In order to find the explicit form of the function of free energy, as seen in expressions (3.34) and (3.41), for quantum systems it is necessary to know the statistical sum (3.35), and for classical systems, the statistical integral (3.43). And to calculate the statistical sum and statistical integral the explicit form of the energy spectrum E_n or the Hamilton function $E(q,p)$ needs to be known. Thus, the solution of the problem is reduced to finding E_n or $E(q,p)$. These problems are solved in quantum and classical mechanics, conformably.

This chain of calculations in statistical physics composes the essence of the *Gibbs method*: in order to find pressure P, entropy S and mean energy E, it is necessary to know free energy F, and to do this it is necessary to compute the statistical sum Z or the statistical integral Z_{kl}, for which it is necessary to know the explicit form of functions E_n or $E(q,p)$. Thus, the complete solution of the statistical problem for concrete systems is reduced to finding E_n or $E(q,p)$.

The essence of the Gibbs method, being on *the top of statistical physics*, can be schematically presented in the form:

$$\begin{pmatrix} E_n \\ E(q,p) \end{pmatrix} \Rightarrow \begin{pmatrix} Z \\ Z_{kl} \end{pmatrix} \Rightarrow F \Rightarrow (P, E, S, C_V, C_P, \ldots). \tag{3.53}$$

Thus, the chain of calculations, beginning with E_n or $E(q,p)$, is accomplished. It should be noted that the basis of this method composes the postulate of statistical physics, i.e. the microcanonical distribution, inasmuch as the canonical distribution, composing the basis of the Gibbs method, also stems from the microcanonical distribution. Consequently, *the microcanonical distribution is the basis of all statistical physics.*

As is seen from the chain of calculations (3.53), first of all, it is necessary to know the energy spectrum E_n or Hamilton function $E(q,p)$. Unfortunately, in the general case, for any system the explicit form of these functions is not known. Therefore, the Gibbs method can be applied only to particular cases.

Energy of interaction between two arbitrary particles in the system is denoted by u_{ik}, and mean kinetic energy of the particle as $\bar{\varepsilon}$. Introduce the dimensionless parameter

$$\eta = \frac{u_{ik}}{\bar{\varepsilon}}. \tag{3.54}$$

Depending on the value of the parameter η all systems surrounding us can be divided into four groups:

1. Ideal gases: $\eta = 0$, i.e $u_{ik} = 0$,
2. Real gases: $\eta \ll 1$, i.e. $u_{ik} \ll \bar{\varepsilon}$,
3. Liquids: $\eta \approx 1$, i.e. $u_{ik} \approx \bar{\varepsilon}$,
4. Crystalline solids: $\eta \gg 1$, i.e. $u_{ik} \gg \bar{\varepsilon}$.

In subsequent chapters we comprehensively consider three of these groups: an ideal gas, real gases and crystalline solids. In the case of an ideal gas, inasmuch as interaction between particles is completely absent ($u_{ik} = 0$), for energy we can write the exact expression and analytically bring the chain of calculations (3.53) to the end. In the case of real gases ($\eta \ll 1$) and crystalline solids ($\eta^{-1} \ll 1$), a small parameter exists. Therefore, we can expand energy in a series and bring calculations (3.53) to the end.

In the case of liquid, rough methods are not applicable owing to the absence of a small parameter, since $\eta \approx 1$. Generally, the analytical statistical theory of liquids is very complex. Therefore, only semi-empirical or phenomenological theories are possible.

3.4 Grand Canonical Distribution for Open Systems

Before passing to the application of the Gibbs method to concrete systems, consider the canonical distribution in the general form, i.e. generalize it for open systems. The thermodynamics of such systems is expounded in Sect. 9.9.

Assume that the considered system with energy E_N, the number of particles N, entropy S, temperature T and the chemical potential μ is a small part of a very large system – the thermostat (Fig. 3.3). Let the corresponding parameters of the thermostat be E'_N, N', S', T', μ'.

Also assume that the volume of the system does not change ($V = \text{const}$); however, at the boundary of division of the system the thermostat transmits particles ($\Delta N \neq 0$) and heat ($\Delta Q \neq 0$). Therefore, in the thermodynamic equilibrium $T = T'$ and $\mu = \mu'$. Thus, the considered system is a small, but

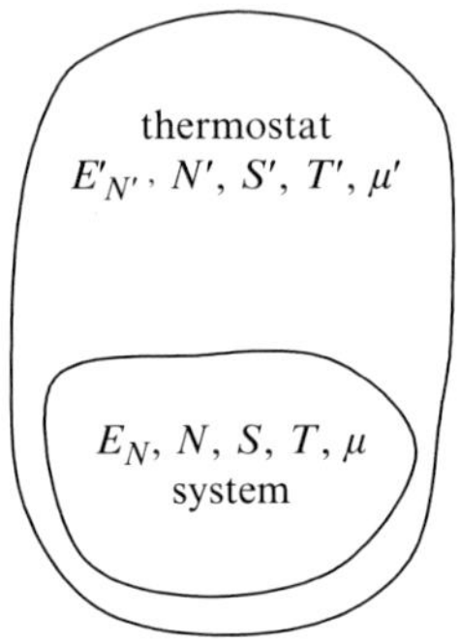

Fig. 3.3. The open system in the thermostat

macroscopic system with a variable number of particles in thermodynamic equilibrium with the thermostat:

$$E_N \ll E'_{N'}, \quad N \ll N', \quad S(E_N, N) \ll S'(E'_{N'}, N'). \tag{3.55}$$

Inasmuch as the system + the thermostat, as a whole, represent a completely isolated system, it can be written as:

$$N + N' = N_0 = \text{const}, \quad E_N + E'_{N'} = E_0 = \text{const}. \tag{3.56}$$

Consider separately quantum and classical systems.

Quantum systems. Inasmuch as in the system the number of particles N changes, energy of microstates ought to depend not only on the totality of the quantum numbers n, but also on N. Let E_{nN} be energy of a microstate of the system with the number of particles N that are found in the quantum state, determined by the totality of quantum numbers n.

It is necessary to find the answer to the question: *What is the probability that the system is found in a concrete microstate with energy E_{nN} and the number of particles N, in order that the thermostat can be found in any of its microstates.*

Denote the probability sought by W_{nN}. In order to find this probability, take into account that the complex system (the system + the thermostat) is completely isolated and apply to it microcanonical distribution (3.8).

Having generalized microcanonical distribution (3.8) for the considered case, we get

$$\mathrm{d}W = \text{const}\, \delta(E'_{N'} + E_N - E_0)\delta_{N'+N,N_0} \mathrm{d}G'_{N'} \mathrm{d}G_N, \tag{3.57}$$

where the δ-function and the δ-symbol show isolatedness of the complex system.

In order to answer the question posed above, it is necessary to accept $E_N = E_{nN}$ and $dG_N = 1$, and thereupon integrate (3.57) over $dG'_{N'}$ and add up with respect to N'. Repeating the reasoning, adduced in Sect. 3.1, the probability sought can be presented in the form analogous to (3.13):

$$W_{nN} = \text{const}\,\exp\left[\frac{1}{k_0}S'(E_0 - E_{nN}, N_0 - N)\right]. \tag{3.58}$$

Inasmuch as the system is significantly less than the thermostat, i.e. condition (3.55) takes place, entropy of the thermostat S' can be expanded with respect to E_{nN} and N. If we restrict ourselves to the linear approximation, we get

$$S'(E_0-E_{nN}, N_0-N) = S'(E_0, N_0)-\left(\frac{\partial S'}{\partial E'}\right)_{V,N} E_{nN}-\left(\frac{\partial S'}{\partial N'}\right)_{V,E} N. \tag{3.59}$$

Thermodynamic relationship for systems with a variable number of particles (2.194) can be presented in the form

$$\mathrm{d}S = \frac{1}{T}\mathrm{d}E + \frac{P}{T}\mathrm{d}V - \frac{\mu}{T}\mathrm{d}N. \tag{3.60}$$

Hence we get

$$\left(\frac{\partial S}{\partial E}\right)_{V,N} = \frac{1}{T};\left(\frac{\partial S}{\partial N}\right)_{V,E} = -\frac{\mu}{T}. \tag{3.61}$$

Taking into account these relationships in (3.59), we get

$$S'(E_0 - E_{nN}, N_0 - N) = S'(E_0, N_0) - \frac{E_{nN}}{T} + \frac{\mu N}{T}. \tag{3.62}$$

From expressions (3.58) and (3.62) we get the probability sought

$$W_{nN} = A\,\mathrm{e}^{(\mu N - E_{nN})/k_0 T}. \tag{3.63}$$

This expression is called *the grand canonical distribution for open systems.*

In order to find the constant A, entering into (3.63), use the expression of entropy (1.113), presented in the form

$$S = -k_0\overline{\ln\, W_{nN}}. \tag{3.64}$$

Then from (3.63) and (3.64) we have

$$S = -k_0\, \ln\, A - \frac{\mu\bar{N}}{T} + \frac{\bar{E}_{nN}}{T}. \tag{3.65}$$

Having denoted mean values by $\bar{N} = N$ and $\bar{E}_{nN} = E$, we get

$$k_0 T\, \ln\, A = E - TS - \mu N = F - \mu N = F - \Phi = \Omega. \tag{3.66}$$

Hence we have

$$A = e^{\Omega/k_0T}. \tag{3.67}$$

As a result, the grand canonical distribution (3.63) finally takes the form

$$W_{nN} = e^{\frac{\Omega+\mu N - E_{nN}}{k_0T}}. \tag{3.68}$$

The grand thermodynamic potential $\boldsymbol{\Omega}$ is found from the normalization condition

$$\sum_N \sum_n W_{nN} = 1. \tag{3.69}$$

Having substituted the expression of the distribution function (3.68) into (3.69), we get

$$\boldsymbol{\Omega} = -k_0T \ln \left[\sum_N e^{\mu N/k_0T} \sum_n e^{-E_{nN}/k_0T} \right]. \tag{3.70}$$

With the aid of (3.70) we can find the dependence of the grand thermodynamic potential on T, μ and V, i.e. the explicit form of the function $\boldsymbol{\Omega} = \boldsymbol{\Omega}(T, \mu, V)$ for a concrete system. Thus, using the relationship $\boldsymbol{\Omega} = -PV$, pressure can be determined as a function of temperature, chemical potential and volume:

$$P = -\boldsymbol{\Omega}/V = P(T, \mu, V). \tag{3.71}$$

Solving (3.71) simultaneously with the equation $N = N(T, \mu, V) = -\ (\partial\boldsymbol{\Omega}/\partial\mu)_{T,V}$, obtained in (2.206), and eliminating μ, we find the equation of the state of the system $f(T, V, P, N) = 0$

Now adduce the generalized form of the canonical distribution for multicomponent systems. To do this the expression of the thermodynamic potential $\boldsymbol{\Phi} = \mu N$ is present in the general form

$$\Phi = \sum_i N_i \frac{\partial \Phi}{\partial N_i} = \sum_i \mu_i N_i, \tag{3.72}$$

where N_i and μ_i are the number of particles and chemical potential of the ith component, respectively. Then in (3.68), having carried out the replacement

$$\Phi = \mu N \to \sum_i \mu_i N_i;\ E_{nN} \to E_{nN_1N_2\ldots}, \tag{3.73}$$

we get the generalized form of the canonical distribution for a multicomponent system

$$W_{nN_1N_2\ldots} = \exp\left[\frac{\boldsymbol{\Omega} + \sum\limits_i \mu_i N_i - E_{nN_1N_2\ldots}}{k_0T} \right]. \tag{3.74}$$

Classical systems. Assume that there is a classical system with a number of particles N. The probability that a microstate of this system falls in an element of volume

$$\mathrm{d}\Gamma_N = \mathrm{d}q^{(N}\mathrm{d}p^{(N)} = \prod_{i=1}^{3N} \mathrm{d}q_i\,\mathrm{d}p_i, \tag{3.75}$$

taken near the point $(q,p) \equiv (q_1, q_2, \ldots, q_{3N}, p_1, p_2, \ldots, p_{3N})$ is

$$\mathrm{d}W_N(q,p) = \rho_N(q,p)\mathrm{d}\Gamma_N, \tag{3.76}$$

where $\rho_N(q,p)$ is the probability of a microstate of the system found in a unit volume of the phase space near the point (q,p), i.e. the distribution function. This function, like expression (3.63), can be presented in the form

$$\rho_N(q,p) = A_{\mathrm{cl}}\mathrm{e}^{\frac{\mu N - E_N(q,p)}{k_0 T}}. \tag{3.77}$$

The constant A_{cl} can be found, using the relationship between entropy and the mean value of the distribution function (1.125) and expression (3.77). As a result, we get

$$k_0 T \ln\left[(2\pi\hbar)^{3N} A_{\mathrm{cl}}\right] = E - TS - \mu N = F - \mathbf{\Phi} = \mathbf{\Omega}, \tag{3.78}$$

where $E = \overline{E_N(q,p)}$ and $N = \overline{N}$ are mean values of energy and the number of particles of the system, respectively. From (3.78) it follows that

$$A_{\mathrm{cl}} = (2\pi\hbar)^{-3N}\mathrm{e}^{\Omega/k_0 T}. \tag{3.79}$$

Having substituted this expression into (3.77), we get the final expression for the distribution function of classical open systems

$$\rho_N(q,p) = (2\pi\hbar)^{-3N}\mathrm{e}^{\frac{\Omega + \mu N - E_N(q,p)}{k_0 T}}. \tag{3.80}$$

In the considered case for open systems, the normalization condition has the appearance

$$\sum_N \int \rho_N(q,p)\mathrm{d}\Gamma_N = 1. \tag{3.81}$$

Using (3.81) and (3.80), the general expression of the grand thermodynamic potential can be written down as[1]

$$\mathbf{\Omega} = -k_0 T \ln\left[\sum_N \mathrm{e}^{\mu N/k_0 T}\frac{1}{N!}\int \mathrm{e}^{-E_N(q,p)/k_0 T}\frac{\mathrm{d}\Gamma_N}{(2\pi\hbar)^{3N}}\right]. \tag{3.82}$$

In conclusion, note the distinctions between three fundamental distributions: microcanonical, canonical and grand canonical.

1. Microcanonical distribution (1.99) or (3.57) is just for completely isolated systems. In this case, we do not take into account the fluctuation in either energy or the number of particles (see Fig. 1.5).

---

[1] Here the appearance of the factor $N!$ is based on reasoning when passing from the expression for statistical integral (2.14) to the expression (2.15).

2. Gibbs canonical distribution (3.16) is just for closed systems in the thermostat. In this case, the fluctuation in energy is taken into account (see Fig. 3.2), but the fluctuation in the number of particles is not, i.e. the mean number of particles $\overline{N}$ coincides with their real value. This distribution with reference to the number of particles can be called "the microcanonical distribution".
3. Gibbs grand canonical distribution (3.63) or (3.68) is just for open systems in the thermostat. In this case, the fluctuation both in energy and the number of particles is taken into account. This distribution with reference to energy and the number of particles can be called "the canonical distribution".

In particular, if we disregard the fluctuation in the number of particles, i.e. we suppose that the mean value of the number of particles $\overline{N}$ and the real N coincide, we get $\boldsymbol{\Omega} + \mu N = \boldsymbol{\Omega} + \mu \bar{N} = \boldsymbol{\Omega} + \boldsymbol{\Phi} = F$. Consequently, the grand canonical distribution (3.68) passes into the canonical distribution (3.33).

4

Ideal Gas

Summary. This chapter is devoted to the application of the Gibbs method to an ideal gas. On the basis of this method, free energy and entropy are calculated, and thermal and caloric equations of a state are found. The exposition of classical and quantum theories of the heat capacity of an ideal gas occupies a large place. A mixture of ideal gases and the Gibbs paradox are considered. Thermodynamic properties of an ideal gas consisting of polar molecules in an external electric field as well as an ideal gas consisting of magnetic dipoles (a paramagnetic in the gaseous phase) in an external magnetic field are also considered. At the end of the chapter a simple example of the thermodynamic state of a system with negative absolute temperature is discussed.

4.1 Free Energy, Entropy and Equation of the State of an Ideal Gas

Apply the Gibbs method to an ideal gas of volume V, consisting of a N number of identical particles (molecules). In the given section, we disregard the internal structure of a molecule and present it as a material point with the mass m. In an ideal gas, inasmuch as interaction between molecules is absent, internal energy can be presented in the form:

$$E(q,p) = \sum_{i=1}^{N} \frac{1}{2m}(p_{ix}^2 + p_{iy}^2 + p_{iz}^2). \tag{4.1}$$

In order to find the free energy of an ideal gas, it is necessary to calculate the statistical integral. To do this, substitute the expression of energy (4.1) into (3.43). Then we get

$$Z_{\mathrm{cl}} = \frac{1}{N!(2\pi\hbar)^{3N}} \left[\int \mathrm{e}^{-(p_x^2+p_y^2+p_z^2)/2mk_0T} \mathrm{d}q_x \mathrm{d}q_y \mathrm{d}q_z \mathrm{d}p_x \mathrm{d}p_y \mathrm{d}p_z \right]^N. \tag{4.2}$$

Take into account that

$$\int \mathrm{d}q_x \mathrm{d}q_y \mathrm{d}q_z = V. \tag{4.3}$$

Moreover, according to Appendix A, integrals, entering into (4.2), are easily calculated

$$\int_{-\infty}^{\infty} \mathrm{e}^{-p_x^2/2mk_0T} \mathrm{d}p_x = 2\int_{0}^{\infty} \mathrm{e}^{-p_x^2/2mk_0T} \mathrm{d}p_x = (2\pi m k_0 T)^{1/2}. \tag{4.4}$$

Note that integrals over $\mathrm{d}p_y$, $\mathrm{d}p_z$ also equal (4.4). Then with regard to (4.3) and (4.4) the statistical integral (4.2) takes the form

$$Z_{\mathrm{cl}} = \frac{1}{N!(2\pi\hbar)^{3N}} \left[V(2\pi m k_0 T)^{3/2}\right]^N. \tag{4.5}$$

Logarithmizing this expression and taking into account that for large N the logarithm $N!$ can be presented in the form

$$\ln N! = \ln 1 + \ln 2 + \cdots + \ln N = \sum_{x=1}^{N} \ln x \approx \int_1^N \ln x \,\mathrm{d}x = N \ln (N/e) \tag{4.6}$$

for $\ln Z_{\mathrm{cl}}$ we get

$$\ln Z_{\mathrm{cl}} = N \ln \left[\frac{eV}{N}\left(\frac{mk_0T}{2\pi\hbar^2}\right)^{3/2}\right]. \tag{4.7}$$

As a result, the free energy of an ideal gas $F = -k_0 T \ln Z_{\mathrm{cl}}$ takes the form

$$F = -k_0 N T \ln \left[\frac{eV}{N}\left(\frac{mk_0T}{2\pi\hbar^2}\right)^{3/2}\right]. \tag{4.8}$$

Using the expression of free energy (4.8) and relationship $P = -\left(\partial F/\partial V\right)_{T,N}$, we get the known thermal equation of the state of an ideal gas, i.e. the Mendeleev–Clapeyron equation

$$P = \frac{k_0 N T}{V}. \tag{4.9}$$

With regard to the expression of free energy (4.8) and the definition $S = -\left(\partial F/\partial T\right)_{V,N}$, entropy of an ideal gas can be expressed by T and V:

$$S(T,V) = \frac{3}{2}k_0 N + k_0 N \ln \left[\frac{eV}{N}\left(\frac{mk_0T}{2\pi\hbar^2}\right)^{3/2}\right]. \tag{4.10}$$

Expression of entropy (4.10) can be rewritten in the form

$$S(T,V) = k_0 N \ln\left(\frac{eV}{N}\right) + \frac{3}{2}k_0 N \ln T + BN, \tag{4.11}$$

where B is the constant for a specified gas, equal to

$$B = \frac{3}{2}k_0\left[1 + \ln\left(\frac{mk_0}{2\pi\hbar^2}\right)\right].$$

Then from (4.11) for the isochoric heat capacity of an ideal gas $C_V = T\left(\partial S/\partial T\right)_V$ we get the known expression

$$C_V = \frac{3}{2}k_0 N. \tag{4.12}$$

If this expression is taken into account in (4.11), entropy takes the form

$$S(T,V) = k_0 N \ln\left(\frac{Ve}{N}\right) + C_V \ln T + BN. \tag{4.13}$$

Use the equation of the state of an ideal gas $V = k_0 TN/P$ in (4.13). Then entropy can be presented as a function of T and P:

$$S(T,P) = (k_0 N + C_V)\ln T - k_0 N \ln P + B_0 N. \tag{4.14}$$

Here $B_0 = B + k_0(1 + \ln k_0)$. Taking into account the expression (4.14) and the definition $C_P = T\left(\partial S/\partial T\right)_P$, for the isobaric heat capacity we get

$$C_P = C_V + k_0 N = C_V + R. \tag{4.15}$$

For the considered model of an ideal gas it is not difficult to determine the caloric equation of a state, i.e. the expression of energy. To do this, take into account (4.8) and (4.10) in the expression $E = F + TS$. As a result, we get

$$E = \frac{3}{2}k_0 NT. \tag{4.16}$$

As is seen, energy of an ideal gas does not depend on volume and is determined only by temperature. Note that having taken the derivative with respect to temperature in (4.16), we get the expression of the isochoric heat capacity (4.12).

If we use the equation of state (4.9), expressions of entropy (4.13) and isobaric heat capacity (4.15), we get equations of the isotherm ($T = \text{const}$) and adiabat ($S = \text{const}$) of an ideal gas:

$$\begin{aligned} &\text{if } T = \text{const}, \text{then } PV = \text{const} \\ \text{and}& \\ &\text{if } S = \text{const}, \text{then } PV^{\gamma} = \text{const}, \end{aligned} \tag{4.17}$$

where $\gamma = C_P/C_V$ is the ratio of heat capacities. Inasmuch as $\gamma > 1$, from (4.17) it follows that the slope of the curve of the adiabat is more like the slope of curve of the isotherm.

Using the equation of state (4.9) and the equation of the adiabat (4.17), known relationships between T and V, as well as T and P, can be found.

$$TV^{\gamma-1} = \text{const}; \quad T^{\gamma}P^{1-\gamma} = \text{const}. \tag{4.18}$$

Note that the latter relationship in the case of an adiabatic process ($S = \text{const}$) can also be obtained from expression (4.14).

4.2 Mixture of Ideal Gases: Gibbs Paradox

Dwell upon one important question of the classical statistics more comprehensively. This question is associated with correction of the statistical integral, i.e. with the presence of the factor $1/N!$ in expression (3.43)[1].

An example of an ideal gas shows that without consideration of the given factor, it is the violated additivity of entropy and free energy which leads to the Gibbs paradox. Indeed, *if the factor* $1/N!$ *in the expression of the statistical integral* (4.2) *is disregarded*, for free energy instead of (4.8) and for entropy instead of (4.13) we get the following expression, conformably,

$$F = -k_0NT \ln \left[V \left(\frac{mk_0T}{2\pi\hbar^2} \right)^{3/2} \right], \tag{4.19}$$

$$S(T, V) = k_0N \ln V + C_V \ln T + BN. \tag{4.20}$$

From these expressions it is seen that neither free energy nor entropy satisfies the condition of additivity. For instance, at pre-assigned conditions (identical temperatures) as the number of particles N increases two times, F and S increase more than two times, i.e. inasmuch as N increases, the heat capacity and volume also increase two times.

The application of expression for entropy (4.20), obtained without consideration of $1/N!$, to a mixture of ideal gases leads to the appearance of the Gibbs paradox. In order to show this, assume that a vessel with an ideal gas is divided by a partition *ab* into two parts with volumes V_1 and V_2. In each part, different ideal gases (with different heat capacities) with a number of molecules N_1 and N_2, respectively, are found. Temperature and pressure of the gases in both parts are identical (Fig. 4.1).

Entropy of the complete system, consisting of two parts divided by the partition, with regard to expression (4.20) can be presented as:

[1] In quantum statistics this problem is solved on the basis of the indistinguishability principle: a wave function of an ideal gas, consisting of identical particles, ought to be either symmetric or antisymmetric (see Chap. 7)

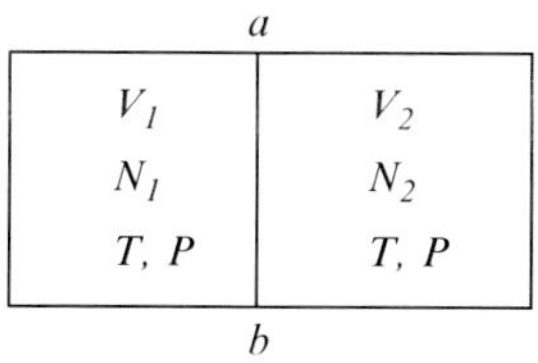

Fig. 4.1. The illustration for the Gibbs Paradox

$$\begin{aligned} S_0 = S_1 + S_2 = k_0 N_1 \ln V_1 + k_0 N_2 \ln V_2 \\ +(C_{V_1} + C_{V_2}) \ln T + B_1 N_1 + B_2 N_2. \end{aligned} \tag{4.21}$$

If the partition ab is taken away, as a result of diffusion of gases a mixture of gases forms. The system with the number of molecules $(N_1 + N_2)$ is found in a state of equilibrium with the identical density over the entire volume $(V_1 + V_2)$. In this final state, entropy of the system can also be presented on the basis of (4.20) in the form

$$S = k_0(N_1 + N_2) \ln(V_1 + V_2) + (C_{V_1} + C_{V_2}) \ln T + B_1 N_1 + B_2 N_2. \tag{4.22}$$

The change in entropy as a result of taking away the partition is

$$\Delta S = S - S_0 = k_0 N_1 \ln \left(\frac{V_1 + V_2}{V_1} \right) + k_0 N_2 \ln \left(\frac{V_1 + V_2}{V_2} \right). \tag{4.23}$$

It is seen that as a result of the diffusion of gases the system passes into a new state with the larger entropy: $\Delta S > 0$, i.e. diffusion of different gases is an irreversible process, and so it ought to be. However, from (4.23) it is seen that in the expression of ΔS no parameter characterizing the distinction of gases enters. Thus it follows that if between two parts of the system with identical gases the partition is taken away, then, according to formula (4.23), entropy increases by a magnitude ΔS as a result of self-diffusion. This alogical result is precisely the Gibbs paradox.

The paradox consists of the fact that as identical gases are mixed the macroscopic state of the system does not change and therefore entropy ought not to change. However, entropy, calculated on the basis of formula (4.20), changes. In the particular case, if $V_1 = V_2 = V$ and $N_1 = N_2 = N$, according to (4.23), the change in entropy, independent of the kind of gas, is a constant quantity, equal to

$$\Delta S = 2R \ln 2, \tag{4.24}$$

where $R = k_0 N$ is the universal gas constant.

Now show that if to a mixture of ideal gases, the expression for entropy (4.13) obtained with consideration of $1/N!$ is applied, the above-indicated Gibbs paradox disappears, i.e. *as different gases are mixed entropy increases, and as identical gases are mixed entropy does not change.*

Indeed, inasmuch as on the right and left parts of the system, presented in Fig. 4.1, different gases are found, entropy of the original state of the system on the basis of formula (4.13) has the appearance:

$$S_0 = S_{01} + S_{02} = k_0 N_1 \ln\left(\frac{eV_1}{N_1}\right) + k_0 N_2 \ln\left(\frac{eV_2}{N_2}\right) \\ +(C_{V_1} + C_{V_2}) \ln T + B_1 N_1 + B_2 N_2, \tag{4.25}$$

and after mixing (taking away the partition), entropy of the system in the final state ought to equal

$$S = S_1 + S_2 = k_0 N_1 \ln\left[\left(\frac{(V_1+V_2)e}{N_1}\right)\right] \\ + k_0 N_2 \ln\left[\left(\frac{(V_1+V_2)e}{N_1}\right)\right] + (C_{V_1} + C_{V_2}) \ln T + B_1 N_1 + B_2 N_2, \tag{4.26}$$

and the change in entropy is

$$\Delta S = S - S_0 = k_0 N_1 \left[\ln\left(\frac{V_1+V_2}{N_1}\right) - \ln\frac{V_1}{N_1}\right] \\ + k_0 N_2 \left[\ln\left(\frac{V_1+V_2}{N_2} - \ln\frac{V_2}{N_2}\right)\right]. \tag{4.27}$$

Inasmuch as each of the expressions entering into (4.27) is positive, then

$$\Delta S > 0. \tag{4.28}$$

Assume that on both sides of the partition an identical ideal gas is found (Fig. 4.1). In the presence of the partition, entropy on the basis of formula (4.13) has the appearance:

$$S_0' = k_0 N_1 \ln\frac{eV_1}{N_1} + k_0 N_2 \ln\frac{eV_2}{N_2} + (C_{V_1} + C_{V_2}) \ln T + B(N_1 + N_2). \tag{4.29}$$

After taking away the partition the system occupies volume $(V_1 + V_2)$ with the number of molecules $(N_1 + N_2)$. Therefore the expression of entropy in this final state, according to (4.13), takes the form

$$S' = k_0(N_1 + N_2) \ln\frac{e(V_1+V_2)}{N_1+N_2} + (C_{V_1} + C_{V_2}) \ln T + B(N_1 + N_2). \tag{4.30}$$

As a result, the change in entropy is

$$\Delta S' = S' - S_0' = k_0 N_1 \left[\ln\frac{V_1+V_2}{N_1+N_2} - \ln\frac{V_1}{N_1}\right] \\ + k_0 N_2 \left[\ln\tfrac{V_1+V_2}{N_1+N_2} - \ln\frac{V_2}{N_2}\right]. \tag{4.31}$$

Inasmuch as in the presence and in the absence of the partition, pressure and temperature are identical, from the equation of state (4.9) follows the identity:

$$\frac{V_1 + V_2}{N_1 + N_2} = \frac{V_1}{N_1} = \frac{V_2}{N_2}. \tag{4.32}$$

Taking into account (4.32) in (4.31) we see that entropy does not change:

$$\Delta S' = 0. \tag{4.33}$$

Thus, in classical statistics, consideration of the factor $1/N!$ leads to the correct calculation of the statistical integral and the circumventment of the Gibbs paradox.

4.3 Law About Equal Distribution of Energy Over Degrees of Freedom: Classical Theory of Heat Capacity of an Ideal Gas

The essence of this law consists in the fact that if the motion of particles of an ideal gas is described by equations of classical mechanics, mean values of energy, accounting for each degree of freedom of a classical particle, are identical and the value of this energy is determined by temperature. Note that this result for the simplest case was obtained in Sect. 4.1, in which an atomar ideal gas was considered and for the mean energy of each particle the expression $E/N = 3k_0T/2$ was obtained [see (4.16)]. In this case, inasmuch as the degree of freedom of a particle (an atom) equals 3, energy corresponding to each degree of freedom equals $E/3N = k_0T/2$.

Here we will consider the more general case. Let N molecules occupy volume V; therewith each molecule consists of a $n \geq 2$ number of atoms. Assume that the translational and the rotational motion of a molecule and also vibrational motions of atoms in reference to each other inside a molecule are classical. Apply the Gibbs method to this ideal gas and find free energy, entropy, mean energy and equation of the state of the system.

Having substituted the expression of the total internal energy of a classical ideal gas

$$E(q,p) = \sum_{i=1}^{N} \varepsilon_i(q,p) \tag{4.34}$$

into the expression of the statistical integral (4.74), we get:

$$Z_{kl} = \frac{(2\pi\hbar)^{-3Nn}}{N!} \left[\int \mathrm{e}^{-\varepsilon(q,p)/k_0T} \mathrm{d}q_1 \mathrm{d}q_2 \ldots \mathrm{d}q_{3n} \mathrm{d}p_1 \mathrm{d}p_2 \ldots \mathrm{d}p_{3n} \right]^N, \tag{4.35}$$

here $\varepsilon(q,p)$ is the energy of an arbitrary molecule, n is the number of atoms in a molecule, $3n$ is the number of degrees of freedom of one molecule, $q_1, q_2, \ldots, q_{3n}$ are generalized coordinates, determining the position of the molecule, and $p_1, p_2, \ldots, p_{3n}$ are generalized impulses.

It is known that the total energy is a quadratic function of coordinates and impulses

$$\varepsilon(q,p) = a_{ik}p_ip_k + b_{ik}q_iq_k, \tag{4.36}$$

where a_{ik} and b_{ik} are intrinsic tensors of the molecule with constant components. Then (4.35) takes the form

$$Z_{kl} = \frac{(2\pi\hbar)^{-3Nn}}{N!}\left[\int \mathrm{e}^{-(a_{ik}p_ip_k+b_{ik}q_iq_k)/k_0T}\,\mathrm{d}q_1\mathrm{d}q_2\ldots\mathrm{d}q_{3n}\mathrm{d}p_1\mathrm{d}p_2\ldots\mathrm{d}p_{3n}\right]^N. \tag{4.37}$$

The position of a molecule as a whole can be determined by $3n$ coordinates, knowing the Cartesian coordinates of each atom. However, for convenience, coordinates determining the position of a molecule can be chosen as follows:

coordinates of the centre of masses of a molecule

$$q_1, q_2, q_3 \Rightarrow x, y, z,$$

and angles of rotation of a molecule around three mutual-perpendicular axes, passing through centres of masses

$$q_4, q_5, q_6 \Rightarrow \varphi_1, \varphi_2, \varphi_3.$$

The remaining $(3n-6)$ coordinates characterize the vibrational motion of atoms in reference to each other inside the molecule.

Note that the above is true only for molecules with the number of atoms $n > 2$ and when all atoms are not found in one straight line (the non-linear configuration of a molecule). For two-atom ($n = 2$) and multi-atom ($n > 2$) molecules, in which atoms are positioned in one straight line (the linear molecule), the number of rotational degrees of freedom equals not 3, but 2; therefore, for the indicated case, the number of vibrational degrees of freedom equals $(3n-5)$.

Consider a gas consisting of multi-atom ($n > 2$) non-linear molecules. It is clear that in the absence of an external field, energy of such molecules $\varepsilon(q,p)$ does not depend on coordinates determining translational (x, y, z) and rotational $(\varphi_1, \varphi_2, \varphi_3)$ motions. Therefore, the integral with respect to six coordinates in (4.37) gives the volume of the gas

$$\int \mathrm{d}q_1\mathrm{d}q_2\mathrm{d}q_3 \Rightarrow \int \mathrm{d}x\,\mathrm{d}y\,\mathrm{d}z = V, \tag{4.38}$$

and the constant angle $(2\pi)^3$

$$\int \mathrm{d}q_4, \mathrm{d}q_5, \mathrm{d}q_6 \Rightarrow \int \mathrm{d}\phi_1, \mathrm{d}\phi_2, \mathrm{d}\phi_3 = (2\pi)^3. \tag{4.39}$$

Taking into account (4.38) and (4.39) in (4.35), we get

$$Z_{\text{cl}} = \frac{(2\pi\hbar)^{-3Nn}}{N!} \times \left[(2\pi)^3 V \int \mathrm{e}^{-a_{ik}p_i p_k + b_{ik} q_i q_k / k_0 T} \underbrace{\mathrm{d}q_7 \mathrm{d}q_8 \ldots \mathrm{d}q_{3n}}_{3n-6} \mathrm{d}p_1 \mathrm{d}p_2 \ldots \mathrm{d}p_{3n} \right]^N . \tag{4.40}$$

Pass to new variables q_i' and p_i':

$$q_i = \sqrt{k_0 T} q_i'; \quad p_i = \sqrt{k_0 T} p_i' \tag{4.41}$$

and denote all quantities, not depending on temperature and volume, by $A =$ const. Then the statistical integral acquires the very simple shape

$$Z_{\text{cl}} = A(VT^{\ell/2})^N. \tag{4.42}$$

Here for multi-atom ($n > 2$) non-liner molecules

$$\ell = 6n - 6. \tag{4.43}$$

In the case of two-atom ($n = 2$) and multi-atom ($n > 2$) linear molecules for the statistical integral Z_{cl} the same expression as (4.42) is obtained only in this case

$$\ell = 6n - 5. \tag{4.44}$$

Taking into account the expression of the statistical integral (4.42) in the relationship $F = -k_0 T \ln Z_{\text{cl}}$, for the free energy we get

$$F = -k_0 T \ln A - k_0 N T \ln V - k_0 N T \frac{\ell}{2} \ln T. \tag{4.45}$$

Hence, the equation of the state of an ideal gas $P = -\left(\partial F / \partial V\right)_T$ can be found:

$$P = \frac{k_0 N T}{V}, \tag{4.46}$$

entropy $S = -\left(\partial F / \partial T\right)_V$

$$S = k_0 \ln A + k_0 N \ln V + \frac{\ell}{2} k_0 N \ln T + \frac{\ell}{2} k_0 N, \tag{4.47}$$

and the total energy $E = F + TS$

$$E = \frac{\ell}{2} k_0 N T. \tag{4.48}$$

Hence, two conclusions can be drawn: firstly, from (4.46) it is seen that the pressure of an ideal gas (the thermal equation of state) does not depend on the degree of complexity and structure of molecules ℓ and is determined only by the number of molecules; secondly, from (4.48) it follows that the mean energy of an ideal gas (the caloric equation) does not depend on volume and is determined by temperature T and complexity of the structure of molecules, i.e. the parameter ℓ.

Expression for the mean energy (4.48) can be presented in the form

$$\frac{E}{N\ell} = \frac{k_0 T}{2}. \tag{4.49}$$

Hence the law about the equal distribution of energy of an ideal gas over degrees of freedom follows: *If all kinds of motion (translational, rotational and vibrational) of molecules forming an ideal gas are classical, for each value of the number ℓ, associated with degrees of freedom of a molecule, the identical energy, equal to $k_0 T/2$, accounts.*

Now elucidate the relation of the number ℓ determined by the expressions (4.43) and (4.44) to the number of degrees of freedom. The number ℓ can be presented in the form

$$\ell = \ell_{\mathrm{t}} + \ell_{\mathrm{r}} + 2\ell_{\mathrm{v}}, \tag{4.50}$$

where ℓ_{t}, ℓ_{r} and ℓ_{v} are numbers of degrees of freedom of translational, rotational and vibrational motions, respectively.

For a one-atom gas $(n = 1):\ \ell_{\mathrm{r}} = \ell_{\mathrm{v}} = 0;\ \ell = \ell_{\mathrm{t}} = 3$

For a two-atom $(n = 2)$ molecule from (4.44) it follows that $\ell = 7$. This number between degrees of freedom is distributed as follows:

$$\ell_{\mathrm{t}} = 3;\ \ell_{\mathrm{r}} = 2;\ \ell_{\mathrm{v}} = 1\,. \tag{4.51}$$

For multi-atom $(n > 2)$ linear molecules, according to (4.44), $\ell = 6n - 5$. For such molecules degrees of freedom equal:

$$\ell_{\mathrm{t}} = 3;\ \ell_{\mathrm{r}} = 2;\ \ell_{\mathrm{v}} = 3n - 5\,. \tag{4.52}$$

For multi-atom $(n > 2)$ non-linear molecules from (4.43) it follows that $\ell = 6n - 6$. In this case

$$\ell_{\mathrm{t}} = 3;\ \ell_{\mathrm{r}} = 3;\ \ell_{\mathrm{v}} = 3n - 6\,. \tag{4.53}$$

In conclusion, note the following. Why is the vibrational degree of freedom doubled, in contradistinction to translational and rotational motions, in the number ℓ in expression (4.50)? This is associated with the fact that in the absence of an external field, the coordinate does not enter into the expression of energy of translational and rotational motions, but both the impulse and the coordinate enter into the expression of energy of the vibrational motion. In other words, energy of the vibrational motion is the sum of kinetic and potential energies.

4.3.1 Classical Theory of Heat Capacity of an Ideal Gas

The expression obtained above for energy of a classical ideal gas (4.48) makes it possible to easily calculate the isochoric heat capacity $C_V = (\partial E/\partial T)_V$:

$$C_V = \frac{\ell}{2} k_0 N = \frac{\ell}{2} R. \tag{4.54}$$

Hence it is seen that according to classical theory, the heat capacity of an ideal gas does not depend on temperature and is determined only by the degree of complexity and structure (ℓ) of a molecule.

1. For an atomar gas ($n = 1$) $\ell = \ell_t = 3$ and $C_V = 3R/2$.
2. For a two-atom ($n = 2$) and a multi-atom ($n > 2$) ideal gas, consisting of linear molecules, the heat capacity is determined by the expression

$$C_V = \frac{6n - 5}{2} R;\ n \geq 2\,. \tag{4.55}$$

3. For a multi-atom ($n > 2$) ideal gas, consisting of non-linear molecules, the heat capacity equals

$$C_V = \frac{6n - 6}{2} R;\ n > 2\,. \tag{4.56}$$

In the particular case, according to (4.55), the heat capacity of an ideal two-atom gas equals $C_V = 7R/2$. However, the test shows that the heat capacity of a two-atom ideal gas remains constant (T_0), only in the region of room temperatures, and the value of C_V equals not $7R/2$, but $5R/2$. For instance, at room temperature for hydrogen H_2, oxygen O_2 and nitrogen N_2 values of the heat capacity C_V/R equal 2.45, 2.51 and 2.51, respectively. However, with the growth of temperature, beginning from the room, the heat capacity C_V increases, and with the lowering of temperature, it lowers. In Fig. 4.2, values of the heat capacity C_V/R, obtained on the basis of classical theory, are indicated and the behaviour of the heat capacity depending on temperature is schematically depicted. From the figure the nonconformity of experimental data and classical theory is seen. Note that only quantum theory circumvents this nonconformity (see Sect. 4.4).

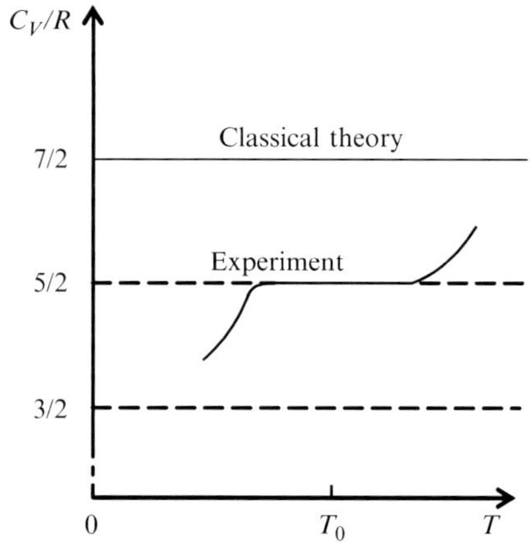

Fig. 4.2. The temperature dependence of the heat capacity C_V/R for a two-atom ideal gas

In conclusion, according to classical theory, determine the ratio of heat capacities. It is known that the difference of heat capacities $C_P - C_V$ is determined by the thermal equation of the state of an ideal gas $P = P(V, T)$. Inasmuch as the equation of a state does not depend on the degree of complexity of molecules forming a gas, and is determined only by their concentration, for a gas consisting of molecules of any complexity we get $C_P - C_V = R$. Taking this into account, and also relationships (4.55) and (4.56), for the ratio C_P/C_V we get the following values:

1. For an atomar gas ($n = 1$)

$$\frac{C_P}{C_V} = 1 + \frac{R}{C_V} = \frac{5}{3}. \tag{4.57}$$

2. For a two-atom ($n = 2$) and a multi-atom ($n > 2$) ideal gas, consisting of linear molecules

$$\frac{C_P}{C_V} = 1 + \frac{R}{C_V} = 1 + \frac{2}{6n-5} = \frac{6n-3}{6n-5}; \quad n \geq 2. \tag{4.58}$$

3. For a multi-atom ($n > 2$) ideal gas, consisting of non-linear molecules

$$\frac{C_P}{C_V} = 1 + \frac{R}{C_V} = 1 + \frac{2}{6n-6} = \frac{3n-2}{3n-3}; \quad n > 2. \tag{4.59}$$

As is seen from relationships (4.58) and (4.59), with the growth in the number of atoms in a molecule, the ratio C_P/C_V tends to unity $\lim\limits_{n\to\infty} (C_P/C_V) = 1$, whereas the difference $C_P - C_V = R$ does not depend on the number of atoms n in a molecule. This is associated with the fact that the difference $C_P - C_V$ is determined only by the equation of state. And the equation of the state (4.46) does not depend on the degree of complexity of molecules.

4.4 Quantum Theory of Heat Capacity of an Ideal Gas: Quantization of Rotational and Vibrational Motions

In Sect. 4.3 we noted the qualitative distinction between the experiment and classical theory in the temperature dependence of the heat capacity of a two-atom ideal gas (Fig. 4.2). The basic cause of this distinction resides in the supposition that all kinds of motion of molecules (translational, rotational and vibrational) which do not correspond to reality are classical.

Here we will consider the more general case, supposing that all motions of molecules are quantum, i.e. the energy spectrum, corresponding to these motions, is quantized. However we will show that at finite temperatures the translational motion is not quantized, i.e. this motion can be considered as a classical one. At the same time, consideration of the quantization of two other motions (rotational and vibrational ones) is very important in the theory of heat capacity. The quantum theory, taking into account this quantization, successfully circumvents the distinction indicated above.

Assume that an ideal gas, consisting of N number of two-atom molecules, occupies volume V. The total energy of an ideal gas has the appearance

$$E_{n'} = \sum_{i=1}^{N} \varepsilon_k^i, \tag{4.60}$$

where ε_k^i is energy of the ith molecule, k is the totality of quantum numbers, determining the state of one molecule, n' is the totality of all quantum numbers, determining one microstate of an ideal gas as a whole. Inasmuch as we suppose that all kinds of motions are independent of each other, the energy of one arbitrary ith molecule can be written as:

$$\varepsilon_k^i = \varepsilon_{\mathrm{t}} + \varepsilon_{\mathrm{r}} + \varepsilon_{\mathrm{v}}, \tag{4.61}$$

where $\varepsilon_{\mathrm{t}}, \varepsilon_{\mathrm{r}}, \varepsilon_{\mathrm{v}}$ is energy of the translational, rotational and vibrational motions, respectively. It is supposed that energy of all kinds of the motion is quantized and is determined by different corresponding quantum numbers. Their explicit form will be given slightly later. And for now the statistical sum present in the general form

$$Z = \sum_{n'} \mathrm{e}^{-E_{n'}/k_0 T} = \frac{1}{N!} \left(\sum_{k} \mathrm{e}^{-\varepsilon_k / k_0 T} \right)^N, \tag{4.62}$$

where the factor $1/N!$, entering into this expression, takes into account the fact that at the transposition of two molecules the state of a gas as a whole does not change, i.e. it is regarded that molecules are identical (see Sects. 3.2 and 4.1).

If we take into account the expression of energy of a molecule (4.61) in (4.62), we get

$$Z = \frac{1}{N!} \left(z_{\mathrm{t}} \cdot z_{\mathrm{r}} \cdot z_{\mathrm{v}} \right)^N, \tag{4.63}$$

where

$$z_{\mathrm{t}} = \sum \mathrm{e}^{-\varepsilon_{\mathrm{t}}/k_0 T}, \tag{4.64}$$

$$z_{\mathrm{r}} = \sum \mathrm{e}^{-\varepsilon_{\mathrm{r}}/k_0 T}, \tag{4.65}$$

$$z_{\mathrm{v}} = \sum \mathrm{e}^{-\varepsilon_{\mathrm{v}}/k_0 T} \tag{4.66}$$

are minor statistical sums, corresponding to different motions of one arbitrary molecule.

Note that in these expressions the summation is conducted with respect to quantum numbers, corresponding to translational, rotational and vibrational motions.

If we take into account $\ln N! \approx N \ln(N/e)$ in (4.63), the free energy of an ideal gas $F = -k_0 T \ln Z$ takes the form

$$F = -k_0 TN \left[\ln\left(\frac{e}{N} z_t\right) + \ln z_r + \ln z_v \right]. \tag{4.67}$$

Then for entropy of the gas from the expression $S = -\left(\partial F/\partial T\right)_V$ and (4.67) we get

$$S = k_0 N \left[\ln\left(\tfrac{e}{N} z_t\right) + \ln z_r + \ln z_v\right] + k_0 NT \frac{\partial}{\partial T}\left[\ln z_t + \ln z_r + \ln z_v\right]. \tag{4.68}$$

The mean energy of the gas $E = F + TS$ is

$$E = k_0 NT^2 \frac{\partial}{\partial T}\left[\ln z_t + \ln z_r + \ln z_v\right] \tag{4.69}$$

or

$$E = E_t + E_r + E_v. \tag{4.70}$$

Here

$$E_t = k_0 NT^2 \frac{\partial}{\partial T} \ln z_t, \tag{4.71}$$

$$E_r = k_0 NT^2 \frac{\partial}{\partial T} \ln z_r, \tag{4.72}$$

$$E_v = k_0 NT^2 \frac{\partial}{\partial T} \ln z_v. \tag{4.73}$$

The isochoric heat capacity of an ideal gas, conformable to the expression of energy (4.69), consists of three items:

$$C_V = C_V^t + C_V^r + C_V^v. \tag{4.70a}$$

It is seen that all problems of the quantum theory of the heat capacity are reduced to calculations of minor statistical sums z_t, z_r, z_v of one molecule. Calculate these sums for each kind of motions separately.

4.4.1 Translational Motion

Assume that in a cube with dimensions of sides L and volume $V = L^3$ N number of molecules is found. When considering the translational motion of molecules, the complexity and construction of a molecule do not matter, inasmuch as it is supposed to behaves as a material point with the reduced mass at the centre of masses of the molecule. According to quantum mechanics, energy of a material point (a particle) with the mass m, translationally moving in volume $V = L^3$, is pre-assigned by the expression

$$\varepsilon_t = \frac{\pi^2 \hbar^2}{2mL^2}\left(n_1^2 + n_2^2 + n_3^2\right), \tag{4.74}$$

here $n_i = 1, 2, 3, \ldots$ are quantum numbers.

In order to calculate the explicit form of the minor statistical sum, characterizing the translational motion, substitute (4.74) into (4.64):

$$z_t = \sum_{n_1 n_2 n_3} e^{-\frac{\pi^2\hbar^2}{2mL^2 k_0 T}(n_1^2+n_2^2+n_3^2)}. \tag{4.75}$$

Introduce the concept *of the characteristic temperature* for the translational motion

$$T_t = \frac{\pi^2\hbar^2}{2mk_0L^2} \tag{4.76}$$

and write down sum (4.75) in the form

$$z_t = \left(\sum_{n=1}^{\infty} e^{-n^2 T_t/T}\right)^3. \tag{4.77}$$

Inasmuch as at arbitrary values of temperature sum (4.77) cannot be presented in the analytical form, consider two limiting cases.

Low temperatures: $T \ll T_t$. In this case in sum (4.77) it can be restricted to the first two terms. Then

$$z_t = e^{-3T_t/T}\left(1 + 3e^{-3T_t/T}\right); \; T \ll T_t. \tag{4.78}$$

Using this expression in (4.71), we get the mean energy of the translational motion

$$E_t = E_0 + 3E_0 e^{-3T_t/T}; \; T \ll T_t \, , \tag{4.79}$$

where

$$E_0 = 3k_0 N T_t = \frac{3\pi^2\hbar^2 N}{2mL^2} \tag{4.80}$$

is the zero energy of an ideal gas, when all molecules are found in the fundamental quantum state ($n_1 = n_2 = n_3 = 1$).

If (4.79) is used, the heat capacity of a gas, associated with the translational motion, at low temperatures has the appearance

$$C_V^t = 27k_0N\left(\frac{T_t}{T}\right)^2 e^{-3T_t/T}; \; T \ll T_t \, . \tag{4.81}$$

It is seen that as $T \to 0$ the translational part of the heat capacity tends to zero.

High temperatures: $T \gg T_t$. In the region of temperatures, satisfying this condition, the discreteness of the translational motion can be neglected and it can be regarded as a classical one. If we use notion (4.76), the condition of classicity of the motion can be presented as

$$\frac{\pi^2\hbar^2}{2mL^2} \ll k_0T \text{ or } \lambda \ll L, \tag{4.82}$$

where $\lambda = h/\sqrt{2mk_0T}$ is the de Broglie wavelength of a molecule. It is evident that (4.82) is the condition that in the energy spectrum of the translational motion the difference of energies of two adjacent levels is far less than that of the thermal energy k_0T. Simultaneously, the condition of classicity (4.82) can be formulated as follows: the motion is classical if the de Broglie wavelength of a molecule (a particle) is far less than the dimensions of the space in which the motion occurs.

Note that the condition of classicity (4.82) can also be presented in the form

$$mvL \gg h \text{ or } S_\mathrm{t} \gg h, \tag{4.83}$$

i.e. the action of the given motion $S_\mathrm{t} = mvL$ ought to be much more than the Planck constant h. When obtaining the condition of classicity (4.83) we supposed that $k_0T \approx mv^2$. Therefore, (4.82) and (4.83) are equivalent conditions of classicity of the translational motion.

The sum entering into the expression of the statistical sum (4.77) can be presented in the form

$$\sum_{n=1}^{\infty} \mathrm{e}^{-n^2T_\mathrm{t}/T} = \sum_{n=0}^{\infty} \mathrm{e}^{-n^2T_\mathrm{t}/T} - 1. \tag{4.84}$$

Inasmuch as in the region of high temperatures ($T \gg T_\mathrm{t}$) two adjacent terms are little distinct from each other, to calculate this sum the Euler summation formula can be used:

$$\sum_{n=0}^{\infty} f(n) = \int_0^{\infty} f(n)\mathrm{d}n + \frac{1}{2}\left[f(0) + f(\infty)\right] + \frac{1}{12}\left[f'(\infty) - f'(0)\right] - \cdots \tag{4.85}$$

Here it is supposed that the function $f(n)$ is integratable in the range of $(0, \infty)$ and all its derivatives at lower and upper limits take on finite values.

In our case $f(n) = \mathrm{e}^{-n^2T_\mathrm{t}/T}$. Taking this into account and also

$$\int_0^{\infty} \mathrm{e}^{-n^2T_\mathrm{t}/T}\mathrm{d}n = \frac{1}{2}\sqrt{\frac{\pi T}{T_\mathrm{t}}} \tag{4.86}$$

in (4.84), sum (4.83) can be presented in the form

$$\sum_{n=1}^{\infty} \mathrm{e}^{-n^2T_\mathrm{t}/T} = \frac{1}{2}\left[\left(\frac{\pi T}{T_\mathrm{t}}\right)^{1/2} - 1\right]. \tag{4.87}$$

Having substituted (4.86) into (4.77), for the statistical sum we get

$$z_\mathrm{t} = \frac{1}{8}\left(\frac{\pi T}{T_\mathrm{t}}\right)^{3/2}\left[1 - 3\left(\frac{T_\mathrm{t}}{\pi T}\right)^{1/2}\right]; \quad T \gg T_\mathrm{t}. \tag{4.88}$$

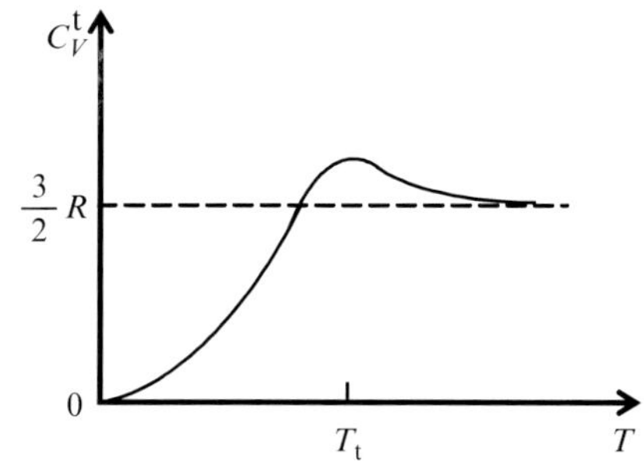

Fig. 4.3. The temperature dependence of the translational part of the heat capacity

Taking into account this expression, and also (4.71), for the mean energy of the translational motion in the region of high temperatures we get

$$E_t = \frac{3}{2}k_0NT\left[1+\left(\frac{T_t}{\pi T}\right)^{1/2}\right]; \quad T \gg T_t. \tag{4.89}$$

and the heat capacity, corresponding to this motion, equals

$$C_V^t = \frac{3}{2}k_0N\left[1+\frac{1}{2}\left(\frac{T_t}{\pi T}\right)^{1/2}\right]; \quad T \gg T_t. \tag{4.90}$$

Second items in the square bracket in (4.89) and (4.90) are quantum corrections to the mean energy, associated with the quantization of the translational motion and the heat capacity. As is seen, this correction insignificantly increases the classical value of the heat capacity $C_V^t = 3R/2$. The schematic dependence $C_V^t(T)$ in all temperature regions with regard to expressions (4.81) and (4.90) is presented in Fig. 4.3, from which it is seen that there is a small maximum.

At ultra-high temperatures ($T \to \infty$) from (4.88) to (4.90) stems purely classical results

$$z_t = \frac{1}{8}\left(\frac{\pi T}{T_t}\right)^{3/2} = \frac{V}{\hbar^3}\left(\frac{mk_0T}{2\pi}\right)^{3/2}, \tag{4.91}$$

$$E_t = \frac{3}{2}k_0NT, \tag{4.92}$$

$$C_V^t = \frac{3}{2}k_0N. \tag{4.93}$$

4.4.2 Rotational Motion

Consider an ideal gas consisting of N number of two-atom molecules; therewith atoms with masses m_1 and m_2 are found at a constant distance r from each other. From the point of view of the rotational motion each molecule can be considered as a rotator.

According to quantum mechanics, energy of a rotator has the appearance

$$\varepsilon_{\mathrm{r}} = \frac{\hbar^2}{2I} l(l+1), \tag{4.94}$$

where $l = 0, 1, 2, \ldots$ is the rotational quantum number, $I = mr^2$ is the moment of inertia of a molecule, $m = m_1 m_2/(m_1 + m_2)$ is the reduced mass of a molecule.

To find the mean value of energy of a system, corresponding to the rotational motion, according to expression (4.72), it is required to calculate the statistical sum (4.65). It is known that energy levels (4.94) of the rotational motion are $(2l+1)$-fold degenerate. Taking this into account, substitute (4.94) into (4.65). Then for the statistical sum, corresponding to the rotation, we get

$$z_{\mathrm{r}} = \sum_{l=0}^{\infty} (2l+1) \mathrm{e}^{-\frac{\hbar^2}{2Ik_0T} l(l+1)}. \tag{4.95}$$

If we introduce the concept of *the characteristic temperature for the rotational motion*

$$T_{\mathrm{r}} = \frac{\hbar^2}{2k_0 I}, \tag{4.96}$$

(4.95) takes the form

$$z_{\mathrm{r}} = \sum_{l=0}^{\infty} (2l+1) \mathrm{e}^{-\frac{T_{\mathrm{r}}}{T} l(l+1)}. \tag{4.97}$$

Inasmuch as at arbitrary values of temperature the sum (4.97) cannot be presented in the analytical form, consider two limiting case.
Low temperatures: $T \ll T_{\mathrm{r}}$. In this case, inasmuch as with the lowering of temperature terms of the sum (4.97) fast decrease, it can be restricted to the first two terms

$$z_{\mathrm{r}} \approx 1 + 3\mathrm{e}^{-2T_{\mathrm{r}}/T}. \tag{4.98}$$

Taking this into account in (4.72), for the mean energy we get

$$E_{\mathrm{r}} = 6k_0 N T_{\mathrm{r}} \mathrm{e}^{-2T_{\mathrm{r}}/T}, \tag{4.99}$$

and for the heat capacity

$$C_V^{\mathrm{r}} = 12k_0 N \left(\tfrac{T_{\mathrm{r}}}{T}\right)^2 \mathrm{e}^{-2T_{\mathrm{r}}/T};\ T \ll T_{\mathrm{r}} \cdot \tag{4.100}$$

Hence it is seen that with the lowering of temperature the heat capacity exponentially drops and tends to zero.

High temperatures: $T \gg T_{\mathrm{r}}$. In the region of temperatures, satisfying this condition, the discreteness of the energy spectrum of a rotator (4.94) can be neglected and the motion can be regarded as a classical one. If notation (4.96) is used, the condition of classicity of the rotational motion $T \gg T_\mathrm{r}$ can be presented as

$$T \gg T_\mathrm{r}; \quad k_0 T \gg \frac{\hbar^2}{2mr^2}; \quad \lambda \ll r; \quad S_\mathrm{r} \gg \hbar, \tag{4.101}$$

where $\lambda = \hbar/\sqrt{2mk_0T}$ is the de Broglie wavelength, $S_\mathrm{r} = mvr$ is the action of the rotational motion, v is the linear velocity of rotation.

In order to compute the sum (4.97) for the considered case, use the Euler formula (4.84) and take into account that

$$\int_0^\infty (2l+1)\mathrm{e}^{-\frac{T_\mathrm{r}}{T}l(l+1)}\mathrm{d}l = \int_0^\infty \mathrm{e}^{-\frac{T_\mathrm{r}}{T}x}\mathrm{d}x = \frac{T}{T_\mathrm{r}}. \tag{4.102}$$

Then we get

$$z_\mathrm{r} = \frac{T}{T_\mathrm{r}}\left[1 + \frac{1}{3}\frac{T_\mathrm{r}}{T} + \frac{1}{12}\left(\frac{T_\mathrm{r}}{T}\right)^2\right]; \quad T \gg T_\mathrm{r}. \tag{4.103}$$

Having substituted (4.103) into (4.72), for the mean energy of the rotational motion we get

$$E_\mathrm{r} = k_0 NT\left[1 - \frac{1}{12}\left(\frac{T_\mathrm{r}}{T}\right)^2\right]; \quad T \gg T_\mathrm{r}, \tag{4.104}$$

and for the heat capacity

$$C_V^\mathrm{r} = k_0 N\left[1 + \frac{1}{12}\left(\frac{T_\mathrm{r}}{T}\right)^2\right]; \quad T \gg T_\mathrm{r}. \tag{4.105}$$

The temperature dependence of the rotational heat capacity, according to expressions (4.100) and (4.105), is schematically shown in Fig. 4.4. As is seen from the figure, the function $C_V^\mathrm{r}(T)$ in the region T_r has a small maximum. On the basis of numerical computations it can be shown that the temperature at which $C_V^\mathrm{r}(T)$ takes on a maximum value equals $T_\mathrm{max} = 0.8T_\mathrm{r}$, and the value of the heat capacity at this temperature $C_V^\mathrm{r}(T_\mathrm{max}) = 1.1k_0N$.

Second items in the square bracket in (4.104) and (4.105) are quantum corrections to the mean energy and heat capacity, associated with the quantization of the rotational motion. As is seen from (4.104) and (4.105), the sign of the correction is positive.

In the limiting case ($T \to \infty$) from expressions (4.104) and (4.105) for energy and heat capacity purely classical results follow: $E_\mathrm{r} = k_0NT$ and $C_V^\mathrm{r} = k_0N$.

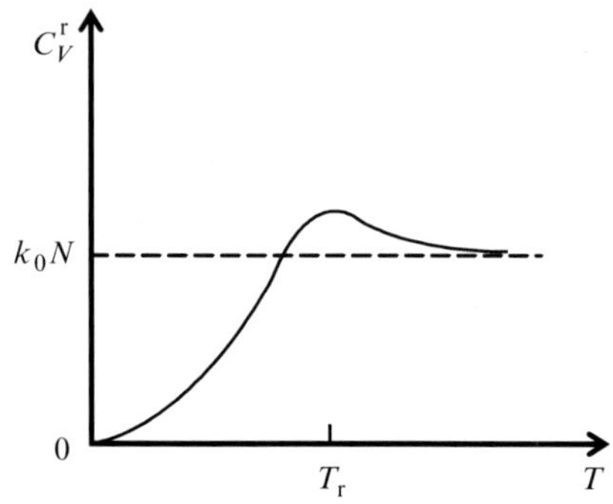

Fig. 4.4. The temperature dependence of the rotational part of the heat capacity

4.4.3 Vibrational Motion

In two-atom molecules atoms perform small intermolecular vibrational motions in reference to each other. Therefore a molecule can be considered as a linear harmonic oscillator. According to quantum mechanics, energy of such an oscillator has the appearance

$$\varepsilon_v = \hbar\omega\left(n + \frac{1}{2}\right), \tag{4.106}$$

where $\omega = \sqrt{\chi/m}$ is the circular frequency of vibrations, χ is the coefficient of elasticity, m is the reduced mass of a molecule, and $n = 0, 1, 2, \ldots$ is the oscillatory quantum number.

Consequently, a two-atom ideal gas can be considered as a gas consisting of N number of non-interacting harmonic oscillators with the frequency ω. Calculate the mean energy and heat capacity of such a gas. To do this substitute (4.106) into (4.66) and add up with respect to the quantum number n. As a result, for the vibrational statistical sum we get

$$z_v = \frac{e^{-\hbar\omega/2k_0T}}{1 - e^{-\hbar\omega/k_0T}} = \left[2sh\frac{\hbar\omega}{2k_0T}\right]^{-1}. \tag{4.106a}$$

Having substituted (4.106a) into (4.73) for the mean energy of the vibrational motion of molecules of a gas we get

$$E_v = \frac{N\hbar\omega}{2} + \frac{N\hbar\omega}{e^{\hbar\omega/k_0T} - 1} = N\varepsilon(\omega, T). \tag{4.107}$$

Hence for the mean energy of one oscillator $\varepsilon(\omega, T) = E_v/N$ we have

$$\varepsilon(\omega, T) = \frac{\hbar\omega}{2} + \frac{\hbar\omega}{e^{\hbar\omega/k_0T} - 1}. \tag{4.108}$$

Here $\varepsilon_0 = \hbar\omega/2$ is energy of the oscillator as $T = 0$, i.e. the zero energy. At temperatures satisfying the condition of classicity $k_0T \gg \hbar\omega$, from (4.108)

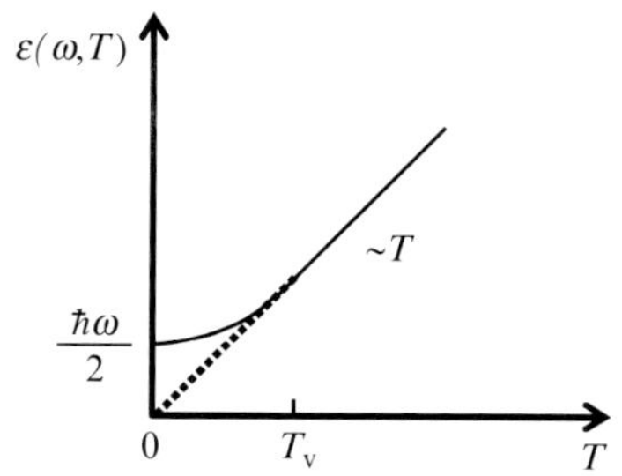

Fig. 4.5. The temperature dependence of the energy of a linear harmonic oscillator

stems the known classical result for energy of an oscillator $\varepsilon(T) \approx k_0 T$ (Fig. 4.5).

Note that the presence of zero energy of an oscillator and the fact that energy of a harmonic oscillator depends on the frequency are purely quantum effects. Using expression (4.107), for the heat capacity $C_V^{\mathrm{v}} = (\partial E_{\mathrm{v}}/\partial T)$, associated with the vibrational motion, we get

$$C_V^{\mathrm{v}} = k_0 N \left(\frac{\hbar\omega}{k_0 T}\right)^2 \frac{\mathrm{e}^{\hbar\omega/k_0 T}}{\left(\mathrm{e}^{\hbar\omega/k_0 T} - 1\right)^2}. \tag{4.109}$$

If we introduce the concept of *the characteristic temperature of the vibrational motion*

$$T_{\mathrm{v}} = \frac{\hbar\omega}{k_0}, \tag{4.110}$$

expression for heat capacity (4.109) takes the form

$$C_V^{\mathrm{v}} = k_0 N \left(\frac{T_{\mathrm{v}}}{T}\right)^2 \frac{\mathrm{e}^{T_{\mathrm{v}}/T}}{\left(\mathrm{e}^{T_{\mathrm{v}}/T} - 1\right)^2}. \tag{4.111}$$

This expression of heat capacity applies for any value of temperature. On this basis consider the limiting cases.

Low temperatures: $T \ll T_{\mathrm{v}}$. If in this case in the denominator of (4.111) we neglect unity compared with $\exp(T_{\mathrm{v}}/T) \gg 1$, (4.111) takes the form

$$C_V^{\mathrm{v}} \approx k_0 N \left(\frac{T_{\mathrm{v}}}{T}\right)^2 \mathrm{e}^{-T_{\mathrm{v}}/T}; \quad T \ll T_{\mathrm{v}}, \tag{4.112}$$

i.e. the heat capacity exponentially depends on temperature and as $T \to 0$ tends to zero.

High temperatures: $T \gg T_{\mathrm{v}}$. In the region of temperatures and frequencies, satisfying this condition, the vibrational motion is classical. This condition, according to (4.110), can be presented in the form

$$\hbar\omega \ll k_0 T, \tag{4.113}$$

i.e. the discreteness (the difference between two adjacent levels) of the energy spectrum of an oscillator ought to be of much less energy than that of the thermal motion.

If we write down the classical law of motion of an oscillator in the form $y = y_{\max} \cos \omega t$ and accept that $\dot{y} = v$ and $mv^2 \approx k_0 T$, the condition of quasi-classicity (4.113) acquires the following equivalent shape:

$$\frac{\hbar}{mv} \ll y \text{ or } \lambda \ll y; \quad \hbar \ll mvy \text{ or } S_{\mathrm{v}} \gg \hbar, \tag{4.114}$$

where $S_{\mathrm{v}} = mvy$ is the action of the vibrational motion and $\lambda = h/mv$ is the de Broglie wavelength.

In order to consider the region of high temperatures, introduce the notation $x = T_{\mathrm{v}}/T$ and rewrite the expression of heat capacity (4.111) in the form

$$C_V^{\mathrm{v}} = k_0 N \varphi(x), \tag{4.115}$$

where

$$\varphi(x) = \left(\frac{x}{\mathrm{e}^x - 1}\right)^2 \mathrm{e}^x = \left[\frac{x/2}{sh(x/2)}\right]^2. \tag{4.116}$$

Inasmuch as in the region of high temperatures $x \ll 1$, $sh(x/2)$ can be expanded into a series, restricting it to the first two terms:

$$sh\frac{x}{2} = \frac{x}{2} + \frac{1}{6}\left(\frac{x}{2}\right)^3. \tag{4.117}$$

If we substitute (4.117) into (4.116), the function $\phi(x)$ takes the form

$$\varphi(x) = \left(1 + x^2/24\right)^{-2} = \left(1 - x^2/12\right); x \ll 1. \tag{4.118}$$

As a result from (4.115) and (4.118) in the limiting case of high temperatures $x = T_{\mathrm{v}}/T \ll 1$ for heat capacity we get

$$C_V^{\mathrm{v}} = k_0 N \left[1 - \frac{1}{12}\left(\frac{T_{\mathrm{v}}}{T}\right)^2\right]; \quad T \gg T_{\mathrm{v}}. \tag{4.119}$$

Here $C_V^{\mathrm{v}} = k_0 N$ is the classical result, and $-k_0 N/12 \left(T_{\mathrm{v}}/T\right)^2$ is the correction to heat capacity, associated with the quantization of the vibrational motion. As is seen, this correction, in contradistinction to the correction of the rotational heat capacity, is negative. Therefore $C_V^{\mathrm{v}}(T)$ has no maximum. The temperature dependence of $C_V^{\mathrm{v}}(T)$ is schematically presented in Fig. 4.6.

Note the following. At the beginning of the section we supposed that all motions of a molecule are independent. Therefore the total energy of molecules can be presented in the form (4.74). However, the supposition in reference to rotational and vibrational motions is true only in the case when the vibrational motion is harmonic. In the case of anharmonic vibrations the mean distance between atoms changes, which leads to the change in the moment of inertia $I = mr^2$, and, thereby, parameters of the rotational motion changes, i.e. the anharmonic vibration can influence the rotational motion.

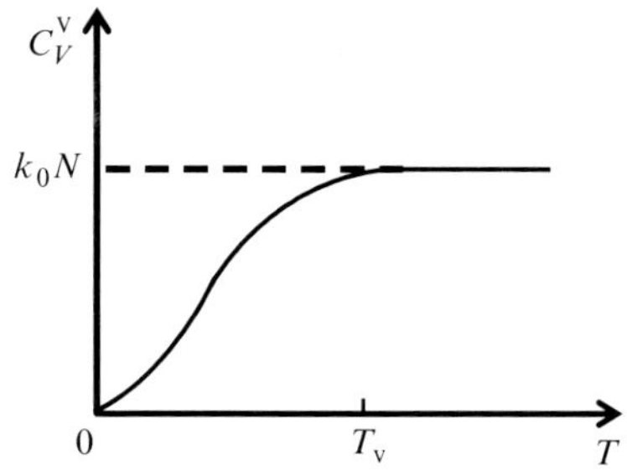

Fig. 4.6. The temperature dependence of the vibrational part of the heat capacity

4.4.4 Total Heat Capacity

Up to now we considered different forms of the motion of two-atom molecules separately and analysed temperature dependences of heat capacities in different temperature ranges. Now consider temperature dependences of a two-atom ideal gas in the wide temperature region. To do this remember that while expounding the quantum theory of heat capacity for each kind of motion (translational, rotational, vibrational) the notion of the characteristic temperature ($T_{\mathrm{t}}, T_{\mathrm{r}}, T_{\mathrm{v}}$) was introduced. Evaluate these temperatures.

If we accept that the reduced mass of a molecule is $m = 10^{-24}$ g, and linear dimensions of the vessel $L = 0.1$ cm, according to (4.76), for T_{t} we get

$$T_{\mathrm{t}} = \frac{\pi^2\hbar^2}{2mk_0L^2} \sim 10^{-12}\,\mathrm{K}. \qquad (4.120)$$

If we accept that the distance between atoms is $r = 10^{-8}$ cm, according to (4.96), for T_{r} we have

$$T_{\mathrm{r}} = \frac{\hbar^2}{2mk_0r^2} \sim 50\,\mathrm{K}. \qquad (4.121)$$

If we accept that the cyclic frequency is equal to $\omega \approx 10^{14}\,\mathrm{s}^{-1}$ (the infrared region), according to (4.110), for T_{v} we get

$$T_{\mathrm{v}} = \frac{\hbar\omega}{k_0} \sim 10^3\,\mathrm{K}. \qquad (4.122)$$

Apart from those indicated, there are two characteristic temperatures: temperature of condensation T_{cond} and temperature of dissociation of molecules T_{d}.

Note that the quantum theory of the heat capacity expounded here is applicable only in the region of temperatures $T_{\mathrm{cond}} < T < T_{\mathrm{d}}$.

Characteristic temperatures for three two-atom gases are adduced in Table 4.1. In this table the value of T_{t} is absent, inasmuch as, according to (4.120), its value almost equals zero.

From the above-adduced values of theoretical evaluations, and also from Table 4.1 it follows that above the condensation temperature in the gaseous

Table 4.1. Characteristic temperatures for three two-atom gases

	H_2	O_2	N_2
T_{cond} (K)	20	90.2	77.5
T_{r} (K)	85.4	2.1	2.9
T_{v} (K)	6.1×10^3	2.2×10^3	3.3×10^3
T_{d} (K)	5.2×10^4	5.9×10^4	8.5×10^4

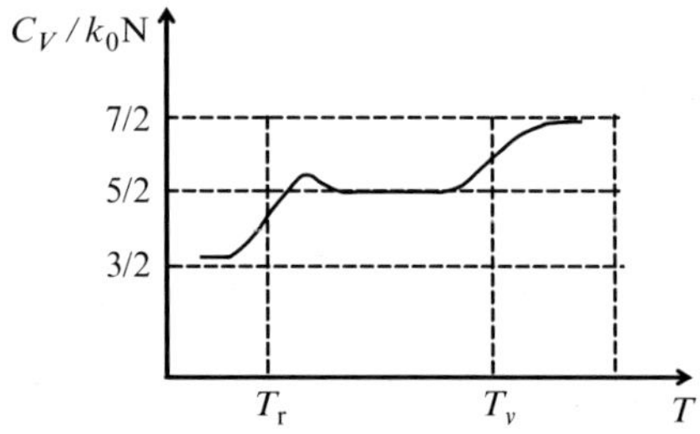

Fig. 4.7. The temperature dependence of the total heat capacity for a two-atom ideal gas

phase the translational motion of two-atom molecules is always classical and the corresponding isochoric heat capacity $C_V^{\mathrm{t}} = 3k_0N/2$.

When approaching the temperature of a gas T_{r}, the rotational motion of molecules gets excited and, thereby, the heat capacity grows. In the region $T \gg T_{\mathrm{r}}$ the rotational motion becomes classical and the total heat capacity, corresponding to translational and rotational motions in this region, equals $5k_0N/2$. At the further raise of temperature, i.e. in the region $T_{\mathrm{r}} < T < T_{\mathrm{v}}$ the vibrational motion also gets excited, which leads to the growth of C_V. In the region of temperatures $T_{\mathrm{d}} > T \gg T_{\mathrm{v}}$ the vibrational motion also becomes classical and the total heat capacity, being determined by all types of motions, theoretically ought to reach the limiting value $C_V = C_V^{\mathrm{t}} + C_V^{\mathrm{r}} + C_V^{\mathrm{v}} = 7k_0N/2$. As is seen from (4.122), the characteristic temperature T_{v} is of rather high magnitude; therefore in practice the limiting value is not reachable. Inasmuch as the characteristic temperature of the translational motion T_{t} is triflingly small, in the gaseous phase ($T_{\mathrm{t}} < T_{\mathrm{r}}$) the value $3k_0N/2$ is not reachable, either. In Fig. 4.7 the temperature dependence of the total heat capacity is schematically presented. If we compare this dependence with experimental results, adduced in Fig. 4.2, it can be noticed that the experimental result coincides with the theoretical one in the region of temperatures $(T_{\mathrm{r}} - \Delta T) < T < (T_{\mathrm{v}} - \Delta T)$, where ΔT is a certain finite range of temperatures. From this comparison it follows that in the indicated region of temperatures translational and rotational motions are classical, and the vibrational motion only just begins to get excited.

So, we come to the following physical conclusions: *the theory of the heat capacity of a two-atom gas in the general case ought to be quantum one; the translational motion of molecules is always classical; the rotational motion in* *the region $T \leq T_{\rm r}$ bears the quantum character, and in the region $T >> T_{\rm r}$ it is classical; the vibrational motion in the region $T \leq T_{\rm v}$ is quantum, and in the region $T >> T_{\rm v}$ it bears the classical character.*

Thus, theoretically and experimentally studying temperature dependences of the heat capacity in the wide temperature range, the nature of the motion (classical or quantum) of molecules, forming a gas, can be determined. It is possible since the heat capacity of a system (in our case, a gas) is very sensitive to the internal structure of particles (molecules). The superiority of the heat capacity over other thermodynamic parameters resides just in this.

4.5 Ideal Gas Consisting of Polar Molecules in an External Electric Field

Earlier when considering an ideal gas it was supposed that particles (molecules) forming it possess only mass (the reduced mass). We also supposed that these particles possess neither an electric charge nor electric dipole moment. Therefore, when such a gas is in a thermodynamic state, an external electric field does not act on it.

Molecules depending on the distribution of an electric charge in them are divided into *polar* and *non-polar* groups. In polar molecules centres of positive and negative charges do not coincide, and therefore even in the absence of an external electric field such molecules possess the intrinsic electric dipole moment $\boldsymbol{p}_0$. If we place such a molecule in an electric field, as a result of the change in the distance between centres of charges a supplementary dipole moment can arise, i.e. $\boldsymbol{p}_0$ can change. However, inasmuch as in weak electric fields this change is much less of the intrinsic dipole moment $\boldsymbol{p}_0$, it can be said that the intrinsic moment $\boldsymbol{p}_0$ remains constant.

In non-polar molecules, centres of negative and positive charges coincide, and the molecule does not possess the electric dipole moment. The dipole moment in such molecules can arise only in an external electric field, since an electric field divides centres of negative and positive charges of the molecule.

In the given section we will consider the thermodynamic properties of an ideal gas consisting of polar molecules. It is supposed that the gas is rarefied and therefore the dipole–dipole interaction can be neglected. At first, we calculate the coefficient of the orientational polarization.

4.5.1 Orientational Polarization

Assume that an ideal gas of volume V, consisting of a N number of polar molecules with the intrinsic dipole moment $\boldsymbol{p}_0$, is found in an external uniform electric field $\boldsymbol{\mathcal{E}}$. It is known that the potential energy of a dipole $\boldsymbol{p}_0$ in an external electric field has the appearance

$$U = -(\boldsymbol{p}_0\boldsymbol{\mathcal{E}}) = -p_0\mathcal{E}\cos\theta, \tag{4.123}$$

where θ is the angle between the dipole moment $\boldsymbol{p}_0$ and the electric field $\boldsymbol{\mathcal{E}}$.

Inasmuch as in the absence of an external electric field molecules in the volume occupied are distributed chaotically, on average the polarization is absent in the gas.

Under action of an external electric field dipoles attempt to line up along the field owing to the condition of the minimality of the potential energy U. As a result of this, the gas as a whole is polarized, i.e. the dipole moment, accounting for the unit volume – *the polarization vector* – becomes distinct from zero.

We now investigate the thermodynamic properties of a polarized ideal gas. In particular, we calculate the mean value of the polarization vector

$$\mathscr{P}_0 = \sum_i \boldsymbol{p}_{0i}, \tag{4.124}$$

where the summation is conducted over all polar molecules found in the unit volume.

To do this it is necessary to know the free energy of the gas in the electric field and on the strength of (2.163) to calculate the mean value of the polarization vector by the formula

$$\mathscr{P}_0 = -\frac{1}{V}\left(\frac{\partial F}{\partial \boldsymbol{\mathcal{E}}}\right)_{T,V}. \tag{4.125}$$

For definiteness consider two-atom polar molecules. Inasmuch as the polarization of a gas occurs at the expense of the rotation of molecules in an electric field, we can restrict ourselves to the finding of the free energy associated with the rotation of molecules: $F = F_{\mathrm{r}}$. The presence of the translational and vibrational motion of a molecule does not exercise influence on the dipole moment p_0.

In the expression of the free energy (4.67) we separate the part associated only with the rotation and assume that the rotational motion is classical. Then we get

$$F_{\mathrm{r}} = -k_0 TN \ln z_{\mathrm{r}}, \tag{4.126}$$

where z_{r} is the statistical integral associated with the rotation of one molecule. Using the polar coordinate system, the statistical integral can be presented in the form

$$z_{\mathrm{r}} = \int \mathrm{e}^{-\varepsilon_{\mathrm{r}}(\theta,\varphi)/k_0 T} \frac{\mathrm{d}\varphi\, \mathrm{d}\theta\, \mathrm{d}p_\varphi\, \mathrm{d}p_\theta}{(2\pi\hbar)^2}. \tag{4.127}$$

Here

$$\varepsilon_{\mathrm{r}}(\theta,\phi) = \frac{1}{2I}\left(p_\theta^2 + \frac{p_\phi^2}{\sin^2\theta}\right) - p_0\mathcal{E}\cos\theta \tag{4.128}$$

is the Hamilton function of a molecule in the polar coordinate system in an electric field, $I = mr^2$ is the moment of inertia of a molecule, $m =$

$m_1 m_2/(m_1+m_2)$ is the reduced mass of a molecule, m_1 and m_2 are masses of atoms forming a molecule.

If we substitute the expression of energy (4.128) into (4.127) and take into account the integration boundaries, we get

$$z_{\mathrm{r}} = \frac{1}{(2\pi\hbar)^2} \int_0^{2\pi} \mathrm{d}\varphi \int_0^{\pi} \mathrm{d}\theta \int_{-\infty}^{\infty} \mathrm{d}p_\theta \int_{-\infty}^{\infty} \mathrm{d}p\,\varphi \\ \times \exp\left[-\frac{1}{2Ik_0T}\left(p_\theta^2 + \frac{p_\varphi^2}{\sin^2\theta}\right) + a\cos\theta\right], \tag{4.129}$$

where

$$a = \frac{p_0\mathcal{E}}{k_0T} \tag{4.130}$$

is the dimensionless parameter.

To integrate over $\mathrm{d}p_\theta$ and $\mathrm{d}p_\varphi$ use Appendix A. Then we get

$$\int_{-\infty}^{\infty} \mathrm{e}^{-p_\theta^2/2Ik_0T}\mathrm{d}p_\theta = (2\pi Ik_0T)^{1/2} \tag{4.131}$$

and

$$\int_{-\infty}^{\infty} \mathrm{e}^{-p_\varphi^2/(2Ik_0T\,\sin^2\theta)}\mathrm{d}p_\varphi = (2\pi Ik_0T)^{1/2}\sin\theta. \tag{4.132}$$

If we substitute (4.131) and (4.132) into (4.129), the statistical integral takes the form

$$z_{\mathrm{r}} = \frac{Ik_0T}{\hbar^2}\int_0^{\pi} \mathrm{e}^{a\cos\theta}\sin\theta\,\mathrm{d}\theta. \tag{4.133}$$

If we introduce the notation $x = \cos\theta$, for z_{r} finally we get

$$z_{\mathrm{r}} = \frac{2Ik_0T}{\hbar^2}\frac{sha}{a}. \tag{4.134}$$

Note that in the absence of an electric field ($\mathcal{E} = 0$ or $a = 0$) for the case of ultra-high temperatures ($T \to \infty$) expression (4.134) passes into (4.103).

Having substituted the expression of the statistical integral (4.134) into (4.126), for the free energy we get the following expression

$$F_{\mathrm{r}} = -k_0TN\left[\ln\left(\frac{sh\,a}{a}\right) + \ln\left(\frac{2Ik_0T}{\hbar^2}\right)\right]. \tag{4.135}$$

Taking into account (4.135) in (4.125), the polarization vector takes the form

$$\mathscr{P}_0 = -\frac{1}{V}\left(\frac{\partial F_{\mathrm{r}}}{\partial \boldsymbol{\mathcal{E}}}\right)_{T,V} = k_0T\frac{N}{V}\frac{\partial}{\partial \boldsymbol{\mathcal{E}}}(\ln\,sh\,a - \ln\,a) \tag{4.136}$$

or

$$\mathscr{P}_0 = k_0 T n \left(\frac{ch\, a}{sh\, a} - \frac{1}{a} \right) \frac{\partial a}{\partial \boldsymbol{\mathcal{E}}}. \tag{4.137}$$

If we take into account that $\partial a/\partial \mathcal{E} = p_0/k_0 T$, for the value of the polarization vector we get

$$\mathscr{P}_0 = p_0 n\, L(a), \tag{4.138}$$

where

$$L(a) = ctha - \frac{1}{a} \tag{4.139}$$

is *the Langevin function*, and $n = N/V$ is the concentration of polar molecules.

In the case of weak electric fields and high temperatures ($a = p_0 \mathcal{E}/k_0 T \ll 1$), the hyperbolic cotangent can be expanded into a series

$$cth\, a = \frac{1}{a} + \frac{a}{3} - \frac{a^3}{45} + \cdots \tag{4.140}$$

Then in the linear approximation

$$L(a) = \frac{a}{3} = \frac{p_0 \mathcal{E}}{3 k_0 T}. \tag{4.141}$$

Taking into account (4.141) in (4.123), for the value of the polarization vector we get

$$\mathscr{P}_0 = \frac{n p_0^2}{3 k_0 T} \boldsymbol{\mathcal{E}} = n \alpha \boldsymbol{\mathcal{E}}, \tag{4.142}$$

where $\alpha = p_0^2/3k_0 T$ is *the coefficient of polarisability of a molecule.*

The electric induction D in this case has the appearance

$$D = \mathcal{E} + 4\pi \mathscr{P}_0 = \left(1 + \frac{4\pi n p_0^2}{3 k_0 T} \right) \mathcal{E} = \chi \mathcal{E}, \tag{4.143}$$

where

$$\chi = 1 + \frac{4\pi n p_0^2}{3 k_0 T} \tag{4.144}$$

is *the dielectric constant* of a gas.

In another limiting case, when $a = p_0 \mathcal{E}/3k_0 T \gg 1$, inasmuch as $\lim_{a\to\infty} cth\, a \to 1$, then $\lim_{a\to\infty} L(a) \to 1$. Thus, in strong electric fields and at low temperatures the polarization is saturated and $\mathscr{P} = n p_0$, which corresponds to the orientation of all dipoles along the field.

The schematic plot of the Langevin function, constructed on the basis of considered limiting cases, is presented in Fig. 4.8.

In the case of weak electric fields ($a \ll 1$) from (4.142) it is seen that the polarization is inversely proportional to temperature: $\mathscr{P}_0 \sim 1/T$. Therefore, investigating the temperature dependence of the polarization, on the basis of (4.142) the intrinsic dipole moment p_0 can be calculated. The majority of molecules possess the intrinsic dipole moment of the order $p_0 \approx 10^{-18}$ of electrostatic units $= 1\,\mathrm{D}$.

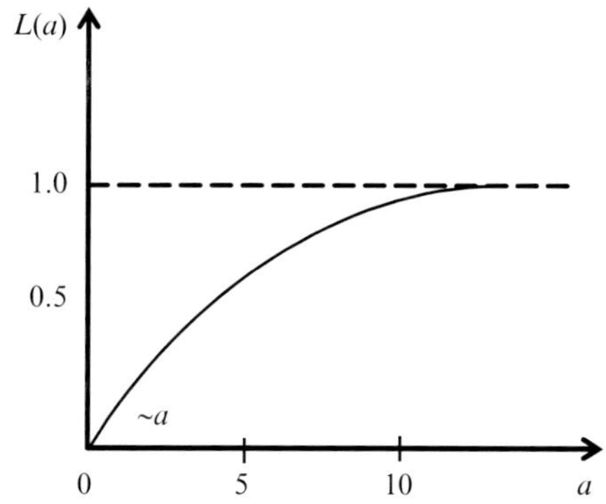

Fig. 4.8. The schematic plot of the Langevin function

4.5.2 Entropy: Electrocaloric Effect

It is known that in an ideal gas a molecule can participate in three kinds of motion: translational, rotational and vibrational. In Sect. 4.4 we showed that to excite the vibrational motion very high temperatures ($T \approx T_{\mathrm{v}}$) are required. Here we consider the region of temperatures $T \ll T_{\mathrm{v}}$. In this region, molecules perform only translational and rotational motions. Then in the expression of free energy (4.67) it is sufficient to keep two terms

$$F = -k_0 NT \left[\ln \left(\frac{e}{N} z_{\mathrm{t}} \right) + \ln\, z_{\mathrm{r}} \right]. \tag{4.145}$$

Let us assume that translational and rotational motions are classical, i.e. restrict ourselves to the region of temperature $T_{\mathrm{r}} \ll T < T_{\mathrm{v}}$. In this case, the expression of the statistical integral (4.91) and (4.134) can be used, taking account the polarization. Then free energy (4.145) takes the form

$$\begin{aligned} F = &-k_0 NT \left[\ln \left(\frac{eV}{N\hbar^3} \right) \left(\frac{mk_0 T}{2\pi} \right)^{3/2} \right] \\ &-k_0 NT \left[\ln \left(\frac{2Ik_0 T}{\hbar^2} \right) + \ln \left(\frac{sh\, a}{a} \right) \right]. \end{aligned} \tag{4.146}$$

In the considered case entropy of a gas $S = -\left(\partial F / \partial T \right)_{V,\mathcal{E}}$ equals

$$S(\mathcal{E}) = S(0) - k_0 N \left[aL(a) - \ln \left(\frac{sh\, a}{a} \right) \right], \tag{4.147}$$

where S_0 is entropy of a gas in the absence of an electric field:

$$S(0) = \frac{5}{2} k_0 N + k_0 N \left[\ln \left(\frac{eV}{N\hbar^3} \right) \left(\frac{mk_0 T}{2\pi} \right)^{3/2} \right] + k_0 N\, \ln \left(\frac{2Ik_0 T}{\hbar^2} \right). \tag{4.148}$$

It can be shown that the function

$$\varphi(a) = aL(a) - \ln \frac{sh\, a}{a}, \tag{4.149}$$

entering into the expression of entropy (4.147) is positive, i.e. as a result of the polarization entropy decreases, which is associated with the rise of the order at the expense of the orientation of dipoles along the field and, thereby, with the decrease in chaoticity.

At small values of the field $a \ll 1$, $L(a) = a/3$, $sh\ a = a + \frac{a^3}{6}$. Then

$$\varphi(a) = \frac{a^2}{6} = \frac{1}{6}\left(\frac{p_0\mathcal{E}}{k_0T}\right)^2 \tag{4.150}$$

and

$$S(\mathcal{E}) = S(0) - \frac{1}{6}k_0N\left(\frac{p_0\mathcal{E}}{k_0T}\right)^2. \tag{4.151}$$

Hence it follows that in ideal polar gases we can observe the electrocaloric effect, analogous to the magnetocaloric one, and use it to lower the temperature of the gas. In order to obviously present this, in Fig. 4.9 the temperature dependence of entropy in the absence ($\mathcal{E} = 0$) and presence ($\mathcal{E} \neq 0$) of an electric field is adduced.

Assume that an ideal gas is found in state 1. Switch on an electric field, and translate the gas from this state into state 2. At that, the gas is polarized and entropy decreases. Further adiabatically translate the gas from state 2 into state 3. At that, temperature drops from T_1 to T_2. By repeating these processes many times, the temperature of the gas can be significantly lowered.

4.5.3 Mean Value of Energy: Caloric Equation of State

From the expression of free energy (4.146) and from the definition of pressure $P = -\left(\partial F/\partial V\right)_T$ it is seen that the thermal equation of the state of a polar ideal gas does not depend on the degree of polarization of the gas and has the usual appearance of an ideal gas $P = k_0NT/V$.

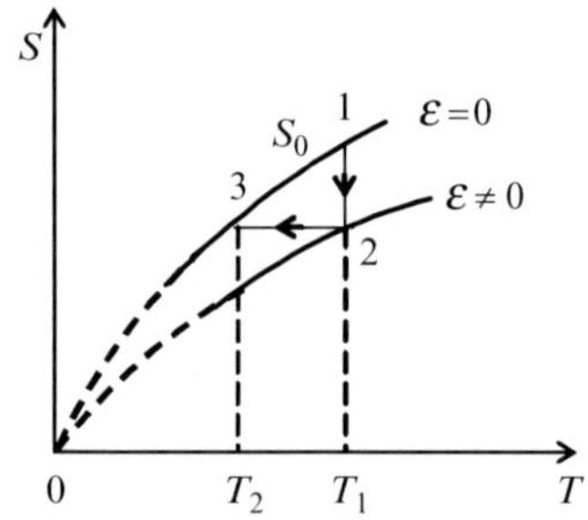

Fig. 4.9. The temperature dependence of entropy in the absence and presence of an electric field

In contradistinction to the thermal equation, the caloric equation of the state of a gas (the mean value of energy) sharply depends on the degree of polarization and an electric field. Indeed, taking into account expressions of the free energy (4.146) and entropy (4.147) in $E = F + TS$, for the mean value of energy (the caloric equation) of a two-atom gas we get

$$E = \frac{5}{2}k_0NT - Np_0\mathcal{E}L(a), \tag{4.152}$$

where $L(a)$ is the Langevin function [see (4.139)].

The first item in this expression is the mean energy of classical translational and rotational motions, and the second item is the mean value of the potential energy of N number of polar molecules in an external electric field $\mathcal{E}$.

Inasmuch as in weak electric fields ($a \ll 1$) the Langevin function $L(a) = a/3 = p_0\mathcal{E}/3k_0T$, then

$$E = \frac{5}{2}k_0NT - \frac{T}{3}k_0NT\left(\frac{p_0\boldsymbol{\mathcal{E}}}{k_0T}\right)^2. \tag{4.153}$$

In strong electric fields ($a \gg 1$), inasmuch as $ctha = 1$ and the Langevin function $L(a) = (1 - 1/a)$, the mean energy has the appearance

$$E = \frac{5}{2}k_0NT - Np_0\boldsymbol{\mathcal{E}}\left(1 - \frac{k_0T}{p_0\boldsymbol{\mathcal{E}}}\right). \tag{4.154}$$

4.5.4 Heat Capacity: Determination of Electric Dipole Moment of Molecule

The heat capacity of a polar ideal gas can be calculated by substituting the expression of entropy of a gas (4.147) into $C_V = T\,(\partial S/\partial T)_V$ or the expression of the mean energy of a gas (4.152) into $C_V = (\partial E/\partial T)_V$. In both cases, naturally, identical results are obtained:

$$C_V = \frac{5}{2}k_0N - Np_0\mathcal{E}\frac{\partial L(a)}{\partial a}\frac{\partial a}{\partial T}. \tag{4.155}$$

Finally we get

$$C_V = \frac{5}{2}k_0N + k_0NA(a), \tag{4.156}$$

where

$$A(a) = 1 + a^2 - (a\,ctha)^2 \equiv 1 - \left(\frac{a}{sha}\right)^2 \tag{4.157}$$

is the dimensionless function.

For weak electric fields ($a \ll 1$) it can be restricted to the first two terms of the series in an expansion of the hyperbolic cotangent (4.140):

$$A(a) = \frac{a^2}{3}. \tag{4.158}$$

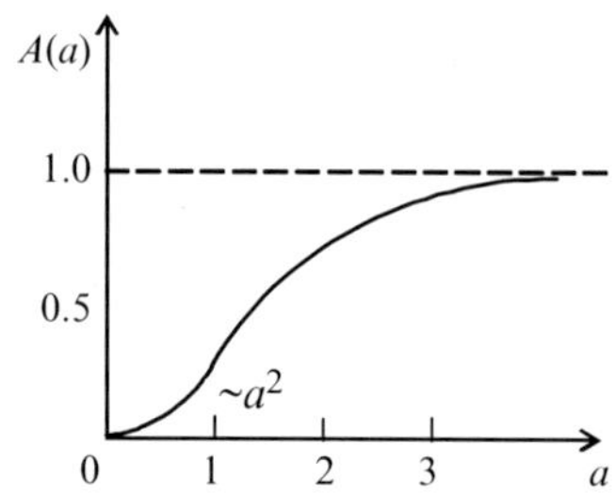

Fig. 4.10. The schematic plot of the $A(a)$ - functions (see (4.157))

In strong electric fields ($a \gg 1$), inasmuch as $cth\, a = 1$, from (4.158) it follows that $A(a) = 1$. Thus, like the function $L(a)$, the function $A(a)$ changes in the range (0–1), with the disparity that at small values of the argument $A(a) \sim a^2$ (compare Figs. 4.8 and 4.10).

Inasmuch as the function $A(a)$ is always positive $A(a) > 0$, from (4.156) it follows that the heat capacity in the presence of an electric field is always larger: $C_V(\mathcal{E}) > C_V(0)$.

This conclusion can be explained as follows. In an electric field dipole moments of molecules line up along the field, i.e. the gas is polarized. As temperature rises dipoles are disordered and, thereby, entropy grows, inasmuch as entropy is a monotonic function of temperature $(\partial S/\partial T) > 0$. To effect disorder supplementary energy is required, as a result of which the heat capacity increases.

In the considered temperature region ($T_\mathrm{r} \ll T \ll T_\mathrm{v}$) translational and rotational motions are classical, and in the absence of an electric field ($\mathcal{E} = 0$) the heat capacity of a gas $C_V(0) = 5/2k_0N$. Taking this into account, expression (4.156) can be presented in the form

$$\frac{C_V(\mathcal{E}) - C_V(0)}{C_V(0)} = \frac{2}{5}A(a). \tag{4.159}$$

Thus, in the region $T_\mathrm{r} \ll T \ll T_\mathrm{v}$, measuring the heat capacity in the absence of an electric field $C_V(0)$ and in its presence $C_V(\mathcal{E})$, the left-hand side of equation (4.159) can be experimentally determined. This experimentally determined value is denoted by $\gamma = [C_V(\mathcal{E}) - C_V(0)]/C_V(0)$. Knowing γ, it can be found that the root $a = a_0$ of the transcendental equation

$$1 + a^2 - (a\, cth\, a)^2 = \frac{5}{2}\gamma, \tag{4.160}$$

with the aid of which the electric dipole moment of a molecule can be immediately determined

$$p_0 = \frac{k_0 T}{\mathcal{E}} a_0. \tag{4.161}$$

Note that the above expressions adduced for energy (4.152) and heat capacity (4.156) are only for two-atom polar gases.

As for multi-atom ($n > 2$) non-linear molecules, the rotational degree of freedom equals 3, and the heat capacity of such an ideal gas is calculated by the formula

$$C_V = 3k_0N + k_0NA(a), \tag{4.162}$$

where $C_V(0) = 3k_0N = 3R$ is the heat capacity of a gas in the absence of an electric field ($a = 0$). Then instead of (4.160) it can be written as

$$1 + a^2 - (a\, cth\, a)^2 = 3\gamma. \tag{4.163}$$

Thus, to determine the dipole moment of a multi-atom molecule ($n > 2$) in (4.161) roots need to be used (4.163), but not in (4.160).

4.6 Paramagnetic Ideal Gas in External Magnetic Field

4.6.1 Classical Case

The statement of the problem in the present section is analogous to the preceding one. It is supposed that an ideal gas of volume V, consisting of N number of particles (molecules) with the intrinsic magnetic dipole moment $\boldsymbol{\mu}$, is found in an external uniform magnetic field $\boldsymbol{H}$. It is known that the potential energy of a dipole $\boldsymbol{\mu}$ in an external magnetic field has the appearance

$$U = -(\boldsymbol{\mu H}) = -\mu H \cos\theta, \tag{4.164}$$

where θ is the angle between the magnetic dipole moment $\boldsymbol{\mu}$ and the magnetic field $\boldsymbol{H}$.

Inasmuch as in the absence of an external magnetic field, magnetic dipoles in the volume occupied are distributed chaotically, as a whole, the gas is not magnetized, i.e. the magnetization vector $\boldsymbol{M} = 0$.

Under action of the field, dipoles line up along the field owing to the condition of minimality of the potential energy. As a result of this, the gas as a whole is magnetized, i.e. the dipole moment accounting for the unit of volume – *the magnetization vector* – becomes distinct from zero $\boldsymbol{M} \neq 0$.

Let us investigate the thermodynamic properties of such a magnetized state of an ideal gas. In particular, let us calculate the mean value of the magnetization vector

$$\boldsymbol{M} = \sum_i \boldsymbol{\mu}_i, \tag{4.165}$$

where the summation is conducted over all magnetic dipoles found in the unit of volume.

It is evident that the magnetization vector is directed along the external magnetic field, and its numerical value equals

$$M = n\overline{\mu \cos\theta}, \tag{4.166}$$

where n is the concentration of magnetic dipoles, and $\overline{\mu\cos\theta}$ is the mean value of the projection of an arbitrary dipole in the direction of the magnetic field.

The statistically mean value of the magnetization vector in a paramagnetic ideal gas was computed by the French physicist Paul Langevin for the first time in 1905. It is supposed that all molecules have an identical magnetic dipole moment, and to calculate M, according to (4.166), it is sufficient to calculate the mean value $\cos\theta$. In 1905, when quantum mechanics as a science did not exist, it was supposed that in a magnetic field dipoles can be found at any angle θ, i.e. $\overline{\cos\theta}$ in the range 0–1 can take on any value. Using the expression of potential energy (4.164), Paul Langevin calculated $\overline{\cos\theta}$ on the strength of the Boltzmann distribution and, thereby, found the mean value of the magnetization vector. At first, let us consider the classical method proposed by Langevin.

In the absence of a magnetic field magnetic dipoles are positioned chaotically, i.e. the probability of finding a magnetic dipole in any direction is identical.

In the presence of an external magnetic field $\boldsymbol{H}$ every possible direction of a magnetic dipole is non-equally probable. In this case, the distribution is determined by the Boltzmann distribution function, corresponding to (4.164). It is evident that the probability of the fact that the direction of the magnetic dipole is found in an element of the solid angle $\mathrm{d}\Omega = \sin\theta\,\mathrm{d}\theta\,\mathrm{d}\varphi$ ought to be proportional to $\mathrm{d}\boldsymbol{\Omega}$. As a result, the probability of the magnetic dipole $\boldsymbol{\mu}$ found in the element of the solid angle $\mathrm{d}\boldsymbol{\Omega}$ is

$$W(\theta)\mathrm{d}\boldsymbol{\Omega}, \tag{4.167}$$

where

$$W(\theta) = A\,\mathrm{e}^{-\frac{U(\theta)}{k_0T}} = A\,\mathrm{e}^{\frac{\mu H}{k_0T}\cos\theta} \tag{4.168}$$

is the probability of formation of an angle θ between the magnetic dipole and a magnetic field. The normalizing constant A is determined from the condition

$$\int W(\theta)\mathrm{d}\boldsymbol{\Omega} = A\int_0^{2\pi} d\varphi \int_0^{\pi} \mathrm{e}^{b\cos\theta}\sin\theta\,\mathrm{d}\theta = 1. \tag{4.169}$$

Hence it can be easily found that

$$A = \frac{1}{4\pi}\cdot\frac{b}{shb}, \tag{4.170}$$

where

$$b = \frac{\mu H}{k_0T}. \tag{4.171}$$

As a result, for the mean value of the projection of the magnetic dipole moment $\boldsymbol{\mu}$ on the direction of a magnetic field $\boldsymbol{H}$ we have

$$\overline{\mu \cos\theta} = \int \mu \cos\theta\, W(\theta) \mathrm{d}\boldsymbol{\Omega} = \frac{\mu b}{2shb} \int_0^\pi \cos\theta\, \mathrm{e}^{b\cos\theta} \sin\theta\, \mathrm{d}\theta. \tag{4.172}$$

Introducing the notation $\cos\theta = x$, we get

$$\int_0^\pi \cos\theta\, \mathrm{e}^{b\cos\theta} \sin\theta\, \mathrm{d}\theta = \frac{\partial}{\partial b} \int_{-1}^{1} \mathrm{e}^{bx} \mathrm{d}x = \frac{\partial}{\partial b}\left(\frac{\mathrm{e}^b - \mathrm{e}^{-b}}{b}\right). \tag{4.173}$$

Then from the two latter expressions we finally get

$$\overline{\mu \cos\theta} = \mu L(b), \tag{4.174}$$

where

$$L(b) = cthb - \frac{1}{b} \tag{4.175}$$

is the Langevin function.

From (4.166) and (4.174) the statistical mean value of the magnetization vector can be found:

$$M = n\mu L(b). \tag{4.176}$$

The limiting value and its dependence on the argument are schematically presented in Fig. 4.8 (see Sect. 4.5). If we take into account that in a weak magnetic field at high temperatures $b = \mu H/k_0 T \ll 1$ and $L(b) \approx b/3$, for the mean value of the magnetization vector we get from (4.176)

$$M = \frac{1}{3} n\mu b = \frac{n\mu^2 H}{3k_0 T}. \tag{4.177}$$

This dependence ($M \sim H/T$) of the mean value of the magnetization on a magnetic field and temperature corresponds to the Curie experimental law, established in 1895.

Note that the mean value of the magnetization vector (4.176) can be obtained also with the aid of the Gibbs method (analogous calculations are adduced in Sect. 4.5), calculating F, and using the relationship

$$M = -\frac{1}{V}\left(\frac{\partial F}{\partial H}\right)_{T,V}. \tag{4.178}$$

4.6.2 Quantum Case

Up to now we reasoned classically, supposing that the magnetic moment of a molecule in an external magnetic field can form any angle with the magnetic field; in other words, we regarded that the projection of the magnetic dipole moment in the direction of the magnetic field can take on any value. However, after the rise of quantum mechanics it became known that the magnetic moment of a molecule (the moment formed at the expense of spin and orbital motions) cannot form any angle; therefore the projection of the magnetic moment in the direction of the field takes on only discrete values.

For simplicity, assume that magnetic moments of molecules forming a gas correspond to the intrinsic moment (the spin) of electrons: $s\hbar$, where $\hbar = h/2\pi, h$ is the Planck constant, and s is an integer or semi-integer number. Along the magnetic field $\boldsymbol{H}$ the possible number of projections of moments $s\hbar$ is $(2s+1)$, i.e. projections along the magnetic field can take on values $j\hbar$, where j takes on values $-s(-s+1), \ldots, (s-1), s$.

From quantum mechanics it is also known that energy corresponds to the magnetic moment of a molecule placed in a magnetic field $\boldsymbol{H}$.

$$\varepsilon_j = -2j\mu_{\mathrm{B}}H, \tag{4.179}$$

where $\mu_{\mathrm{B}} = e\hbar/2mc = 0.927 \times 10^{-20}$ erg/G is the Borh magneton.

Consequently, the energy spectrum of a molecule having the spin $s\hbar$ consists of $(2s+1)$ levels.

In order to determine the magnetization of a gas, with regard to discreteness of the energy spectrum (4.179), use the Gibbs method. To do this it is necessary to calculate the free energy F entering into (4.178) on the basis of the spectrum (4.179).

The free energy of an ideal gas consisting of N molecules has the appearance

$$F = -k_0TN \ln z, \tag{4.180}$$

where z is the statistical integral or statistical sum of one molecule which can be presented in the form

$$z = z_0 \cdot z_H, \tag{4.181}$$

where z_0 is the statistical integral of a molecule in the absence of a magnetic field, z_H is the statistical sum associated with the discreteness of spectrum (4.179) in a magnetic field. The computation of z_0 is adduced in Sect. 4.4. Therefore it is sufficient to calculate z_H, i.e. the statistical sum:

$$z_H = \sum_{j=-s}^{s} \mathrm{e}^{-\varepsilon_j/k_0T} = \sum_{j=-s}^{s} \mathrm{e}^{2j\mu_{\mathrm{B}}H/k_0T}. \tag{4.182}$$

If we take into account (4.181) in (4.180), we get

$$F = F_0 + F_H, \tag{4.183}$$

where F_0 is the free energy in the absence of a magnetic field $(H = 0)$, F_H is the free energy, being determining by the statistical sum z_H and depending on a magnetic field:

$$F_H = -k_0TN \ln z_H. \tag{4.184}$$

Magnetization

Having substituted (4.183) and (4.184) into the definition of magnetization (4.178) and having taken into account that F_0 does not depend on a magnetic field, for the magnetization we get

$$M = nk_0T\left(\frac{\partial}{\partial H}\ln z_H\right)_{T,V}, \tag{4.185}$$

where $n = N/V$ is the concentration of molecules of a gas.

To calculate the statistical sum (4.182) introduce the notation

$$a = \frac{\mu_\mathrm{B}H}{k_0T}. \tag{4.186}$$

Then sum (4.182) can be presented in the open form:

$$\begin{aligned} z_H = \sum_{j=-s}^{s} \mathrm{e}^{2aj} &= \mathrm{e}^{-2as} + \mathrm{e}^{-2a(s-1)} + \mathrm{e}^{-2a(s-2)} + \cdots \\ &+\mathrm{e}^{-2a} + 1 + \mathrm{e}^{2a} + \mathrm{e}^{+4a} + \cdots + \mathrm{e}^{2as} = \mathrm{e}^{-2as}\Big[1 + \mathrm{e}^{2a} + \mathrm{e}^{4a} + \cdots \\ &+\mathrm{e}^{2a(s-1)} + \mathrm{e}^{2as} + \mathrm{e}^{2a(s+1)} + \mathrm{e}^{2a(s+2)} + \cdots + \mathrm{e}^{4as}\Big]. \end{aligned} \tag{4.187}$$

To calculate the sum of the geometric progression, entering into (4.187) and consisting of $(2s+1)$ terms, make use of the formula

$$\sum_{i=1}^{n} a_1q^{i-1} = a_1\left(1 + q + q^2 + \cdots + q^{n-1}\right) = a_1\frac{q^n - 1}{q - 1}. \tag{4.188}$$

From comparison (4.188) with (4.187) it is seen that

$$a_1 = \mathrm{e}^{-2as};\ q = \mathrm{e}^{2a};\ n = 2s + 1. \tag{4.189}$$

Then we get

$$z_H = \mathrm{e}^{-2as}\frac{\mathrm{e}^{2a(2s+1)} - 1}{\mathrm{e}^{2a} - 1} = \frac{sh\left[(2s+1)a\right]}{sha}. \tag{4.190}$$

If we take into account (4.184) and (4.186), the magnetic part of the free energy takes the form:

$$F_H = -k_0TN\ln\left[\frac{sh\left[(2s+1)a\right]}{sha}\right] = -k_0TN\ln\left[\frac{sh\left[(2s+1)(\mu_\mathrm{B}H)/k_0T\right]}{sh\left(\mu_\mathrm{B}H/k_0T\right)}\right]. \tag{4.191}$$

Taking into account (4.190) in (4.185), for the magnetization of a gas we get

$$M = n(2s+1)\mu_\mathrm{B}B_s(H), \tag{4.192}$$

where

$$B_s(H) = cth\left[(2s+1)\frac{\mu_B H}{k_0 T}\right] - \frac{1}{(2s+1)} cth\left(\frac{\mu_B H}{k_0 T}\right) \tag{4.193}$$

is the generalized Brillouin function, which is the quantum analogue of the Langevin function.

Show that in the quasi-classical case function (4.193) passes into (4.175), and magnetization (4.192), conformably, into (4.176). Indeed, in the quasi-classical approximation, $s \to \infty$, $\hbar \to 0$ and $\mu_B \to 0$ are the conditions of continuity of the energy spectrum. If we introduce the notation $(2s+1)\,\mu_B \to \mu$ and take into account that $cth\,(\mu_B H/k_0 T) = k_0 T/\mu_B H$, the Brillouin function (4.193) passes into the Langevin function (4.175) and expressions (4.192) and (4.176) coincide.

In the other limiting case ($s = 1/2$), the expression (4.193) can be presented in the form

$$B_{1/2}(H) = cth\left(2\frac{\mu_B H}{k_0 T}\right) - \frac{1}{2} cth\left(\frac{\mu_B H}{k_0 T}\right). \tag{4.194}$$

If we use the known trigonometric formula $cth 2x = (1 + cth^2 x)/2 cth\, x$, (4.194) takes the form

$$B_{1/2}(H) = \frac{1}{2} th\left(\frac{\mu_B H}{k_0 T}\right), \tag{4.195}$$

and the magnetization equals

$$M = n\mu_B th\left(\frac{\mu_B H}{k_0 T}\right). \tag{4.196}$$

In the region of high temperatures and weak magnetic fields, when the condition $\mu_B H/k_0 T \ll 1$ is fulfilled, the Brillouin function (4.193) takes the form

$$B_s(H) = \frac{4s(s+1)}{3(2s+1)} \frac{\mu_B H}{k_0 T}, \tag{4.197}$$

and magnetization (4.192) equals

$$M = \frac{4s(s+1)}{3} \frac{n\mu_B^2 H}{k_0 T} \sim \frac{H}{T}. \tag{4.198}$$

At $s = 1/2$ magnetization has the appearance

$$M = \frac{n\mu_B^2 H}{k_0 T}, \tag{4.199}$$

which coincides with the result obtained from (4.196) for the considered case.

The dependence of magnetization on a magnetic field and temperature in the form (4.198) or (4.199) was experimentally corroborated by the French physicist Pierre Curie in 1895 and bears the name *the Curie law.*

In the region of low temperatures and strong magnetic fields, when the condition $\mu_B H/k_0 T \gg 1$ is fulfilled, Brillouin function (4.193) does not depend on a magnetic field, and takes the form

$$B_s(H) = \frac{2s}{2s+1}. \tag{4.200}$$

In this case, as is seen from (4.192), magnetization is saturated to the value

$$M = 2sn\mu_B. \tag{4.201}$$

The plot of the Brillouin function (4.193) in a wide region of the parameter $a = \mu_B H/k_0 T$ for different values of s is adduced in Fig. 4.11.

In conclusion, note that using the expression of the magnetic part of the free energy (4.191), mean energy, entropy and heat capacity of a paramagnetic gas in a magnetic field can also be calculated.

Entropy, Mean Energy and Heat Capacity

Using the expression for the total free energy (4.183), with regard to (4.191) for entropy of a paramagnetic gas $S = -\left(\partial F/\partial T\right)_{V,H}$, we get:

$$S(H) = S(0) - k_0 N\varphi(H), \tag{4.202}$$

where $S(0) = -\left(\partial F_0/\partial T\right)_{V,H}$ is entropy of a paramagnetic gas in the absence of a magnetic field, and the function $\varphi(H)$ equals

$$\varphi(H) = (2s+1)aB_s(H,T) - \ln\left[\frac{sh(2s+1)a}{(2s+1)sha}\right]. \tag{4.203}$$

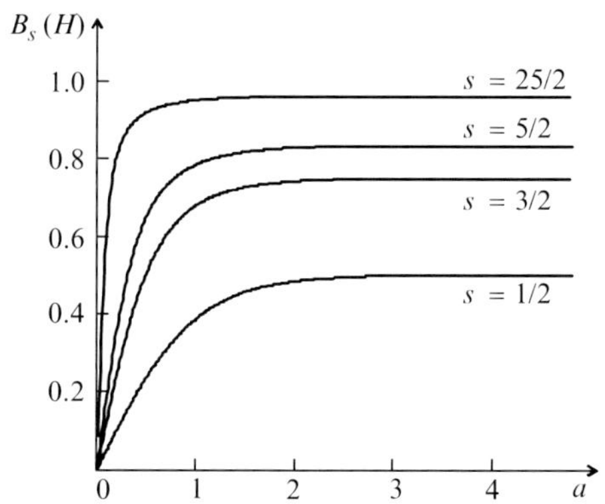

Fig. 4.11. The plot of the Brillouin function

It can be shown that the function $\phi(H)$, entering into the expression of entropy (4.202), is positive; therefore, at the expense of magnetization, entropy of a gas in a magnetic field decreases, since dipoles in a magnetic field orientising themselves create a certain order and the chaoticity decreases.

At small values of the magnetic field and at high temperatures, when $a \ll 1$, using (4.197) and expanding the hyperbolic sine, for entropy we get

$$S(H) = S(0) - k_0 N \frac{4s(s+1)}{3} \left(\frac{\mu_{\mathrm{B}} H}{k_0 T} \right)^2 . \tag{4.204}$$

In particular, at $s = 1/2$

$$S(H) = S(0) - k_0 N \left(\frac{\mu_{\mathrm{B}} H}{k_0 T} \right)^2 . \tag{4.205}$$

The decrease in entropy in the magnetic field leads to the rise of *the magnetocaloric effect*, which is used to produce ultra-low temperatures (see Sect. 2.8).

Using the expression of free energy (4.191) and entropy (4.202), for the mean energy of a paramagnetic gas in a magnetic field we get:

$$E(H) = E_0 - (2s+1) N \mu_{\mathrm{B}} H B_s(H), \tag{4.206}$$

where $E(0)$ is energy of the gas in the absence of a magnetic field.

In the case of weak magnetic fields ($a \ll 1$), according to (4.197), expression (4.206) takes the form

$$E(H) = E_0 - \frac{4s(s+1)}{3} N k_0 T \left(\frac{\mu_{\mathrm{B}} H}{k_0 T} \right)^2 . \tag{4.207}$$

In the region of low temperatures and strong magnetic fields, when the condition $\mu_{\mathrm{B}} H / k_0 T \gg 1$ is fulfilled, according to (4.200) and (4.206), the magnetic part of energy does not depend on temperature

$$E(H) - E_0 = -2s N \mu_{\mathrm{B}} H, \tag{4.208}$$

If we use the expression of mean energy (4.206), for heat capacity of a gas in a magnetic field we get:

$$C_V(H) = C_V(0) - (2s+1) N \mu_{\mathrm{B}} H \frac{\partial}{\partial a} \left[B_s(H,T) \right] \frac{\partial a}{\partial T}, \tag{4.209}$$

where $C_V(0)$ is heat capacity in the absence of a magnetic field. If the explicit form of the Brillouin function (4.193) is taken into account, the expression for heat capacity (4.209) takes the form

$$C_V(H) = C_V(0) + k_0 N A(a), \tag{4.210}$$

where

$$A(a) = \left(\frac{a}{sha}\right)^2 \left\{1 - \left[\frac{(2s+1)\,sha}{sh\,[(2s+1)a]}\right]^2\right\} \tag{4.211}$$

is the dimensionless function.

It can be shown that the function $A(a)$ is positive. In particular, at weak magnetic fields ($a \ll 1$), expanding the hyperbolic sine, we get

$$A(a) = \frac{4s(s+1)}{3}a^2. \tag{4.212}$$

Then in this approximation heat capacity of a paramagnetic ideal gas has the appearance

$$C_V(H) = C_V(0) + \frac{4s(s+1)}{3}k_0 N \left(\frac{\mu_B H}{k_0 T}\right)^2. \tag{4.213}$$

Hence it is seen that the magnetic part of heat capacity of an ideal paramagnetic gas in a weak magnetic field ($a \ll 1$) quadratically grows: $C_V(H) > C_V(0)$.

This result is explained by the fact that magnetic dipoles are oriented along the magnetic field and as temperature rises this orientation is partially destroyed. A supplementary energy is expended for the process of destruction of the order and, thereby, heat capacity grows.

In the other limiting case of *strong magnetic fields and low temperatures* ($a \gg 1$), as is seen from (4.210), the function $A(a)$ tends to zero.

Plots of the function $A(a)$ in a wide region of the parameter $a = \mu_B H/k_0 T$ for different values of s are adduced in Fig. 4.12. It is seen that this function, determining the magnetic part of heat capacity, has a maximum threat with the growth of s and the maximum displacement is towards the small values of the parameter a, i.e. towards the weak magnetic fields.

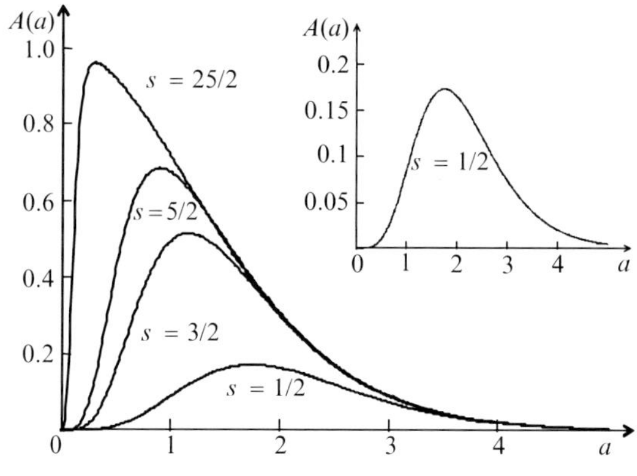

Fig. 4.12. Plots of the function $A(a)$ in a wide region of the parameter a for different values of s (see (4.211))

Thus, the magnetic part of heat capacity of a paramagnetic gas with the growth of the magnetic field (at the pre-assigned value of temperature) grows quadratically first and, passing through the maximum, in very strong magnetic fields tends to zero.

The non-monotonic dependence of the magnetic part of heat capacity is a purely quantum effect and is easily explained by the quantization of energy of the magnetic dipole in a magnetic field [see (4.179)].

4.7 Systems with Negative Absolute Temperature

We introduced the concept of absolute temperature in Sect. 1.8 of the first chapter [see (1.145)]. There it was shown that in the thermodynamically steady state the absolute temperature ought not to be negative (Sect. 1.8, point 4).

This result for systems called "normal systems" follows also from expressions of the statistical integral or statistical sum composing the basis of the Gibbs method. "Normal systems" are systems in which energy of the particles forming them is restricted from below but not from above, i.e energy can take on infinitely large values:

$$\varepsilon_{\min} \le \varepsilon < \infty. \tag{4.214}$$

In the classical case, energy of all three types of motion (translational, rotational and vibrational) of particles forming a system equals zero: $\varepsilon_{\min} = 0$. In the quantum case, conformable to expressions (4.74), (4.94) and (4.106), energy takes on the following values:

$$\varepsilon^t_{\min} = \frac{\pi^2\hbar^2}{2mL^2}; \; \varepsilon^r_{\min} = 0; \; \varepsilon^v_{\min} = \frac{\hbar\omega}{2}. \tag{4.215}$$

In both cases, energy is not restricted from above ($\varepsilon \to \infty$).

Indeed, for "normal systems", satisfying condition (4.214), in order that the statistical integral

$$Z_{\text{cl}} = \int \mathrm{e}^{-E(q,p)/k_0T} \mathrm{d}q\,\mathrm{d}p \tag{4.216}$$

and the statistical sum

$$Z = \sum_n \mathrm{e}^{-E_n/k_0T} \tag{4.217}$$

at large values of energy do not diverge, temperature ought to be only positive ($T > 0$).

However, systems exist in which energy of particles corresponding to some degrees of freedom changes in the finite range, i.e. energy is restricted both from above and from below:

$$\varepsilon_{\min} \le \varepsilon \le \varepsilon_{\max}. \tag{4.218}$$

Such systems are called "anomalous systems". For such systems the statistical integral (4.216) and the statistical sum (4.217) at any value $(-\infty < T < +\infty)$ of temperature take on finite values. Consequently, the state of a system with negative absolute temperature $(T < 0)$ is possible.

Consider a simplest system satisfying condition (4.218). Assume that the considered system, consisting of N number of molecules with the magnetic dipole moment $\boldsymbol{\mu}$, is placed in an external uniform magnetic field $\boldsymbol{H}$. It is supposed that the interaction between molecules (simultaneously, between dipoles) is rather weak.

In the classical case, as is seen from (4.214), energy of a magnetic dipole (a molecule) $\boldsymbol{\mu}$ depending on the angle θ, formed between the dipole and the magnetic field $\boldsymbol{H}$, continuously changes from $-\mu H$ to $+\mu H$

$$-\mu H \leq \varepsilon \leq \mu H. \tag{4.219}$$

In the quantum case, as is seen from (4.179), energy of a magnetic dipole (a molecule) depending on the value of s

$$-2s\mu_{\mathrm{B}}H \leq \varepsilon \leq 2s\mu_{\mathrm{B}}H \tag{4.220}$$

takes on $(2s+1)$ number of discrete values. Consequently, the magnetic part of energy of the considered system is restricted both from below $\varepsilon_{\min} = -2s\mu_{\mathrm{B}}NH = -N\mu H$ and from above $\varepsilon_{\max} = +2s\mu_{\mathrm{B}}NH = N\mu H$.

Consider the thermodynamic properties of a system, the energy spectrum of which is restricted by condition (4.220). To do this, at first find the free energy. The expression of the free energy associated with the magnetic field is determined by formula (4.184) with regard to (4.190). Having combined these expressions, for free energy we get

$$F_H = -k_0 TN \left\{ \ln sh \left[\frac{(2s+1)\mu H}{k_0 T} \right] - \ln sh \left(\frac{\mu H}{k_0 T} \right) \right\}, \tag{4.221}$$

where the notation $\mu \equiv \mu_{\mathrm{B}}$ is introduced.

For simplicity, consider only two-level systems, i.e. systems with $s = 1/2$ (Fig. 4.13). In this case, the expression of the free energy takes the simple form

$$F_H = -k_0 TN \ln \left[2ch \left(\frac{\mu H}{k_0 T} \right) \right], \tag{4.222}$$

where the known formula $sh2x = 2shx \cdot chx$ is used. On the strength of (4.222) with $S = -\left(\partial F_H / \partial T\right)$, mean energy $E = F_H + TS$ and heat capacity $C_V = -\left(\partial E / \partial T\right)_V$, entropy can be calculated as

$$S = k_0 N \left[\ln 2\, ch \left(\frac{\mu H}{k_0 T} \right) - \frac{\mu H}{k_0 T} th \frac{\mu H}{k_0 T} \right], \tag{4.223}$$

$$E = -\mu HNth \frac{\mu H}{k_0 T}, \tag{4.224}$$

$$C_V = k_0 N \left[\frac{\mu H / k_0 T}{ch\left(\mu H / k_0 T\right)} \right]^2. \tag{4.225}$$

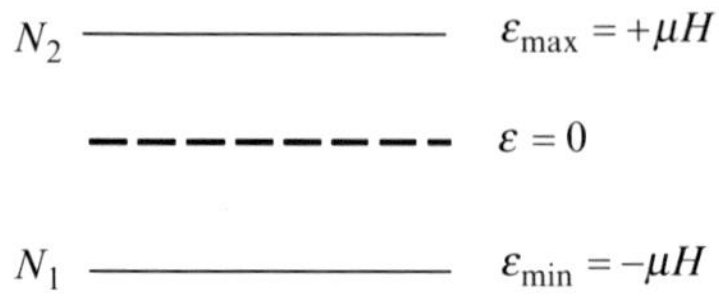

Fig. 4.13. The splitting of the energy levels in a magnetic field

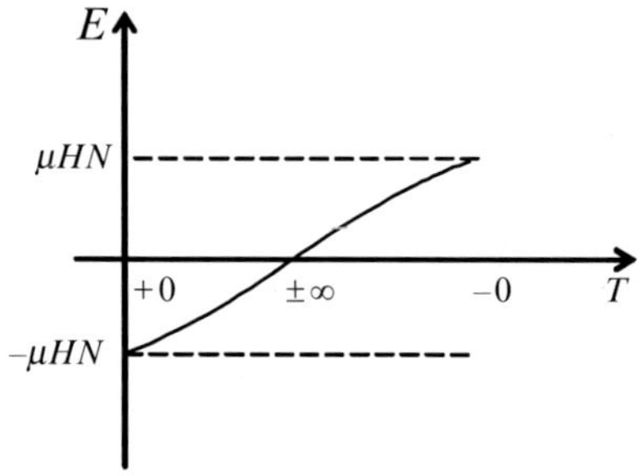

Fig. 4.14. The temperature dependence of the two-level system energy

Before analyzing the expressions of entropy and mean energy, consider the distribution of N number of particles over levels shown in Fig. 4.13.

For the number of particles N_1 and N_2 at the lower and upper levels, respectively, it can be written

$$\begin{aligned} N_1 &= N\,\mathrm{e}^{-\varepsilon_{\min}/k_0T} = N\,\mathrm{e}^{\mu H/k_0T} \\ N_2 &= N\,\mathrm{e}^{-\varepsilon_{\max}/k_0T} = N\,\mathrm{e}^{-\mu H/k_0T}. \end{aligned} \tag{4.226}$$

Hence, for N_1/N_2 we get $N_1/N_2 = \mathrm{e}^{2\mu H/k_0T}$. If we solve this equation in conjunction with the condition $N_1 + N_2 = N$, we get

$$N_1 = \frac{N\,\mathrm{e}^{\mu H/k_0T}}{2ch\mu H/k_0T};\; N_2 = \frac{N\,\mathrm{e}^{-\mu H/k_0T}}{2ch\mu H/k_0T}. \tag{4.227}$$

Inasmuch as the energy spectrum of a system is restricted from below and above according to (4.220), parameters characterizing the system (E, S, C_V, N_1 and N_2) take on finite values in all the temperature ranges $-\infty < T < +\infty$.

For the sake of clarity, we graphically present the temperature dependence of indicated parameters.

In Fig. 4.14 temperature dependence of energy is presented (4.224), from which it is seen that the minimum energy ($-\mu HN$) corresponds to temperature $T \to +0$, and the maximum energy ($+\mu HN$) corresponds to temperature $T \to -0$. Hence, it follows that higher values of energy correspond to the negative region of temperature (Fig. 4.15).

From the energy point of view, limiting states corresponding to temperatures $T = +\infty$ and $T = -\infty$ are equivalent. In both cases, the mean energy of

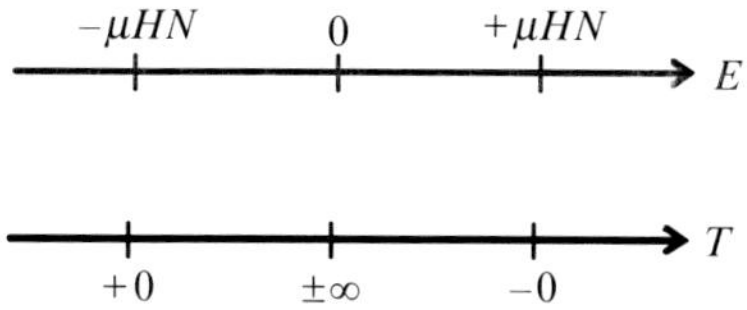

Fig. 4.15. Comparison of energy and temperature scales of two-level system

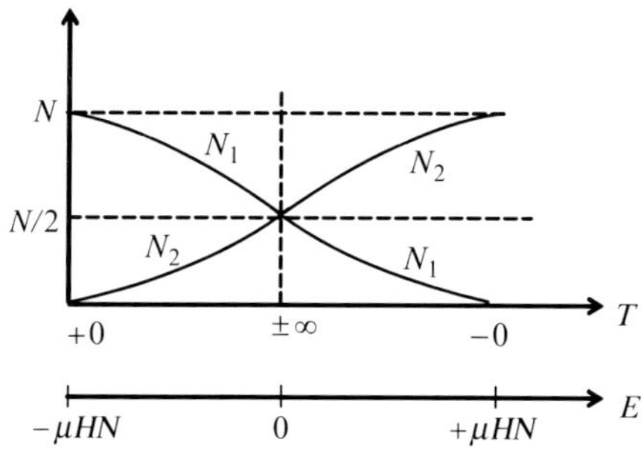

Fig. 4.16. The distribution of particles on levels depending on energy and temperature

the system $E = 0$, inasmuch as in this case half of the particles $N/2$ are found at the level $-\mu H$ and the other half $N/2$ at the level $+\mu H$. In Fig. 4.16 the distribution of particles depending on energy and temperature is presented. As is seen in the negative temperature region the number of particles N_2 at the upper level is larger than the number of particles N_1 at the lower level, i.e. inversion occurs in the distribution.

In order to pass from the positive region of temperature into the negative one, it is necessary not to decrease energy, but, on the contrary, to increase it, in order that temperature increases $T \to +\infty$. Limiting states of temperature $+\infty$ and $-\infty$ correspond to equivalent states, and $T \to +0$ and $T \to -0$, from the energy point of view, correspond to extreme limiting states. In other words, inasmuch as the absolute zero ($T \to +0$) is impossible to produce (the Nernst theorem), it is also impossible to immediately pass into the state with $T \to -0$. Thus, in order to produce the state with $T \to -0$, it is necessary to pass through the state with $T \to \pm\infty$ (Fig. 4.17).

As is seen from this figure, it is impossible to approach the point $T = 0$ either from the right or from the left (the point $T = 0$ is the special point); as energy rises, beginning from the state $-\infty$, which is equavalent to $+\infty$, it can approach $T \to -0$.

We now analyse the dependence of entropy of the system on temperature and energy. A plot of the dependence, constructed on the basis of expression (4.223), is presented in Fig. 4.18. Inasmuch as energy rises from $-\mu HN$ to 0, entropy S also grows and $(\partial S/\partial E) = 1/T > 0$, and as energy rises from 0 to $+\mu HN$, entropy S decreases and $(\partial S/\partial E) = 1/T < 0$.

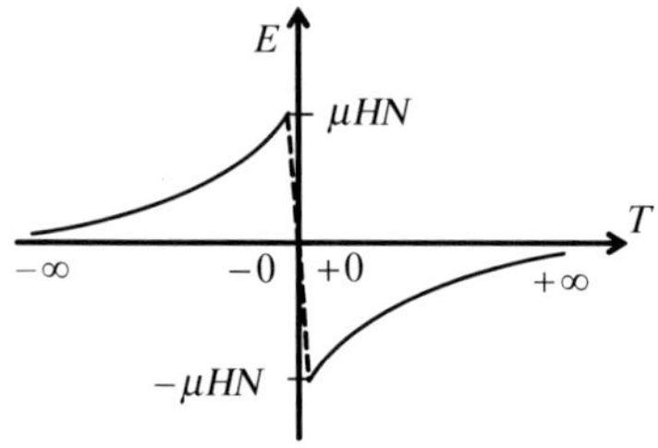

Fig. 4.17. The temperature dependence of the two-level system energy

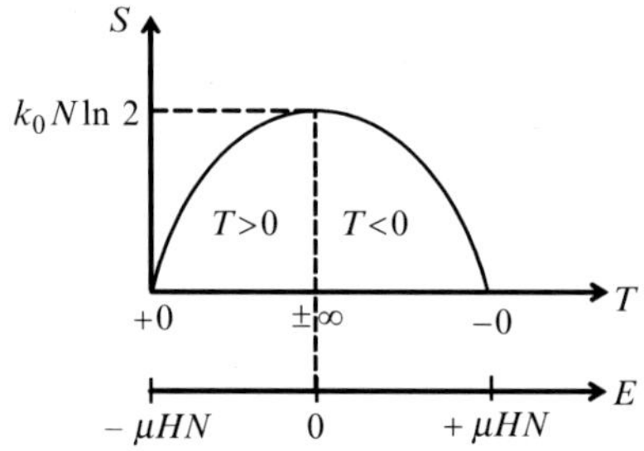

Fig. 4.18. The entropy dependence on the energy and temperature

Note that in the state $T \to \pm 0$ entropy equals zero, inasmuch as in the system there exists order: as $T \to +0$ all magnetic dipoles are directed along the magnetic field, and as $T \to -0$ they all are directed against the field (see Fig. 4.16). As $T \to \pm\infty$ half of the dipoles are directed along the field and the other half are against the field (Fig. 4.16), i.e. the system is completely disordered (chaotic) and entropy takes on values $S = S_{\max} = k_0 N \ln 2$ (Fig. 4.18).

An analysis of the expression of heat capacity (4.225) shows that at $\mu H/k_0 T = \pm 0$ $(T \to \pm\infty)$ the heat capacity C_V equals zero. Outside this point C_V exponentially grows, and passing through the maximum, finally decreases. From the condition of maximum heat capacity C_V

$$ch\left(\frac{\mu H}{k_0 T}\right) - \left(\frac{\mu H}{k_0 T}\right) sh\left(\frac{\mu H}{k_0 T}\right) = 0, \tag{4.228}$$

it can be found that

$$\frac{\mu H}{k_0 T} = \pm 1.2 \, \text{or} \, T_0 = \pm \frac{\mu H}{1.2 k_0}. \tag{4.229}$$

Calculations show that the heat capacity at the point of the maximum takes on the value

$$C_V(T_0) = 0.44\, k_0 N. \tag{4.230}$$

In conclusion, note that the states corresponding to the negative absolute temperature are not steady. If subsystems of this system interact between themselves, the system fast passes into the equilibrium state ($T > 0$). Therefore, states with $T < 0$ are not long-lasting. In order to produce a state with negative absolute temperature, it is nedeed to act as follows. Assume that all magnetic moments of the system are directed along the magnetic field. This corresponds to the state with $T = +0$. The direction of the magnetic field can be instantly changed to the opposite direction so that magnetic moments would have no time to change their direction. Till magnetic dipole moments that are relaxing do not pass into the steady state with $T > 0$, the system in the course of a very small time interval is found in a state with negative absolute temperature.

This process, i.e. "the inversion" of the distribution of moments, is used when creating lasers. This state of "inversion" is frequently realized optically.

5

Non-Ideals Gases

Summary. This chapter is devoted to the application of the Gibbs method to non-ideal – *real* – molecular gases. At first, the equation of the state of rarefied gases, molecules of which interact weakly among themselves, is derived in the general form. Thereupon, thermodynamics of a non-ideal gas of the van der Waals type is considered. In concluding the chapter, plasma – a neutral gas in which Coulomb interaction exists between particles – is considered. In this case, the method of Debye screening is used.

5.1 Equation of State of Rarefied Real Gases

Consider a gas of volume V consisting of N number of molecules. Assume that the motion of molecules of the gas obeys the laws of classical mechanics, and their concentration N/V is sufficiently small (a rarefied gas) so that the energy of the interaction u_{ik} between two molecules is much less than the mean kinetic energy of the molecules $\bar{\varepsilon}$:

$$u_{ik}/\bar{\varepsilon} \ll 1. \tag{5.1}$$

By applying the Gibbs method in approximation (5.1), we find the equation of the state of such a gas. The essence of the Gibbs method is given in Chap. 3. According to this method, the equation of the state of a system (a gas) can be found by knowing the explicit form of the free energy:

$$P = -\left(\frac{\partial F}{\partial V}\right)_T. \tag{5.2}$$

The free energy is related to the statistical integral Z by the relationship

$$F = -k_0 T \ln Z. \tag{5.3}$$

Therefore, it is necessary to calculate the statistical integral

$$Z = \frac{1}{N!} \int \mathrm{e}^{-E(\boldsymbol{q},\boldsymbol{p})/k_0 T} \frac{(\mathrm{d}\boldsymbol{q}\,\mathrm{d}\boldsymbol{p})}{(2\pi\hbar)^{3N}}, \tag{5.4}$$

where $E(\boldsymbol{q}, \boldsymbol{p})$ is the total internal energy of the system, and $(\mathrm{d}\boldsymbol{q}\mathrm{d}\boldsymbol{p})$ is an element of volume in the phase space. In the classical case, the total energy can be presented as the sum of kinetic and potential energies:

$$E(\boldsymbol{p},\boldsymbol{r}) = \sum_{i=1}^{N} \frac{1}{2m} \left(p_{ix}^2 + p_{iy}^2 + p_{iz}^2\right) + U(\boldsymbol{r}_1, \boldsymbol{r}_2, \ldots, \boldsymbol{r}_N), \tag{5.5}$$

where m is the reduced mass of a molecule, $U(\boldsymbol{r}_1, \boldsymbol{r}_2, \ldots, \boldsymbol{r}_N)$ is the potential energy of interaction of molecules of a gas as a whole, and $\boldsymbol{p}_i$ and $\boldsymbol{r}_i$ are the impulse and radius vectors of the centre of mass of the ith molecule. Here and henceforth, the generalized coordinates of a molecule are denoted by $\boldsymbol{r}$. It is assumed that an external field is absent.

On substituting (5.5) into (5.4), we get

$$Z = \frac{1}{N!(2\pi\hbar)^{3N}} \int \mathrm{e}^{-\frac{1}{2mk_0T}\sum\limits_{i=1}^{N}\left(p_{ix}^2+p_{iy}^2+p_{iz}^2\right)} (\mathrm{d}\boldsymbol{p}) \int \mathrm{e}^{-\frac{1}{k_0T}U(\boldsymbol{r}_1,\boldsymbol{r}_2,\ldots,\boldsymbol{r}_N)} (\mathrm{d}\boldsymbol{r}), \tag{5.6}$$

where it is taken that $(\mathrm{d}\boldsymbol{p}) \Rightarrow \mathrm{d}\boldsymbol{p}_1\mathrm{d}\boldsymbol{p}_2\ldots\mathrm{d}\boldsymbol{p}_N$ and $(\mathrm{d}\boldsymbol{r}) \Rightarrow \mathrm{d}\boldsymbol{r}_1\mathrm{d}\boldsymbol{r}_2\ldots\mathrm{d}\boldsymbol{r}_N$.

We introduce the concept of *the configurational integral*

$$Z_N = \frac{1}{V^N} \int \exp\left[-\frac{U(\boldsymbol{r}_1, \boldsymbol{r}_2, \ldots, \boldsymbol{r}_N)}{k_0 T}\right] \mathrm{d}\boldsymbol{r}_1 \mathrm{d}\boldsymbol{r}_2, \ldots, \mathrm{d}\boldsymbol{r}_N, \tag{5.7}$$

which is determined by energy of interaction between molecules $U(\boldsymbol{r}_1, \boldsymbol{r}_2, \ldots, \boldsymbol{r}_N)$ depending on the configuration of particles of a gas, i.e. on the radius vector $\boldsymbol{r}_i$ of each molecule.

Then, the statistical integral (5.6) can be presented in the form

$$Z = Z_{\mathrm{id}} \cdot Z_N. \tag{5.8}$$

where Z_{id} is the statistical integral of an ideal gas (see Sect. 4.1). In the absence of interaction between molecules, when $U = 0$, then $Z_N = 1$ and $Z = Z_{\mathrm{id}}$.

If we insert (5.8) in (5.3), for the free energy we get

$$F = F_{\mathrm{id}} - k_0 T \ln Z_N, \tag{5.9}$$

where $F_{\mathrm{id}} = -k_0 T \ln Z_{\mathrm{id}}$ is the free energy of an ideal gas. From (5.2) and (5.9), we get the equation of state of a non-ideal gas:

$$P = P_{\mathrm{id}} + k_0 T \left(\frac{\partial}{\partial V} \ln Z_N\right)_T, \tag{5.10}$$

where

$$P_{\mathrm{id}} = -\left(\frac{\partial F_{\mathrm{id}}}{\partial V}\right)_T = \frac{Nk_0T}{V} \tag{5.11}$$

is the known equation of state of an ideal gas.

Thus, the problem of finding the explicit form of the equation of state of a real gas is reduced to the calculation of the configurational integral Z_N. As is seen from (5.7), to calculate the configurational integral it is necessary to know the explicit form of the energy of interaction U. However, it is known that a universal form of the function U applicable to all systems does not exist. The explicit form of the function U depends on the nature of interaction of particles forming the system. Therefore, in order to calculate Z_N it is necessary to make supplementary assumptions about the function $U(\boldsymbol{r}_1, \boldsymbol{r}_2, \ldots, \boldsymbol{r}_N)$, i.e. about the interaction between particles.

Assume that the total energy of interaction U can be presented as the sum of energies u_{ik} of the pair-wise interactions of molecules:

$$U = \sum_{i<k}^{1,N} u_{ik}. \tag{5.12}$$

On the basis of this assumption, Z_N can be presented in the form

$$Z_N = \frac{1}{V^N}\int \mathrm{e}^{-\sum\limits_{i<k}^{1,N} u_{ik} \Big/ k_0T} \mathrm{d}\boldsymbol{r}_1, \mathrm{d}\boldsymbol{r}_2, \ldots, \mathrm{d}\boldsymbol{r}_N \tag{5.13}$$

or

$$\begin{aligned} Z_N = \frac{1}{V^N}\int \mathrm{d}\boldsymbol{r}_1 \int \mathrm{d}\boldsymbol{r}_2 \mathrm{e}^{-u_{12}/k_0T} \int \mathrm{d}\boldsymbol{r}_3 \mathrm{e}^{-(u_{13}+u_{23})/k_0T} \\ \ldots \int \mathrm{d}\boldsymbol{r}_N \mathrm{e}^{-(u_{1N}+u_{2N}+\cdots+u_{N-1,N})/k_0T}. \end{aligned} \tag{5.14}$$

If the integral corresponding to an arbitrary kth molecule is denoted by

$$J_k = \int \mathrm{d}\boldsymbol{r}_k \exp\left[-\frac{u_{1k}+u_{2k}+\ldots+u_{k-1,k}}{k_0T}\right], \tag{5.15}$$

we get

$$Z_N = \frac{1}{V^N}\prod_{k=1}^{N} J_k. \tag{5.16}$$

Thus, in order to calculate Z_N it is necessary to find the integral J_k. To do this, we introduce the supplementary function

$$\gamma_{ik} = \mathrm{e}^{-u_{ik}/k_0T} - 1, \tag{5.17}$$

which characterises the interaction between particles, too. Indeed, if $u_{ik} = 0$, then $\gamma_{ik} = 0$, and if $u_{ik} \neq 0$, then $\gamma_{ik} \neq 0$. The integral J_k can be expressed by the function γ_{ik} as follows:

$$J_k = \int \mathrm{d}\boldsymbol{r}_k (1+\gamma_{1k})(1+\gamma_{2k})\dots(1+\gamma_{k-1,k}). \tag{5.18}$$

On multiplication of the brackets, we get

$$J_k = \int \mathrm{d}\boldsymbol{r}_k \left(1 + \sum_{i=1}^{k-1} \gamma_{ik} + \sum_{i \neq l}^{1,k-1} \gamma_{ik}\gamma_{lk} + \cdots\right). \tag{5.19}$$

The first integral corresponds to an ideal gas ($\gamma_{ik} = 0$) and is equal to the volume of a gas:

$$J_k^{(1)} = \int \mathrm{d}\boldsymbol{r}_k = V \Rightarrow \tag{5.20}$$

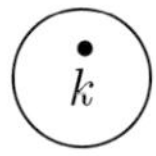

In this case, in the sphere of action of a molecule other molecules are absent. By inserting (5.20) in (5.16), we get $Z_N = 1$.

The second integral in (5.19), i.e.

$$J_k^{(2)} = \int \left(\sum_{i=1}^{k-1} \gamma_{ik}\right) \mathrm{d}\boldsymbol{r}_k \Rightarrow \tag{5.21}$$

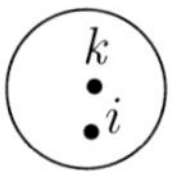

corresponds to the case when an arbitrary kth molecule interacts with only one molecule, i.e. in the sphere of action of the kth molecule only one molecule is found.

The third integral in (5.19), i.e.

$$J_k^{(3)} = \int \left(\sum_{i,l}^{1,k-1} \gamma_{ik}\gamma_{lk}\right) \mathrm{d}\boldsymbol{r}_k \Rightarrow \tag{5.22}$$

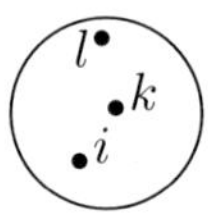

corresponds to a state in which an arbitrary molecule k simultaneously interacts with only two molecules, i.e. in the sphere of action of the kth molecule two molecules are found.

Subsequent integrals correspond to states in which three, four and more molecules are found in the sphere of action of an arbitrary kth molecule at the same time.

Here we consider a rarefied real gas and assume that the concentration of the gas is so small that in the sphere of action of each molecule only one molecule is found. Then, we can restrict (5.19) to only the first two terms:

$$J_k = \int \left(1 + \sum_{i=1}^{k-1} \gamma_{ik}\right) \mathrm{d}\boldsymbol{r}_k, \tag{5.23}$$

because, according to the assumptions, if $\gamma_{ik} \neq 0$, then $\gamma_{lk} = 0$.

In order to continue the calculations, i.e. to calculate the integral (5.23), it is necessary to introduce a supplementary assumption of the function γ_{ik}, i.e. the potential of interaction u_{ik}. Assume that the potential u_{ik} does not depend on the position of a molecule in the space $\boldsymbol{r}_i$ and $\boldsymbol{r}_k$ separately and is determined only by the distance between them, i.e. a spherically symmetric interaction takes place:

$$u_{ik} = u_{ik}(|\boldsymbol{r}_k - \boldsymbol{r}_i|) = u_{ik}(r), \tag{5.24}$$

and, correspondingly, $\gamma_{ik} = \gamma_{ik}(r)$.

In (5.23), we can change over to the spherical coordinate system and integrate over the angles. As a result, in the expression (5.23) we keep the integral which depends only on the distance between two arbitrary molecules (indices i and k are omitted)

$$J_k = V - 2(k-1)B(T), \tag{5.25}$$

where

$$B(T) = -\frac{1}{2}\int \gamma(r) 4\pi r^2 \mathrm{d}r \tag{5.26}$$

is a function that depends only on temperature; the factor $(-1/2)$ is introduced for convenience. The value of $B(T)$ is determined by the explicit form of the function $\gamma(r)$ and, consequently, by the dependence of the function $u(r)$ on the distance between the molecules:

$$\gamma(r) = \mathrm{e}^{-u(r)/k_0T} - 1. \tag{5.27}$$

Not considering the explicit form of the function $u(r)$, substitute (5.25) into (5.16). Then we get

$$Z_N = \frac{1}{V^N}\prod_{k=1}^{N}[V - 2(k-1)B(T)] = \prod_{k=1}^{N}\left[1 - 2(k-1)\frac{B(T)}{V}\right]. \tag{5.28}$$

Hence

$$\ln Z_N = \sum_{k=1}^{N}\ln\left[1 - 2(k-1)\frac{B(T)}{V}\right]. \tag{5.29}$$

In the case of an ideal gas, $\gamma(r) = 0$, and therefore $B(T) = 0$ and $\ln Z_N = 0$. Because the gas is rarefied, the function $B(T)$ is a sufficiently small quantity and therefore it can be taken that

$$\frac{NB(T)}{V} \ll 1. \tag{5.30}$$

Then, in (5.29) the following approximation can be used:

$$\ln\left[1 - 2(k-1)\frac{B(T)}{V}\right] \approx -2(k-1)\frac{B(T)}{V}. \tag{5.31}$$

As a result, (5.29) takes the form

$$\ln Z_N = -\frac{2B(T)}{V}\sum_{k=1}^{N}(k-1) = -\frac{2B(T)}{V}\left[\frac{N(N+1)}{2} - N\right] \tag{5.32}$$

or, for very large N

$$\ln Z_N = -N(N-1)\frac{B(T)}{V} \approx -N^2\frac{B(T)}{V}. \tag{5.33}$$

On substituting (5.33) in (5.9), for the free energy of a real rarefied gas we get the simple expression

$$F = F_{\text{id}} + k_0 T N^2 \frac{B(T)}{V}. \tag{5.34}$$

By substituting (5.34) into (5.2) or (5.33) into (5.10), we get the expression for pressure of a real gas, i.e. *the thermal equation of state*:

$$P = k_0 T\frac{N}{V} + k_0 T\left(\frac{N}{V}\right)^2 B(T). \tag{5.35}$$

If we present (5.35) in the form

$$P = k_0 T\left[\frac{N}{V} + B(T)\left(\frac{N}{V}\right)^2\right], \tag{5.36}$$

we notice that pressure of a gas is obtained as an expansion of the function in powers of the concentration of a gas N/V.

The first item corresponds to an ideal gas, and the second item corresponds to the case in which in the sphere of action of a molecule only one molecule is found. If in (5.19) we keep the terms corresponding to the fact that in the sphere of action of each molecule two, three and more molecules are found, intuitively the following infinite series can be written for pressure:

$$P = k_0 T\left[\frac{N}{V} + B(T)\left(\frac{N}{V}\right)^2 + C(T)\left(\frac{N}{V}\right)^3 + \cdots\right] \tag{5.37}$$

Here, the first term corresponds to an ideal gas, and the second and third terms describe the case in which in the sphere of action of each molecule one or two molecules, respectively, are found. Coefficients $B(T)$ and $C(T)$ entering into the equation of state and characterizing the interaction of molecules are called *second* and *third virial coefficients*, respectively.

Thus, if we calculate all virial coefficients, we can find the equation of state of a real gas in the form of a series in powers of the concentration N/V.

Here and in subsequent sections, we restrict ourselves only to the second virial coefficient in the equation of state (5.37), i.e. we make use of the approximate expressions (5.34) and (5.36).

In the given approximation, the entropy of a real gas, the caloric equation of state and the heat capacity can be expressed by the second virial coefficient $B(T)$:

$$S = S_{\text{id}} - k_0 N^2 \frac{B(T)}{V} - k_0 T \frac{N^2}{V} \left(\frac{\partial B}{\partial T} \right), \tag{5.38}$$

$$E = E_{\text{id}} - k_0 T^2 \frac{N^2}{V} \left(\frac{\partial B}{\partial T} \right), \tag{5.39}$$

$$C_V = C_V^{\text{id}} - 2k_0 T \frac{N^2}{V} \left(\frac{\partial B}{\partial T} \right) - k_0 T^2 \frac{N^2}{V} \left(\frac{\partial^2 B}{\partial T^2} \right), \tag{5.40}$$

where S_{id} is the entropy, $E_{\text{id}} = 3k_0NT/2$ is the energy and $C_V^{\text{id}} = 3k_0N/2$ is the heat capacity of an ideal gas. It is seen that to construct the thermodynamics of a non-ideal gas, it is necessary to know the explicit form of the function $B(T)$.

In conclusion, we once more recall the conditions under which the obtained results, i.e. expressions (5.34), (5.36), (5.38), (5.39) and (5.40), are justified. They are as follows:

1. The motion of the molecules forming a gas ought to obey classical laws.
2. The energy of interaction between molecules ought to be much less than the kinetic energy of motion (5.1).
3. An external field is absent (5.5).
4. The total potential energy of a gas can be presented as the sum of energies of pair-wise interactions of molecules (5.12).
5. A gas ought to be sufficiently rarefied so that each molecule would interact with only one other molecule, i.e. in the sphere of action only one molecule would be found (5.23).
6. The energy of interaction between two molecules depends only on the distance between them (5.24).

Note that even with the six simplifying assumptions introduced above, it is impossible to solve the problem completely, i.e. it is impossible to determine the dependence of the pressure, mean energy, entropy and heat capacity on temperature. To do this, it is necessary to know the explicit form of the function $B(T)$. Section 5.2 is devoted to this problem in the case of a non-ideal van der Waals gas.

At a specified temperature where $B(T_{\text{B}}) = 0$, the thermodynamic properties of a real gas (in particular, its isotherm) coincide with those of an ideal gas. The temperature T_{B} satisfying the equation

$$B(T_{\text{B}}) = 0, \tag{5.41}$$

is called *the Boyle temperature*.

5.2 Second Virial Coefficient and Thermodynamics of Van Der Waals Gas

In Sect. 5.1 we showed that to construct the thermodynamics of a non-ideal gas it is necessary to calculate the function $B(T)$. To do this, as is seen from (5.26) and (5.27), it is required to know the explicit form of the potential of interaction $u(r)$ between molecules:

$$B(T) = \frac{1}{2}\int_0^\infty \left(1 - \mathrm{e}^{u(r)/k_0T}\right) 4\pi r^2 \mathrm{d}r. \tag{5.42}$$

For the potential of interaction of two molecules $u(r)$, different models exist. Among them the van der Waals model is the most commonly used. This interaction can be described as follows: each molecule of a real gas is presented as a spherically symmetric neutral particle. At very close distances, when the distance between the centres of the molecules is equal to their diameter, and electron clouds touch, a strong repulsion arises as a result of the elastic deformation between molecules. At distances larger than their diameter, a force of attraction – the van der Waals force (Fig. 5.1) – arises between them.

The rise of the van der Waals force can be presented as follows: in a molecule that does not possess an electric dipole moment, as a result of fluctuations, the symmetry of distribution of electrical charges inside the molecule can be instantaneously violated and an instantaneous electric dipole moment arises in the molecule. This electric dipole moment can induce a similar dipole moment in an adjacent molecule. As a result, a dipole–dipole interaction arises between adjacent molecules.

The potential in the case where the molecules interact via the medium of van der Waals forces (Fig. 5.1) can be presented in the form

$$u(r) = \begin{cases} +\infty & 0 \le r < d \\ -u_0 \left(d/r\right)^m & d \le r < \infty, \end{cases} \tag{5.43}$$

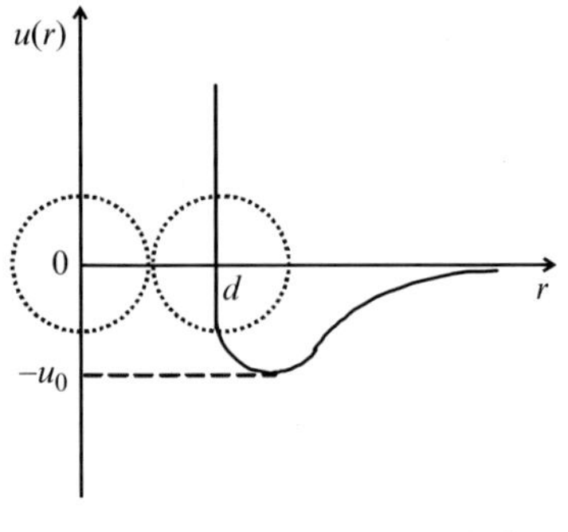

Fig. 5.1. The van der Waals interaction potential dependence on the distances between of two molecules

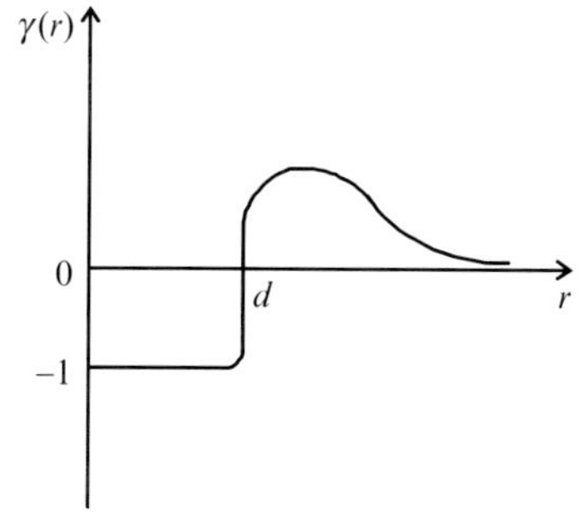

Fig. 5.2. A plot of the function $\gamma(r)$ [see (5.27)]

where d is the diameter of a molecule, r is the distance between the centres of adjacent molecules, u_0 is the maximum potential of attraction, and m is an integer greater than 3 ($m > 3$).

A plot of the function $\gamma(r)$ [see (5.27)] corresponding to the model presented in Fig. 5.1 is schematically shown in Fig. 5.2.

We use this model to calculate the second virial coefficient $B(T)$. A real gas corresponding to this model is called a *van der Waals gas.*

In conformity with Fig. 5.1 and formula (5.43), boundaries of integration in (5.42) to calculate $B(T)$ can be subdivided into two regions:

$$B(T) = \frac{1}{2}\int_0^d 4\pi r^2 \mathrm{d}r + \frac{1}{2}\int_d^\infty \left(1 - \mathrm{e}^{|u(r)|/k_0 T}\right) 4\pi r^2 \mathrm{d}r. \tag{5.44}$$

Here, the first integral corresponds to the repulsion between molecules, and the second integral corresponds to their attraction. The first integral gives the quadrupled volume of a molecule. The exponent entering into the second integral can be expanded into a series, since in the range $d \leq r < \infty$ the inequality $u \ll k_0 T$ is fulfilled. As a result, we get

$$B(T) = 4v_0 - \frac{2\pi}{k_0 T}\int_d^\infty |u(r)|\, r^2 \mathrm{d}r, \tag{5.45}$$

where $v_0 = \frac{4\pi}{3}\left(\frac{d}{2}\right)^3$ is the intrinsic volume of a spherical molecule. In order to derive the van der Waals equation of state for real gases, we substitute (5.45) into (5.35). Then we get

$$P = k_0 T\frac{N}{V} + k_0 T\left(\frac{N}{V}\right)^2 4v_0 - 2\pi\left(\frac{N}{V}\right)^2 \int_d^\infty |u(r)|\, r^2 \mathrm{d}r. \tag{5.46}$$

For convenience, we introduce the constant

$$b = 4Nv_0, \tag{5.47}$$

characterising the repulsion between molecules, and

$$a = \frac{N^2}{2}\int_d^\infty |u(r)|\, 4\pi r^2 \mathrm{d}r \approx \frac{N(N-1)}{2}\int_d^\infty |u(r)|\, 4\pi r^2 \mathrm{d}r, \tag{5.48}$$

characterising the attraction between molecules. Then, the thermal equation of the state of real gases (5.46) takes the form

$$P = k_0 T\frac{N}{V} + k_0 T\frac{N}{V^2}b - \frac{a}{V^2} \tag{5.49}$$

or

$$P = k_0 T\frac{N}{V} + k_0 T\left(\frac{N}{V}\right)^2\left(\frac{b}{N} - \frac{a}{k_0 T N^2}\right). \tag{5.50}$$

If we compare (5.50) with (5.35), the second virial coefficient can be expressed by the constants a and b:

$$B(T) = \frac{b}{N} - \frac{a}{k_0 T N^2}. \tag{5.51}$$

The equation of the state of a rarefied gas (5.49) can also be presented as

$$P = k_0 T\frac{N}{V}\left(1 + \frac{b}{V}\right) - \frac{a}{V^2}. \tag{5.52}$$

The specific volume of molecules b is far less than the volume of gas V, i.e. because $b \ll V$, $(1 + b/V) = (1 - b/V)^{-1}$. As a result, the equation of state of a non-ideal gas takes the form

$$P = \frac{k_0 T N}{V - b} - \frac{a}{V^2}. \tag{5.53}$$

Hence, we get the well-known van der Waals equation

$$\left(P + \frac{a}{V^2}\right)(V - b) = k_0 T N. \tag{5.54}$$

For any real gas, a temperature T_B exists at which $B(T_\mathrm{B}) = 0$. In our case, from (5.51) it follows that

$$T_\mathrm{B} = \frac{a}{k_0 N b}. \tag{5.55}$$

As an example, we present the values of a and b for a diatomic gas of nitrogen N_2 obtained experimentally: $a = 0.14\,\mathrm{J}^3/\mathrm{mol}^2$, $b = 3.92 \times 10^{-6}\,\mathrm{m}^3/\mathrm{mol}$, and according to (5.55), $T_\mathrm{B} \approx 600\,\mathrm{K}$.

At the temperature (5.55) called *the Boyle temperature*, as is seen from (5.50), the equation of state of a real gas (the isotherm) coincides with that of an ideal gas (the isotherm). The equation of state of a real gas (5.50) can be also presented in the form

$$P = k_0 T \frac{N}{V} + k_0 T \frac{Nb}{V^2} \left(1 - \frac{T_B}{T} \right). \tag{5.56}$$

Hence it follows that if $T = T_B$, then $P_{real} = P_{id}$; if $T < T_B$, then $P_{real} < P_{id}$; if $T > T_B$, then $P_{real} > P_{id}$.

Physically it will be clearer if we remember that these conclusions immediately follow from (5.50). Indeed, from (5.50) it is seen that when fulfilling the condition

$$\frac{a}{k_0 NT} > b, \tag{5.57}$$

the pressure of real gases ought to be less than the pressure of ideal gases at the same temperature: $P_{real} < P_{id}$. This fact can be explained as follows: if the attraction in gases (the parameter a) dominates over the repulsion (the parameter b), as is indicated by the condition (5.57), molecules that are found close to the walls of a vessel are attracted by other molecules that are found in the interior of the vessel and, thereby, their impact on the wall weakens; as a result, the pressure of a real gas decreases compared to that of an ideal gas. If the reverse condition to (5.57) is fulfilled, i.e. the repulsion dominates over the attraction, then $P_{real} > P_{id}$.

We analyse also the other thermodynamic properties of the van der Waals gas. To do this, we substitute the expression for $B(T)$ from (5.51) into (5.34), (5.38) and (5.40). As a result, we get

for the free energy

$$F = F_{id} + k_0 T \frac{N}{V} \left(b - \frac{a}{k_0 NT} \right), \tag{5.58}$$

for entropy

$$S = S_{id} - \frac{k_0 N}{V} b, \tag{5.59}$$

for the mean energy

$$E = E_{id} - \frac{a}{V}, \tag{5.60}$$

and for the heat capacity

$$C_V = C_V^{id}. \tag{5.61}$$

Note that the constants of interaction a and b enter into the above thermodynamic functions in different ways: both constants a and b enter into the expression of the free energy and the equation of state; only the repulsion constant b enters into the expression of entropy; and only the attraction constant a enters into the expression of energy. The isochoric heat capacity C_V contains none of the constants: $C_V = C_V^{id}$.

We show that the isobaric heat capacity C_P, on the contrary, depends on a and b. Indeed, using the known relationship

$$C_P = C_V - T\frac{(\partial P/\partial T)_V^2}{(\partial P/\partial V)_T}, \tag{5.62}$$

and also relationships

$$\left(\frac{\partial P}{\partial T}\right)_V = \frac{k_0 N}{V-b}, \quad \left(\frac{\partial P}{\partial V}\right)_T = -\frac{k_0 NT}{(V-b)^2} + \frac{2a}{V^3}, \tag{5.63}$$

obtained from the equation of state (5.53), for the isobaric heat capacity we get

$$C_P = C_V + R\left(1 + \frac{2a(V-b)^2}{RTV^3}\right), \tag{5.64}$$

where $R = k_0 N$ is the universal gas constant.

In the case of ideal gases, $(C_P - C_V)_{\text{id}} = R$, and in the case of real gases, from (5.64), it is seen that $(C_P - C_V)_{\text{real}} > R$. Because $C_V^{\text{real}} = C_V^{\text{id}}$, the isobaric heat capacity of real gases is larger than that of ideal gases: $C_P^{\text{real}} > C_P^{\text{id}}$. From the expression of heat capacity (5.64), it is also seen that the repulsion constant b, though slightly, weakens the inequality $C_P^{\text{real}} > C_P^{\text{id}}$, whereas the attraction constant a sufficiently strengthens the given inequality.

If the repulsion ($b \ll V$) is disregarded, (5.64) takes the form

$$C_P = C_V + R\left(1 + \frac{2a}{RTV}\right). \tag{5.65}$$

The fact that in real gases $C_V^{\text{real}} = C_V^{\text{id}}$ and $C_P^{\text{real}} > C_P^{\text{id}}$ can be explained as follows:

1. Independence of the isochoric heat capacity of interaction between molecules $\left(C_V^{\text{real}} = C_V^{\text{id}}\right)$ is explained by the fact that the heat capacity is measured at constant volume ($V = \text{const}$), and therefore the mean distance between molecules does not change; thereby, the mean value of energy of interaction remains constant and the heat capacity associated with the change in this interaction equals zero.
2. The fact that the isobaric heat capacity of real gases is larger than that of ideal gases $\left(C_P^{\text{real}} > C_P^{\text{id}}\right)$ is explained by the fact that at $P = \text{const}$ the volume of the gas ought to increase (only in this case P remains constant). Therefore, the mean distance between molecules grows. In order for this distance to increase, it is necessary to expend supplementary energy – heat. At the expense of this energy, the heat capacity of a real gas becomes larger.

5.3 Neutral Gas Consisting of Charged Particles: Plasma

In this section, we consider a special kind of real gas – plasma. In contrast to the van der Waals gas particles, those forming plasma possess an electrical charge. Assume that a gas of volume V and the total number of particles $N = \sum\limits_i N_i$ consists of positively and negatively charged ions. Here, N_i is the number of ions of the ith type, $Z_i e$ is the electrical charge of ions of the same type, e is the absolute magnitude of the charge of an electron and Z_i takes on values $Z_i = \pm 1, \pm 2, \ldots$

In the particular case of high temperatures and for a completely ionized atomic gas, Z_i can take on values $Z_1 = Z_{\text{ion}} = +1, Z_2 = Z_{\text{el}} = -1$, i.e. in this case plasma consists of positively charged ions and electrons.

As a whole, a gas ought to be neutral. The condition of neutrality can be written down in the form

$$\sum_i Z_i e N_i = 0 \text{ or } \sum_i Z_i e n_{0i} = 0, \tag{5.66}$$

where $n_{0i} = N_i/V$ is the mean concentration of the ith ion in the case where the interaction between particles is disregarded and it is supposed that all particles of the ith type are uniformly distributed over the entire volume.

Also assume that a completely ionized gas – plasma – is sufficiently rarefied, i.e. the mean distance r between ions is such that the energy of the Coulomb interaction is much less than the energy of the thermal motion $k_0 T$:

$$\frac{Z_i^2 e^2}{r} \ll k_0 T. \tag{5.67}$$

If the concentration of all ions in the plasma is denoted by $n_0 = \sum\limits_i n_{0i}$, then $r \approx n_0^{-1/3}$. Hence, the condition of rarefaction (5.67) can be rewritten in the form

$$n_0 \ll \left(\frac{k_0 T}{Z_i^2 e^2} \right)^3. \tag{5.68}$$

We derive the equation of state of rarefied plasma and calculate some thermodynamic quantities satisfying conditions (5.67) and (5.68). However, it is necessary to note that the method applied to the van der Waals gas in Sect. 5.2 cannot be applied to plasma. This is due to the fact that in plasma the long-range-acting Coulomb interaction exists between ions:

$$u_{ik} = \frac{Z_i Z_k e^2}{r} \sim \frac{1}{r}. \tag{5.69}$$

From this fact two conclusions follow: first, because the radius of action of each ion is large, and plasma is rarefied, not one but many ions are found in the sphere of action of each ion; second, in the case of such an interaction, the integral entering into the expression of the second virial coefficient (5.45) diverges in the upper limit:

$$\int^{\infty} |u_{ik}|\, r^2 \mathrm{d}r = Z_i Z_k e^2 \int^{\infty} r\, \mathrm{d}r \to \infty. \tag{5.70}$$

Therefore, in order to find the free energy and the equation of state of plasma, we use the Helmholtz relationship

$$E = -T^2 \frac{\partial}{\partial T}\left(\frac{F}{T}\right). \tag{5.71}$$

Hence, by knowing the total energy, the free energy can be calculated by the formula

$$F = -T \int \frac{E}{T^2} \mathrm{d}T. \tag{5.72}$$

The total energy of plasma can be presented as the sum

$$E = E_{\mathrm{id}} + E_{\mathrm{coul}}. \tag{5.73}$$

Here, E_{id} is the energy without considering the interaction between ions (an ideal gas), and E_{coul} is the energy of the Coulomb interaction

$$E_{\mathrm{coul}} = \frac{1}{2} \sum_i Z_i e N_i \varphi_i(r), \tag{5.74}$$

where $\varphi_i(r)$ is the potential of the Coulomb field around the ith ion at the point r created by all the ions except the given one.

If we substitute the expression of energy (5.73) into (5.72), we get

$$F = F_{\mathrm{id}} + F_{\mathrm{coul}}, \tag{5.75}$$

where

$$F_{\mathrm{coul}} = -T \int \frac{E_{\mathrm{coul}}}{T^2} \mathrm{d}T \tag{5.76}$$

is the free energy corresponding to the Coulomb interaction.

Thus, the problem of finding the free energy is reduced to finding E_{coul} or the potential $\varphi_i(r)$. In order to determine the explicit form of $\varphi_i(r)$, we use the screening method proposed by Debye and Hueckel in 1923 in the theoretical studies of the properties of electrolytes.

The essence of this method is the fact that in a system of charged particles (in our case, plasma) the potential created at the point of an arbitrary ion being found can be presented in the form

$$\varphi_i(r) = \lim_{r \to 0} \left[\varphi(r) - \frac{Z_i e}{r}\right], \tag{5.77}$$

where $\varphi(r)$ is the potential created by all charges at the point r, and $Z_i e/r$ is the potential created by the given ion at the point r. The potential $\varphi(r)$ can be found from the known Poisson equation

$$\nabla^2\varphi(r) = -4\pi\rho(r), \tag{5.78}$$

where

$$\rho(r) = \sum_i Z_i e n_i(r) \tag{5.79}$$

is the density of charges at the point r, and $n_i(r)$ is the concentration of all ions of the ith type around the point r. From the two latter expressions, we get for the Poisson equation

$$\nabla^2\varphi(r) = -4\pi\sum_i Z_i e n_i(r). \tag{5.80}$$

The concentration $n_i(r)$ on the right-hand side of this equation is distinct from n_{0i}, because $n_i(r)$ is the concentration with regard to the interaction between ions. As a result of the interaction between ions, they are non-uniformly distributed over the entire volume, and each ion is surrounded by an ion with the opposite charge, and the distribution resembles a mosaic. The cause of distinction of $n_i(r)$ from n_{0i} lies in the fact that each ion possesses the potential energy $Z_i e\varphi(r)$ and the distribution of charges in the potential field $\varphi(r)$ can be described with the aid of the Boltzmann statistics:

$$n_i(r) = n_{0i}\exp\left[-\frac{Z_i e\varphi(r)}{k_0 T}\right]. \tag{5.81}$$

From Poisson equation (5.80) and (5.81), it follows that to calculate the potential $\varphi(r)$ it is necessary to know the concentration $n_i(r)$, and, conversely, in order to calculate $n_i(r)$ it is necessary to know $\varphi(r)$. Thus, a certain self-contradictory problem arises. Therefore, it is necessary to simultaneously solve equations (5.80) and (5.81).

If we take into account the inequality $Z_i e\varphi(r) \ll k_0 T$, equivalent to the condition of the rarefied nature of the plasma (5.67), in (5.81), we get

$$n_i(r) = n_{0i}\left[1 - \frac{Z_i e\varphi(r)}{k_0 T}\right]. \tag{5.82}$$

After substituting (5.82) into (5.80), the Poisson equation takes the form

$$\nabla^2\varphi\ (r) = -4\pi\sum_i Z_i e n_{0i} + \frac{4\pi e^2}{k_0 T}\left(\sum_i Z_i^2 n_{0i}\right)\varphi(r). \tag{5.83}$$

Taking into account the condition of neutrality of the plasma (5.66) and introducing the notation

$$r_0^{-2} = \frac{4\pi e^2}{k_0 T}\left(\sum_i Z_i^2 n_{0i}\right), \tag{5.84}$$

the Poisson equation (5.83) can be simplified to

$$\nabla^2\varphi(r) - r_0^{-2}\varphi(r) = 0. \tag{5.85}$$

Because in plasma the electrostatic potential close to each ion possesses spherical symmetry, it is more convenient to write equation (5.85) down in the spherical coordinate system. In this case, the potential depends only on the quantity of the radius vector and it is sufficient to consider the radial part of the operator ∇^2:

$$\frac{1}{r^2}\frac{\mathrm{d}}{\mathrm{d}r}\left(r^2\frac{\mathrm{d}\varphi}{\mathrm{d}r}\right) = r_0^{-2}\varphi(r). \tag{5.86}$$

It is easy to show that the function

$$\phi(r) = \frac{A}{r}\exp\left(-\frac{r}{r_0}\right), \tag{5.87}$$

satisfies this equation.

The constant A entering here can be found from the condition

$$\lim_{r\to 0}\varphi(r) = \frac{Z_i e}{r}, \tag{5.88}$$

to fulfil which it is necessary to have $A = Z_i e$. Then the potential created by any point charge $Z_i e$ at the distance r is

$$\varphi(r) = \frac{Z_i e}{r}\exp\left(-\frac{r}{r_0}\right). \tag{5.89}$$

The potential $\varphi(r)$, pre-assigned by formula (5.89), is called *the screening Coulomb potential* of a point charge, and r_0 is *the Debye screening radius.* If in the volume only one isolated ion is found, the potential created by it at a distance r is

$$\varphi_0(r) = \frac{Z_i e}{r}. \tag{5.90}$$

If a point charge is surrounded (screened) by ions with opposite charges, its field weakens, and at the point $r = r_0$ the potential decreases $e = 2.74$ times (Fig. 5.3).

Thus, if we substitute expression (5.89) into (5.77), for the potential $\varphi_i(r)$ we get

$$\varphi_i(r) = \lim_{r\to 0}\left[\frac{Z_i e}{r}\exp\left(-\frac{r}{r_0}\right) - \frac{Z_i e}{r}\right]. \tag{5.91}$$

If in this limiting case we expand the exponent in powers of $r/r_0 \ll 1$ and restrict ourselves to the first two terms, the potential sought has the appearance

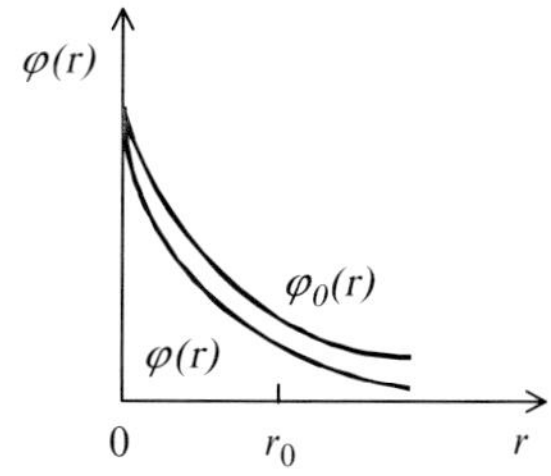

Fig. 5.3. Screening [see (5.89)] and non-screening [see (5.90)] Coulomb potentials of a point charge

$$\varphi_i(r) = \frac{Z_i e}{r_0}. \tag{5.92}$$

Substituting this expression as well as (5.84) in (5.74), we get for the energy of plasma at the expense of the Coulomb interaction

$$E_{\text{coul}} = -e^3 \left(\frac{\pi}{k_0 T}\right)^{1/2} \frac{1}{V^{1/2}} \left(\sum_i Z_i^2 N_i\right)^{3/2}. \tag{5.93}$$

If we introduce the notation

$$q_0 = \left(\sum_i Z_i^2 e^2 N_i\right)^{1/2} \tag{5.94}$$

with the dimensionality of an electrical charge, the energy of the Coulomb interaction of plasma takes the form

$$E_{\text{coul}} = -\left(\frac{\pi}{k_0 T V}\right)^{1/2} q_0^3. \tag{5.95}$$

For the total energy of plasma, from (5.73) and (5.95) we get

$$E = E_{\text{id}} - \left(\frac{\pi}{k_0 T V}\right)^{1/2} q_0^3, \tag{5.96}$$

where $E_{\text{id}} = 3k_0 TN/2$, and $N = \sum_i N_i$ is the total number of all the ions in plasma.

If we take into account the energy of the Coulomb interaction (5.95) in (5.76) and integrate with respect to T, and then insert the obtained expression in (5.75), we have for the free energy of the plasma

$$F = F_{\text{id}} - \frac{2}{3}\left(\frac{\pi}{k_0 T V}\right)^{1/2} q_0^3. \tag{5.97}$$

This expression enables the calculation of all thermodynamic functions and coefficients.

For the *equation of state* $P = -(\partial F/\partial V)_T$, using (5.97), we have

$$P = P_{\text{id}} - \frac{q_0^3}{3V^{3/2}} \left(\frac{\pi}{k_0 T} \right)^{1/2}, \tag{5.98}$$

where $P_{\text{id}} = k_0 TN/V$ is the pressure without considering the Coulomb interaction.

For *entropy* $S = -(\partial F/\partial T)_V$, from (5.97) we get

$$S = S_{\text{id}} - \frac{q_0^3}{3T^{3/2}} \left(\frac{\pi}{k_0 V} \right)^{1/2}. \tag{5.99}$$

The *isochoric heat capacity* $C_V = (\partial E/\partial T)_V$ with regard to the expression (5.96) takes the form

$$C_V = C_V^{\text{id}} + \frac{q_0^3}{2T^{3/2}} \left(\frac{\pi}{k_0 V} \right)^{1/2}, \tag{5.100}$$

where $C_V^{\text{id}} = 3k_0 N/2$. Formula (5.100) can be also obtained from the relationship $C_V = T(\partial S/\partial T)_V$ using (5.99).

Finally, note two conclusions following from the obtained expressions: first, from (5.96) to (5.100), it follows that in plasma the contribution associated with the consideration of the Coulomb interaction decreases with increase in temperature and volume; second, consideration of the Coulomb interaction decreases energy, free energy, pressure and entropy, but increases heat capacity. This result for heat capacity is explained by the fact that in plasma each ion attracts an ion with an opposite charge to itself. As a result, the Coulomb field of each ion is screened and its potential weakens. Consequently, the distribution of ions in plasma resembles a mosaic – a certain ordering. Therefore, entropy decreases, and heat capacity increases, because in order to destroy the order, it is necessary to expend additional energy.

6

Solids

Summary. This chapter is devoted to the classical and quantum theories of thermodynamic properties of non-conducting crystalline solids. At the beginning of the chapter the Hamilton function of a vibrating simple crystalline lattice with normal coordinates is described. Further, in the classical and quantum cases, applying the Gibbs method, the free energy, total energy, equation of state, heat capacity and coefficients of thermal expansion of a solid are calculated. In the classical case, the Hamilton function is accepted as a basis, and in the quantum case the quantized energy spectrum of vibrations of a lattice, corresponding to the Hamilton operator, is accepted.

6.1 Vibration and Waves in a Simple Crystalline Lattice

In this section we will consider vibrations and waves in a simple cubic crystalline lattice, in which only one atom or ion accounts for each elementary cell. However, we should first determine the potential of interaction between particles (atoms or ions) that are found at the crystalline lattice points. In solids there exist different kinds of interactions: ionic, covalent, metallic and van der Waals. The nature of these interactions is usually studied in quantum mechanics. Here, note the following important circumstance. All known mechanisms of interaction in solids have one common peculiarity: *At large distances the interaction between atoms or ions bears a character of attraction and at small distances, that of repulsion.*

In the equilibrium state, atoms or ions positioned at crystalline lattice points are found at a distance of the lattice constant a from each other and all interactions between them are compensated. The non-compensated interaction between particles arises when they deviate from the equilibrium position and the distance between them is distinct from the lattice constant. The potential of interaction between two adjacent atoms of the lattice is schematically shown in Fig. 6.1 (one of the atoms is positioned at the origin of the

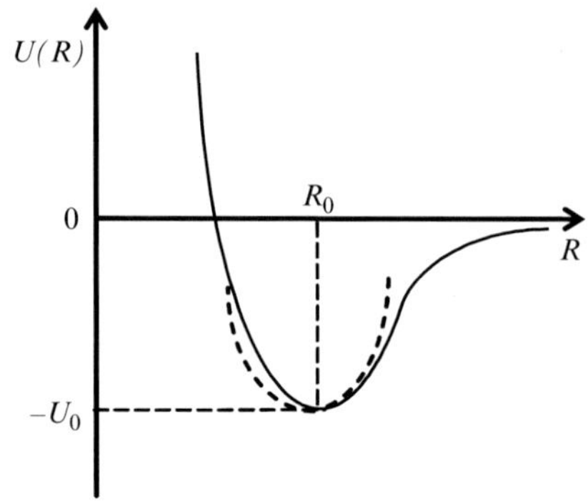

Fig. 6.1. The potential of interaction between two adjacent atoms

coordinates). As it is seen from this figure, $\lim_{R\to\infty} U(R) = 0$ and $U(R_0) = -U_0$ is the minimum of the potential energy, where R_0 is the distance between two particles in the equilibrium state. As $R > R_0$ there arises the force $\boldsymbol{F} = -\text{grad}\, U(R) = -\left(\partial U/\partial R\right) \boldsymbol{n}$, bearing the character of attraction, and as $R < R_0$ the force bears the character of repulsion, here $\boldsymbol{n} = \boldsymbol{R}/R$ is the orth-vector.

In the equilibrium state ($R = R_0$) the force of interaction between atoms or ions does not arise, inasmuch as the potential energy of interaction ($-U_0$) is the constant quantity. As a result of the thermal motion, atoms or ions deviate from their equilibrium position and the distance between them is distinct from R_0. Here the potential energy depends on the distance between atoms, i.e. $U(R) \neq -U_0$. The potential energy $U(R)$ takes on the minimum value at the point R_0; therefore at small deviations $x = (R - R_0) \ll R_0$ can be expanded into a series in powers of $(R - R_0)$:

$$U(R) = U(R_0) + \left(\frac{\partial U}{\partial R}\right)_{R_0} (R - R_0) + \frac{1}{2}\left(\frac{\partial^2 U}{\partial R^2}\right)_{R_0} (R - R_0)^2 + \frac{1}{6}\left(\frac{\partial^3 U}{\partial R^3}\right)_{R_0} (R - R_0)^3 + \cdots \qquad (6.1)$$

Inasmuch as the potential energy at the point $R = 9R_0$ is at a minimum, the ratio $(\partial U/\partial R)_{R_0} = 0$. If we introduce the notations $\left(\partial^2 U/\partial R^2\right)_{R_0} = \beta > 0$, $\left(\partial^3 U/\partial R^3\right)_{R_0} = -2\gamma < 0$ and restrict ourselves to the cubic approximation for constants, entering into (6.1), the potential energy takes the form

$$U(x) = -U_0 + \frac{1}{2}\beta x^2 - \frac{1}{3}\gamma x^3. \qquad (6.2)$$

In order that as $R < R_0$ $(x < 0)$ forces of repulsion dominate over forces of attraction it ought to be $\left(\partial^3 U/\partial R^3\right)_{R_0} < 0$, i.e. $\gamma > 0$.

From (6.2) for the force of interaction between atoms we get

$$F = -\frac{\partial U}{\partial R} = -\frac{\partial U}{\partial x} = -\beta x + \gamma x^2, \qquad (6.3)$$

where β is the coefficient of elasticity, and γ is the parameter characterizing anharmonicity of interaction. Both coefficients are determined by the nature of interaction between atoms and are related between themselves by the relationship $\beta \sim R_0\gamma$. Indeed, from (6.3) it is seen that only in this case the ratio of the anharmonic item to the harmonic one satisfies the condition $\gamma x/\beta \sim x/R_0 \ll 1$.

At crystalline lattice points, at finite temperatures atoms (ions) perform small vibrations with the specified frequency (translational and rotational motions are impossible), which propagate all over the crystal in the form of an acoustic wave. As temperature rises, vibrations with higher frequency are excited, and simultaneously the intensity (the amplitude) of already excited vibrations grows.

In the given crystal, at specified temperature $T = \theta$, all possible frequencies are excited. If we denote the highest frequency by $\omega_{\max}$, this temperature is determined as $\theta = \hbar\omega_{\max}/k_0$, where $\hbar = h/2\pi$, h is the Planck constant, k_0 is the Boltzmann constant, θ is *the Debye temperature* (for more comprehensive details about the Debye temperature, see Sect. 6.3) In a crystal, at temperatures higher than Debye's $T > \theta$, new frequencies are not excited, and only the amplitude of vibrations that have arisen grows.

We are interested in the nature of vibrations of atoms (ions) at crystalline lattice points, i.e. the vibrational motion bears classical or quantum character. In other words, to describe the vibrational motion it is necessary to use classical mechanics of Newton or quantum mechanics of Schrödinger.

In order to answer the question posed, determine the criterion of classical or quantum nature of the motion. It is said that the motion is classical when its action s is much larger than that of the Planck constant $\hbar$:

$$s \gg \hbar, \tag{6.4}$$

where $s \approx Mvx$ is the action of the vibrational motion, M is the mass of the vibrating atom, v is its linear velocity, x is the displacement from the equilibrium position. For simplicity, take $x = A\cos\omega t$, then $v \sim \dot{x} \sim \omega x$, where $x \sim v/\omega$. Then

$$s \sim \frac{Mv^2}{\omega}. \tag{6.5}$$

If we take into account that $Mv^2 \approx k_0 T$, we get

$$s \sim \frac{k_0 T}{\omega}. \tag{6.6}$$

If we rewrite condition (6.4) for the highest frequency $\omega_{\max}$, the condition of classicity of the vibrational motion takes the form

$$\frac{k_0 T}{\omega_{\max}} \gg \hbar \text{ or } k_0 T \gg \hbar\omega_{\max}. \tag{6.7}$$

Consequently, the vibrational motion of a lattice can be regarded as classical only at temperatures satisfying the condition $T \gg \hbar\omega_{\max}/k_0$ or $T \gg \theta$, i.e.

in the region of temperatures higher than the Debye's. At low temperatures $T \leq \theta$, the condition of classicity (6.7) is not fulfilled and the vibrational motion bears the quantum character.

It is known that to construct the statistical theory of thermodynamic properties of a system in the classical case it is necessary to know the Hamilton function, and in the quantum case to do the energy spectrum. We at first consider vibrations and waves in a crystalline lattice at high temperatures ($T \gg \theta$) and find its Hamilton function. Thereupon, using the correspondence principle, we pass from the Hamilton function to the Hamilton operator, i.e. pass to quantum mechanics, determine the energy spectrum and, in this way, embrace the whole temperature range.

At first consider a one-dimensional simple lattice, and then generalize the results obtained to a simple three-dimensional lattice.

6.1.1 One-Dimensional Simple Lattice

Consider a one-dimensional simple crystalline lattice with the lattice constant a, at points of which neutral atoms of the mass M are found. As an initial point of reference accept any point and enumerate other points (Fig. 6.2).

Regard the displacement of atoms to the right as positive, those to the left as negative. Therefore, $u_n(t) > 0$; $u_{n-1}(t) > 0$; $u_{n+1}(t) < 0$.

Inasmuch as each atom is connected with the adjacent one, the displacement (vibrations) over the crystal propagates in the form of an elastic wave. To describe the vibrational motion of an arbitrary point n, use the classical equation of motion:

$$M \frac{\mathrm{d}^2 u_n(t)}{\mathrm{d}t^2} = F_n. \tag{6.8}$$

Here F_n is the force acting on the point n from the sides of other points. In other cases, to find the explicit form of the force F_n accept two suppositions:

1. The potential of interaction between atoms has the parabolic form $U(x) = -U_0 + \beta x^2/2$ (the dashed curve in Fig. 6.1), i.e., the force of interaction between atoms bears the elastic character $F = -\beta x$.
2. Each atom interacts only with the adjacent one, i.e., the atom with number n interacts with atoms with numbers $(n-1)$ and $(n+1)$. Displacements of atoms at points with numbers $(n \pm 2)$, $(n \pm 3), \ldots$ do not influence the displacement of the considered atom n.

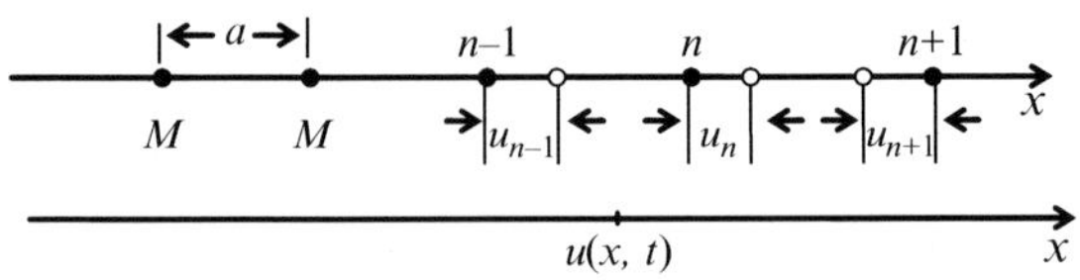

Fig. 6.2. A one-dimensional simple crystalline lattice

On the basis of these suppositions, the force can be written as:

$$F_n = F_{n,n-1} + F_{n,n+1} = -\beta(u_n - u_{n-1}) - \beta(u_n - u_{n+1})$$

or

$$F_n = -\beta(2u_n - u_{n-1} - u_{n+1}). \tag{6.9}$$

Thus, for the displacement of the n-atom we get the following classical equation of motion

$$M\frac{\mathrm{d}^2 u_n(t)}{\mathrm{d}t^2} = -\beta(2u_n - u_{n-1} - u_{n+1}). \tag{6.10}$$

Hence it is seen that to determine the function $u_n(t)$ it is necessary to know the displacement of adjacent atoms $u_{n\pm1}$ and to write the equation of motion. And to do this it is necessary to know the displacement $u_{n\pm2}$. Thus, (6.10) converts into a system of infinitive number of equations, which is practically not possible to solve.

To circumvent this difficulty replace a one-dimensional crystalline lattice (a chain) with a continuous string (Fig. 6.2). Such an approximation is just for long waves. The wave equation for the displacement $u(x,t)$ at the point x at the instance of time t has the known form:

$$\frac{\partial^2 u(x,t)}{\partial t^2} = v_0^2\frac{\partial^2 u(x,t)}{\partial x^2}. \tag{6.11}$$

Here v_0 is the speed of propagation of the elastic (acoustic) wave. It is known that the solution of (6.11) has the appearance

$$u(x,t) = A\,\mathrm{e}^{\mathrm{i}(qx-\omega t)}, \tag{6.12}$$

where A is the amplitude of the elastic wave, q is the wave number, and ω is the frequency. If we substitute (6.12) into (6.11), we get a simple relationship between the wave number and frequency

$$\omega(q) = v_0 q. \tag{6.13}$$

Note that ω and q change in the limits:

$$0 \le q < \infty; \quad 0 \le \omega < \infty.$$

Return to (6.10). Its solution can be presented in the form (6.12), with the contradistinction that in the case of the chain x takes on only discrete values $x = na$. Thus, the solution of (6.10) can be written down as

$$u_n(t) = A\,\mathrm{e}^{\mathrm{i}(qan-\omega t)}. \tag{6.14}$$

If we substitute (6.14) into (6.10), for the frequency we get

$$-M\omega^2 = -\beta(2 - \mathrm{e}^{-iaq} - \mathrm{e}^{iaq}). \tag{6.15}$$

Hence we have the dispersion relationship

$$\omega^2 = 2\frac{\beta}{M}(1 - \cos aq) = 4\frac{\beta}{M}\sin^2\frac{aq}{2} \tag{6.16}$$

or

$$\omega(q) = \omega_0 \left|\sin\frac{aq}{2}\right|, \tag{6.17}$$

where

$$\omega_0 = 2\sqrt{\frac{\beta}{M}} \tag{6.18}$$

is the maximum frequency of the vibration for a one-dimensional simple lattice. For small values of the wave number ($aq \ll 1$) or for long waves ($\lambda \gg 2\pi a$), (6.17) can be expanded into a series and in the first approximation we get

$$\omega(q) \approx \omega_0\frac{aq}{2} = \sqrt{\frac{\beta}{M}}aq. \tag{6.19}$$

It is known that in the elastic string the speed of sound $v_0 = \sqrt{E/\rho}$, where E is the Jung modulus, and ρ is the linear density of the string. For a one-dimensional simple lattice $\rho = M/a$, and the Jung modulus

$$E = \frac{\text{force}}{\text{relative displacement}} = \frac{|f_{n,n-1}|}{|u_n - u_{n-1}|}a = \beta a. \tag{6.20}$$

Consequently, the speed of sound:

$$v_0 = \sqrt{\frac{\beta}{M}}a. \tag{6.21}$$

Having combined (6.19) and (6.21), for the elastic string we get $\omega(q) = v_0 q$. Thus, in the approximation of long waves we substantiated the possibility of replacing a one-dimensional crystalline lattice with the elastic string.

Plots corresponding to the elastic string (6.13) and the expression of the dispersion (6.17) for a one-dimensional lattice are adduced in Fig. 6.3. From the figure it is seen that the frequency of the string changes in the limits $0 \le \omega \le \infty$, whereas the frequency of a one-dimensional lattice changes in the restricted range $0 \le \omega \le \omega_0$. From (6.17) and Fig. 6.3 it is also seen that the frequency is a periodic function of the wave number:

$$\omega(q) = \omega\left(q + \frac{2\pi}{a}g\right), \tag{6.22}$$

where $g = 0, \pm 1, \pm 2, \ldots$ are integer numbers.

If in the wave function (6.14), we replace q with $q' = q + b_g$ (here $b_g = (2\pi/a)\,g$ is the one-dimensional vector of the reciprocal lattice), we get

$$u'_n(t) = A\,\mathrm{e}^{\mathrm{i}(q'an - \omega t)} = A\,\mathrm{e}^{\mathrm{i}(qan - \omega t)} \cdot \mathrm{e}^{\mathrm{i}2\pi g n} = u_n(t), \tag{6.23}$$

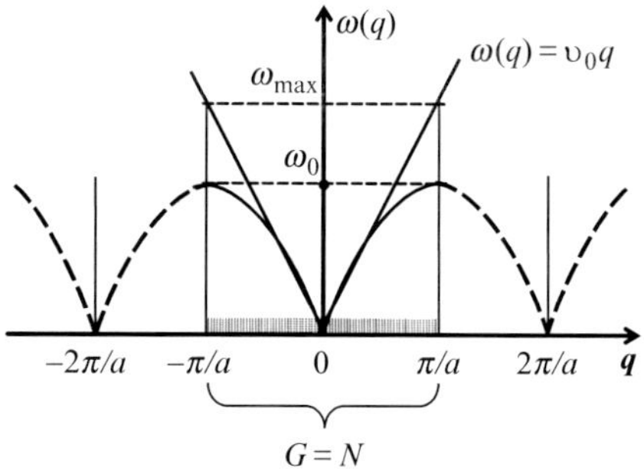

Fig. 6.3. The frequency dependence on the wave number for a one-dimensional lattice

where it is taken into account that $\exp(\mathrm{i}2\pi gn) = \exp(2\pi\mathrm{i}\cdot\text{integer number}) = 1$.

Hence it follows that wave numbers q and $q + (2\pi/a)\,g$ are equivalent, since displacements corresponding to them are identical. In other words, for q it is sufficient to consider its value in the range

$$-\frac{\pi}{a} \leq q \leq +\frac{\pi}{a}. \tag{6.24}$$

This range, in which the wave number takes on independent values, is called *the first Brillouin zone* of a one-dimensional lattice.

Consider the number of possible values of the wave number q, changing in range (6.24). To do this make use of the condition of the Born–Carman cyclicity

$$u_{n+G}(t) = u_n(t), \tag{6.25}$$

where G is a large number, equal to the number of points in the fundamental region of the lattice. In order that the wave (6.14) satisfy the boundary conditions (6.25), $\exp(\pm\mathrm{i}qaG) = 1$, i.e. $qaG = 2\pi g$ ought to take place. Hence we get the following possible values of the wave number:

$$q = \frac{2\pi}{aG}g; \quad g = 0, \pm 1, \pm 2, \ldots \tag{6.26}$$

If these values of q are substituted into (6.24), we get the range of the change in the integer number G

$$-\frac{G}{2} \leq g \leq +\frac{G}{2}. \tag{6.27}$$

It is seen that g takes on values $g = 0, \pm 1, \pm 2, \ldots, \pm G/2$, i.e. in all G values. Inasmuch as one value of q corresponds to each value of g [see (6.26)], and one value of the frequency $\omega(q)$ corresponds to each value of q [see (6.17)], the wave number q in the range $-\pi/a \leq q \leq \pi/a$, and also the frequency in the range $0 \leq \omega \leq \omega_0$ take on the G number of values.

Conclusion: *Inasmuch as G is the number of degrees of freedom (the number of atoms) of a one-dimensional crystalline lattice, it can be asserted that the number of possible values of the frequency in such a crystal is finite and equals the number of degrees of freedom of a one-dimensional chain.*

6.1.2 Three-Dimensional Simple Crystalline Lattice

Generalize the results obtained for a one-dimensional lattice to the case of a three-dimensional lattice. Consider a simple cubic lattice. Let dimensions of the crystal along coordinate axes be $L_x = aG_1$, $L_y = aG_2$, $L_z = aG_3$, where a is the lattice constant, and G_i is a large integer number. If we write the equation of motion, analogous to (6.10), for components of the displacement u_x, u_y, u_z we get the characteristic equation, analogous to (6.15), which in reference to ω^2 is cubic. A solution of this equation gives three roots $\omega_1(\boldsymbol{q})$, $\omega_2(\boldsymbol{q})$ and $\omega_3(\boldsymbol{q})$. It can be shown that each of these frequencies in the $\boldsymbol{q}$-space is a periodic function,

$$\omega_j(\boldsymbol{q}) = \omega_j\left(\boldsymbol{q} + \boldsymbol{q}_0\left(\frac{2\pi}{a}\right)g\right), \tag{6.28}$$

analogous to (6.22), where $\boldsymbol{q}_0$ is the unit vector in the $\boldsymbol{q}$-space, and g is an integer number. Therefore it is sufficient to consider the frequency in the $\boldsymbol{q}$-space in the range

$$-\pi/a \leq \boldsymbol{q} \leq +\pi/a. \tag{6.29}$$

Thus, in contradistinction to a one-dimensional case, in a three-dimensional lattice to each value of the wave number $\boldsymbol{q}$, three values of the frequency $\omega_j(\boldsymbol{q})$ correspond, where $j = 1, 2, 3$, i.e. there exist three branches of frequencies. The dependence of the frequency on $\boldsymbol{q}$ in the range $-\pi/a \leq \boldsymbol{q} \leq +\pi/a$ for the three branches is schematically presented in Fig. 6.4.

Conditions (6.29) in the $\boldsymbol{q}$-space separate a region restricted by six mutually perpendicular planes $q_x = \pm\pi/a$, $q_y = \pm\pi/a$, $q_z = \pm\pi/a$. This region represents a cube of volume $(2\pi)^3/a^3 = (2\pi)^3/\Omega_0$, where $\Omega_0 = a^3$ is the volume of an elementary cell of the crystalline lattice and is called *the first Brillouin zone.*

Now, even if we do not know the explicit form of the function $\omega_j(\boldsymbol{q})$, we can determine the number of possible values of $\boldsymbol{q}$ and frequencies in the first Brillouin zone, i.e. in the range $-\pi/a \leq \boldsymbol{q} \leq +\pi/a$. For possible values of the wave vector $\boldsymbol{q}$ in conformity with (6.26) it can be written as:

$$q_x = \frac{2\pi}{aG_1}g_1, \quad q_y = \frac{2\pi}{aG_2}g_2, \quad q_z = \frac{2\pi}{aG_3}g_3, \tag{6.30}$$

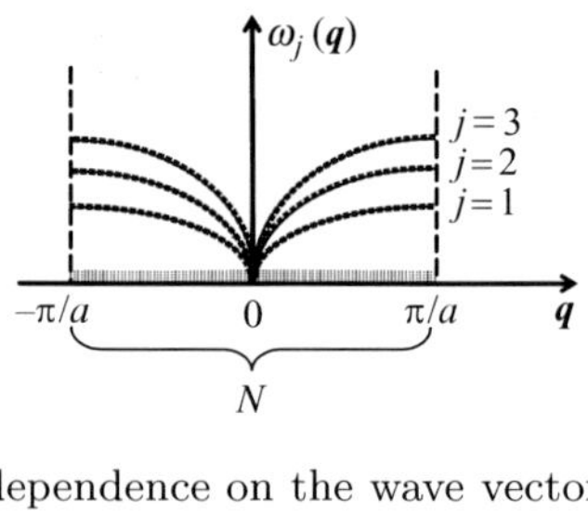

Fig. 6.4. The frequency dependence on the wave vector in the first Brillouin zone for a three-dimensional simple crystalline lattice

where $g_i = 0, \pm 1, \pm 2, \ldots$ are integer numbers. In order to determine in what range g_i changes, we substitute (6.30) into (6.29). Then we get

$$-\frac{G_i}{2} \leq g_i \leq +\frac{G_i}{2}. \tag{6.31}$$

From (6.30) and (6.31) it follows that in the first Brillouin zone components q_x, q_y, q_z take on values G_1, G_2, G_3, respectively.

Thus, we come to the conclusion that in the first Brillouin zone the wave vector $\boldsymbol{q}$ *takes on the finite* $N = G_1 \cdot G_2 \cdot G_3$ *number of values, equal to the number of points (atoms) in the fundamental region of a crystal.*

Inasmuch as in the first Brillouin zone to each value of $\boldsymbol{q}$, *three values of the frequency* $\omega_j(\boldsymbol{q})$ *correspond* (Fig. 6.4), *the total number of frequencies in the first Brillouin zone is* $3N$, *i.e. the number of possible frequencies equals the number of degrees of freedom in the fundamental region of the three-dimensional simple lattice.*

In conclusion, determine the volume that corresponds to each value of the wave vector $\boldsymbol{q}$ in the first Brillouin zone. In the case of the crystalline lattice the wave vector takes on discrete values, the number of which is finite and therefore to each wave vector in the $\boldsymbol{q}$-space, a small but finite volume $\Delta\boldsymbol{q}$ corresponds. It is evident that this small volume equals the ratio of volume of the first Brillouin zone $(2\pi)^3/a^3$ to the possible number of values of the wave vector N

$$\Delta\boldsymbol{q} = \frac{(2\pi)^3}{a^3 \cdot N} = \frac{(2\pi)^3}{V}, \tag{6.32}$$

where $V = Na^3 = N\Omega_0$ is the volume of the fundamental region of the cubic crystal.

Note that (6.32) can be obtained also from (6.30). Indeed, if we take into account that to two adjacent values q_x, q_y, q_z, the range $\Delta g_i = 1$ corresponds, we have

$$\Delta\boldsymbol{q} = \Delta q_x \cdot \Delta q_y \cdot \Delta q_z = \frac{(2\pi)^3}{a^3 \cdot G_1 G_2 G_3} = \frac{(2\pi)^3}{V}. \tag{6.33}$$

Knowing that volume (6.32), corresponds to one value of the wave-vector $\boldsymbol{q}$, the number of possible values of the wave-vector in the element of volume $d\boldsymbol{q}$ can be determined:

$$\frac{\mathrm{d}\boldsymbol{q}}{(2\pi)^3/V} = \frac{V}{(2\pi)^3}\mathrm{d}\boldsymbol{q}. \tag{6.34}$$

Inasmuch as the volume of the fundamental region $V = Na^3$ is proportional to the large number of points N in this region, the volume $(2\pi)^3/V$, corresponding to one value of the wave-vector $\boldsymbol{q}$, is sufficiently small. Therefore $\boldsymbol{q}$ is a quasi-continuous quantity and in computations it can be passed from the sum with respect to $\boldsymbol{q}$ to the integral over $\mathrm{d}\boldsymbol{q}$. For instance, in the finite volume of the $\boldsymbol{q}$-space the number of values of $\boldsymbol{q}$ can be presented in the form

$$\sum_q 1 \Rightarrow \frac{V}{(2\pi)^3} \int \mathrm{d}\boldsymbol{q}. \tag{6.35}$$

For any function $\phi(\boldsymbol{q})$ it can be passed from the sum to the integral as follows:

$$\sum_{q} \phi(\boldsymbol{q}) \Rightarrow \frac{V}{(2\pi)^3} \int \phi(\boldsymbol{q}) \mathrm{d}\boldsymbol{q}. \tag{6.36}$$

6.2 Hamilton Function of Vibrating Crystalline Lattice: Normal Coordinates

It is known that in the classical case in order to apply the Gibbs method to solids, it is necessary to know the explicit form of its Hamilton function. Therefore, here we find the Hamilton function, i.e. the total energy of a vibrating lattice. The total energy E of a vibrating crystal in the classical case equals the sum of kinetic K and the potential energies U:

$$E = K + U. \tag{6.37}$$

For simplicity, we find the total energy of a one-dimensional simple lattice (Fig. 6.2), and thereupon generalize the results to the case of a three-dimensional lattice. Kinetic K and potential U energies of a one-dimensional simple lattice, presented in Fig. 6.2, in the approximation of the nearest neighbors can be presented in the form:

$$K = \frac{M}{2} \sum_{n=1}^{G} \dot{u}_n^2, \tag{6.38}$$

$$U = \frac{\beta}{2} \sum_{n=1}^{G} (u_n - u_{n-1})^2, \tag{6.39}$$

where G is the number of elementary cells in the fundamental region, and $\dot{u}_n$ is the time derivative of the displacement. The fact that the expression of the potential energy (6.39) is true is seen from the fact that the derivative with respect to u_n with the opposite sign $F_n = -\partial U/\partial u_n = -\beta(2u_n - u_{n-1} - u_{n+1})$ gives the force acting on the nth atom [see (6.9)].

Inasmuch as $u_n(t)$ is a periodic function, it can be expanded in harmonics. Then the real displacement can be written down in the form:

$$u_n = \sum_{q} [A_q \mathrm{e}^{\mathrm{i}(qan - \omega_q t)} + A_q^* \mathrm{e}^{-\mathrm{i}(qan - \omega_q t)}] = \frac{1}{\sqrt{G}} \sum_{q} \{a_q \mathrm{e}^{\mathrm{i}qan} + a_q^* \mathrm{e}^{-\mathrm{i}qan}\}, \tag{6.40}$$

where the notation

$$a_q = \sqrt{G} A_q \, \mathrm{e}^{-\mathrm{i}\omega_q t}. \tag{6.41}$$

is introduced. If we take into account that $\dot{a}_q = -\mathrm{i}\omega_q a_q$ and $\dot{a}_q^* = \mathrm{i}\omega_q a_q^*$, and substitute (6.40) into (6.38), the expression of the kinetic energy takes the form

$$K = \frac{M}{2}\sum_{n=1}^{G}\dot{u}_n\dot{u}_n = \frac{M}{2}\sum_{n=1}^{G}\frac{1}{\sqrt{G}}\sum_{q}\left\{\dot{a}_q e^{iqan} + \dot{a}_q^* e^{-iqan}\right\}$$
$$\times\frac{1}{\sqrt{G}}\sum_{q'}\left\{\dot{a}_{q'}e^{iq'an} + \dot{a}_{q'}^* e^{-iq'an}\right\} = -\frac{M}{2G}\sum_{qq'}\sum_{n=1}^{G}\omega_q\omega_{q'}$$
$$\times\left\{a_q a_{q'}e^{i(q+q')an} - a_q a_{q'}^* e^{i(q-q')an} - a_q^* a_{q'}e^{-i(q-q')an} + a_q^* a_{q'}^* e^{-i(q+q')an}\right\}. \tag{6.42}$$

Hence it is seen that to determine K it is necessary to calculate the sum of the type $L = \sum_{n=1}^{G} e^{iqan}$. If we take into account the value of the wave number (6.26), for the sum sought we get

$$L = \sum_{n=1}^{G} e^{iqan} = \sum_{n=1}^{G} e^{i\frac{2\pi}{G}gn} = \sum_{n=1}^{G} l^n, \tag{6.43}$$

where the notation

$$l = e^{i\frac{2\pi}{G}g}. \tag{6.44}$$

is introduced.

Consider two cases:

1. $q \neq 0$, i.e. $g \neq 0$. In this case $l \neq 1$ and inasmuch as $l^G = e^{i2\pi g} = 1$, then

$$L = \sum_{n=1}^{G} l^n = l + l^2 + \cdots + l^G = \frac{l(1-l^G)}{1-l} = 0 \tag{6.45}$$

or

$$L = \sum_{n=1}^{G} e^{iqan} = 0, q \neq 0. \tag{6.46}$$

2. $q = 0$, i.e. $g = 0$. In this case from (6.43) it is seen that $l = 1$ and

$$L = \sum_{n=1}^{G} l^n = \sum_{n=1}^{G} 1 = G. \tag{6.47}$$

Thus, we have

$$\sum_{n=1}^{G} e^{iqan} = \begin{cases} 0, \text{ at } q \neq 0 \\ G, \text{ at } q = 0. \end{cases} \tag{6.48}$$

Analogously it can be written as:

$$\sum_{n=1}^{G} e^{i(q\pm q')an} = \begin{cases} 0, \text{ at } q \pm q' \neq 0 \\ G, \text{ at } q \pm q' = 0. \end{cases} \tag{6.49}$$

If we apply the indicated rules of summation in (6.42) and take into account that $\omega_q = \omega_{-q}$, for the kinetic energy we get:

$$K = \frac{M}{2} \sum_q \omega_q^2 (2a_q a_q^* - a_q a_{-q} - a_q^* a_{-q}^*). \tag{6.50}$$

Now we translate the expression of the potential energy (6.39). If we substitute (6.40) into (6.39), for the potential energy, we get:

$$\begin{aligned} U = \frac{\beta}{2} \sum_{n=1}^{G} (u_n - u_{n-1})(u_n - u_{n-1}) &= \frac{\beta}{2G} \sum_{qq'} \sum_{n=1}^{G} \\ &\times [a_q \mathrm{e}^{\mathrm{i}qan} + a_q^* \mathrm{e}^{-\mathrm{i}qan} - a_q \mathrm{e}^{-\mathrm{i}qa} \mathrm{e}^{\mathrm{i}qan} - a_q^* \mathrm{e}^{\mathrm{i}qa} \mathrm{e}^{-\mathrm{i}qan}] \\ &\times \left[a_{q'} \mathrm{e}^{\mathrm{i}q'an} + a_{q'}^* \mathrm{e}^{-\mathrm{i}q'an} - a_{q'} \mathrm{e}^{-\mathrm{i}q'a} \mathrm{e}^{\mathrm{i}q'an} - a_{q'}^* \mathrm{e}^{\mathrm{i}q'a} \mathrm{e}^{-\mathrm{i}q'an}\right]. \end{aligned} \tag{6.51}$$

Having multiplied the brackets and with the aid of (6.49) having added up in respect to n and q', we get:

$$U = \frac{M}{2} \sum_q \omega_q^2 \left(2a_q a_q^* + a_q a_{-q} + a_q^* a_{-q}^*\right). \tag{6.52}$$

Here it was considered that $\mathrm{e}^{\mathrm{i}qa} + \mathrm{e}^{-\mathrm{i}qa} = 2\cos aq$; $1 - \cos aq = 2\sin^2 aq/2$, and also $\sin^2 aq/2 = M\omega_q^2/4\beta$ was used (6.16).

If we substitute (6.50) and (6.52) into (6.37), for the total energy of a one-dimensional crystalline lattice we get:

$$E = K + U = 2M \sum_q \omega_q^2 a_q a_q^*. \tag{6.53}$$

As is seen, the total energy of a vibrating one-dimensional lattice, expressed by quantities a_q, has a very simple form. However, it is expedient to pass from complex coordinates a_q to real ones X_q and P_q as follows:

$$\begin{aligned} X_q &= a_q + a_q^* = 2\,\mathrm{Re}\, a_q \\ P_q &= M\dot{X}_q = -\mathrm{i}\, M\omega_q (a_q - a_q^*) = -\mathrm{i}\, 2M\omega_q\, \mathrm{Im}\, a_q. \end{aligned} \tag{6.54}$$

Hence we get

$$\begin{aligned} a_q &= \frac{1}{2}\left(X_q + \mathrm{i}\frac{P_q}{M\omega_q}\right), \\ a_q^* &= \frac{1}{2}\left(X_q - \mathrm{i}\frac{P_q}{M\omega_q}\right). \end{aligned} \tag{6.55}$$

If we substitute (6.55) into (6.53), for the total energy or Hamilton function we finally get:

$$E = \sum_q \left\{\frac{1}{2M} P_q^2 + \frac{1}{2} M\omega_q^2 X_q^2\right\} = H\left(X_q, P_q\right). \tag{6.56}$$

Here the expression within brackets under the sum sign is the energy of a harmonic oscillator with the frequency ω_q and mass M. X_q and P_q are called *normal coordinates*, and vibrations expressed by these coordinates are called *normal vibrations* or *modes.*

Inasmuch as the wave number of a one-dimensional lattice and the frequency according to (6.26) take on a $G = N$ number of values, in expression (6.56) there enters the energy of the same number of oscillators.

Thus, expressing the Hamilton function by normal coordinates X_q and P_q, we come to a very important conclusion: *The total energy of a vibrating one-dimensional simple crystalline lattice equals the sum of energies of non-interacting harmonic oscillators, the number of which equals the number of possible frequencies, i.e. the number of degrees of freedom* $G = N$.

From this conclusion stems the advantage of the normal coordinates compared with the displacement u_n.

Expression of the total energy (6.56) can be generalized also to the case of a three-dimensional lattice. In this case, the number of frequencies $\omega_j(\boldsymbol{q})$ and, conformably, the number of oscillators equals $3N$ (Fig. 6.4). As a result, the total energy (the Hamilton function) of a three-dimensional simple lattice can be written as:

$$E = \sum_{\boldsymbol{q}} \sum_{j=1}^{3} \left\{ \frac{1}{2M} P_{\boldsymbol{q}j}^2 + \frac{M}{2} \omega_{\boldsymbol{q}j}^2 X_{\boldsymbol{q}j}^2 \right\} = H(X_{\boldsymbol{q}j}, P_{\boldsymbol{q}j}), \qquad (6.57)$$

where $X_{\boldsymbol{q}j}$ and $P_{\boldsymbol{q}j}$ are normal coordinates of a three-dimensional lattice, and $H(X_{\boldsymbol{q}j}, P_{\boldsymbol{q}j})$ is the Hamilton function.

The simple form of the Hamilton function obtained here is very important to construct the classical theory of thermodynamic properties of solids, to which Sect. 6.3 is devoted.

6.3 Classical Theory of Thermodynamic Properties of Solids

In Sect. 6.1 it was shown that at temperatures higher than that of Debye's ($T \gg \theta$), the vibrational motion of a crystalline lattice is classical. In this section we will consider the classical theory of the thermodynamic properties of solids with a simple crystalline lattice in the region of high temperatures $T \gg \theta$. Consequently, all expressions for thermodynamic functions and thermodynamic quantities obtained in this section hold only at high temperatures $T \gg \theta$.

1. *Free energy.* At first apply the Gibbs method to the considered case and calculate the free energy

$$F = -k_0 T \ln Z_{\text{cl}}, \qquad (6.58)$$

where Z_{cl} is the statistical integral.

For the given case the statistical integral can be presented in the form

$$Z_{\rm cl} = \frac{1}{(2\pi\hbar)^{3N}} \int {\rm e}^{-E(X_{qj},P_{qj})/k_0T} ({\rm d}X_{qj}{\rm d}P_{qj}), \tag{6.59}$$

where the symbolic notation

$$({\rm d}X_{\boldsymbol{q}j}{\rm d}P_{\boldsymbol{q}j}) = \prod_{\boldsymbol{q}j} {\rm d}X_{qj}{\rm d}P_{qj}. \tag{6.60}$$

is introduced. If we substitute the expression of the total energy of a three-dimensional crystal (6.57) into (6.59), the statistical integral takes the form:

$$Z_{\rm cl} = \prod_{q=1}^{N} \left[\int_{-\infty}^{+\infty} {\rm e}^{-P_q^2/2Mk_0T} {\rm d}P_q \int_{-\infty}^{+\infty} {\rm e}^{-M\omega_q^2 X_q^2/2k_0T} \frac{{\rm d}X_q}{2\pi\hbar} \right]^3. \tag{6.61}$$

To integrate over ${\rm d}P_q$ and ${\rm d}X_q$ we make use of the formulae adduced in Appendix A. As a result, we get

$$Z_{\rm cl} = \prod_q \left(\frac{k_0T}{\hbar\omega_q}\right)^3. \tag{6.62}$$

Substituting (6.62) into (6.58), we get:

$$F = -3k_0T \sum_q \ln\left(\frac{k_0T}{\hbar\omega_q}\right). \tag{6.63}$$

Passing from the summation with respect to $\boldsymbol{q}$ to the integration, according to (6.36), for F we get:

$$F = \frac{3k_0TV}{(2\pi)^3} \int \ln\left(\frac{\hbar\omega_q}{k_0T}\right) {\rm d}\boldsymbol{q}. \tag{6.64}$$

To calculate this integral use the continuum approximation $\omega_q = \upsilon_0 q$ and pass to the spherical coordinate system. The integration over angles gives 4π and as a result we have:

$$F = \frac{3k_0TV}{2\pi^2} \int_0^{q_{\rm max}} \ln\left(\frac{\hbar\omega_q}{k_0T}\right) q^2{\rm d}q, \tag{6.65}$$

where $q_{\rm max}$ is the maximally possible value of the wave number of a crystal, which we will determine slightly later.

Introduce the dimensionless variable:

$$x = \frac{\hbar\omega_q}{k_0T} = \frac{\hbar\upsilon_0 q}{k_0T}. \tag{6.66}$$

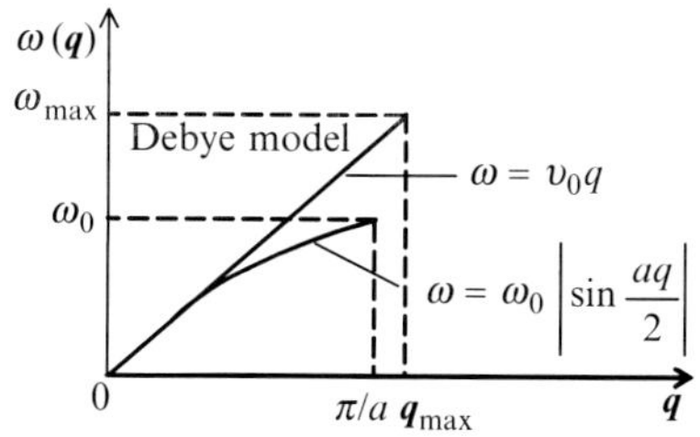

Fig. 6.5. The frequency dependence on the wave vector in the Debye model

Then free energy (6.65) takes the form:

$$F = \frac{3k_0TV}{2\pi^2}\left(\frac{k_0T}{\hbar\upsilon_0}\right)^3 \int\limits_0^{x_{\max}} x^2 \ln x \,\mathrm{d}x, \tag{6.67}$$

where

$$x_{\max} = \frac{\hbar\upsilon_0}{k_0T} q_{\max} = \frac{\hbar\omega_{\max}}{k_0T}, \tag{6.68}$$

and $\omega_{\max} = \upsilon_0 q_{\max}$ in the continuum approximation is the maximally possible frequency of a crystal (Fig. 6.5).

Integrating (6.67) in parts and using (6.68), for the free energy we finally have

$$F = \frac{k_0TV}{6\pi^2}\left(\frac{\omega_{\max}}{\upsilon_0}\right)^3 \left[3 \ln\left(\frac{\hbar\omega_{\max}}{k_0T}\right) - 1\right]. \tag{6.69}$$

2. *Debye temperature.* We mentioned earlier in Sect. 7.1 about the Debye temperature. Characteristic to each solid, it is defined as follows:

$$\theta = \frac{\hbar\omega_{\max}}{k_0}. \tag{6.70}$$

If we take into account this definition in (6.69), for the free energy we get

$$F = \frac{k_0TV}{6\pi^2}\left(\frac{k_0\theta}{\hbar\upsilon_0}\right)^3 \left[3 \ln\left(\frac{\theta}{T}\right) - 1\right]. \tag{6.71}$$

The concept of the Debye temperature was introduced by the Dutch physicist–theorist Debye in 1912 when considering the quantum theory of the heat capacity of solids. According to (6.70), the Debye temperature has the following physical sense: θ *is the temperature at which in the given crystal all possible frequencies* ($\omega \leq \omega_{\max}$) are excited. Each solid has the intrinsic maximum frequency $\omega_{\max}$ and, consequently, the Debye temperature. The magnitude of the Debye temperature depends on the elastic properties of a crystal, and concretely, on the speed of propagation of sound in the crystal and the lattice constant (6.79). To determine these dependences we use the following known condition: *The number of possible values of the wave vector* $\boldsymbol{q}$ *in the first Brillouin zone equals the number of elementary cells – in our case the number of atoms* N. If we consider that in a three-dimensional simple lattice the number of branches of frequencies equals three, the indicated condition can be written in the form:

$$\sum_{q}\sum_{j=1}^{3} 1 = 3N, \tag{6.72}$$

where the summation with respect to $\boldsymbol{q}$ is carried out in the limits of the first Brillouin zone. According to the rule of passing from the sum over the wave vector to the integral (6.35), condition (6.72) has the appearance:

$$\frac{3V}{(2\pi)^3}\int \mathrm{d}\boldsymbol{q} = 3N \tag{6.73}$$

or

$$\frac{3V}{(2\pi)^3}\int\limits_0^{q_{\max}} 4\pi q^2 \mathrm{d}q = 3N. \tag{6.74}$$

Using the Debye model (Fig. 6.5) and the relationship $\omega = v_0 q$, and also passing from the integral over $\mathrm{d}q$ to the integral over $\mathrm{d}\omega$, we get

$$\frac{3V}{2\pi^2 v_0^3}\int\limits_0^{\omega_{\max}} \omega^2 \mathrm{d}\omega = 3N, \tag{6.75}$$

where $\omega_{\max} = v_0 q_{\max}$.

The number of frequencies found in the range $0 - \omega_{\max}$ can also be presented in the form:

$$\int\limits_0^{\omega_{\max}} g(\omega)\mathrm{d}\omega = 3N, \tag{6.76}$$

where $g(\omega)$ is the number of frequencies in the unit range around the frequency ω, i.e. *the function of the density of frequencies*. Comparing (6.76) and (6.75), for this function we get:

$$g(\omega) = \frac{3V}{2\pi^2 v_0^3}\omega^2 \sim \omega^2 \tag{6.77}$$

(Fig. 6.6). It is seen that in the continuum approximation $\omega = v_0 q$ the density of the number of frequencies grows quadratically with the growth of the frequency: $g(\omega) \sim \omega^2$.

From (6.75) it is easy to determine the maximum frequency $\omega_{\max}$.

$$\omega_{\max} = v_0\left(\frac{6\pi^2 N}{V}\right)^{1/3}. \tag{6.78}$$

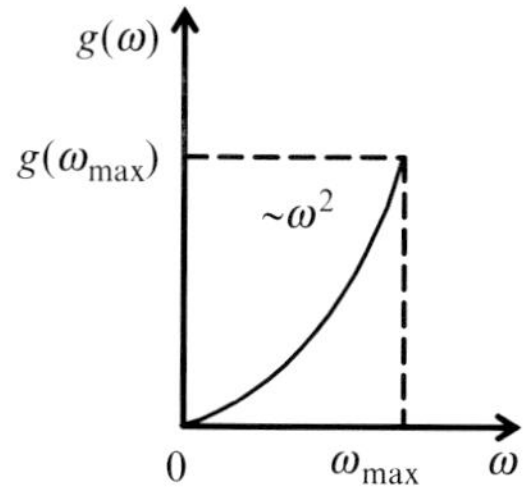

Fig. 6.6. The function of the density of frequencies

Then, according to definition (6.70), the Debye temperature θ has the appearance:

$$\theta = \frac{\hbar v_0}{k_0}\left(\frac{6\pi^2 N}{V}\right)^{1/3}. \tag{6.79}$$

If (6.79) in substituted into (6.71), for the free energy at high temperatures ($T \gg \theta$), it can be written as:

$$F = 3k_0 TN \ln(\theta/T) - k_0 TN; T \gg \theta. \tag{6.80}$$

Knowing the expression of the free energy, we can calculate the basic thermodynamic quantities of a solid.

3. *Entropy.* From the definition $S = -(\partial F/\partial T)_V$ and (6.80) we have

$$S = -3k_0 N \ln(\theta/T) + 4k_0 N. \tag{6.81}$$

4. *Mean energy - caloric equation of state.* From the relationship $E = F + TS$, and also from (6.80) and (6.81) we get

$$E = 3k_0 TN; T \gg \theta. \tag{6.82}$$

   It is seen that in the region of high temperatures the mean energy, i.e. the caloric equation of the state of a solid depends only on temperature. Note that this result also follows from the theorem about the equal distribution of energy over degrees of freedom, obtained on the basis of the Boltzmann statistics.

5. *Thermal equation of state. Grüneisen parameter.* The dependence of the free energy F on volume V in (6.80) is not seen explicitly. It is determined by the dependence of the Debye temperature on volume $\theta(V)$ [see (6.79)]. Therefore the thermal equation of the state of a solid $P = -(\partial F/\partial V)_T$ can be presented in the form:

$$P = -3k_0 TN \frac{1}{\theta}\frac{\mathrm{d}\theta}{\mathrm{d}V}. \tag{6.83}$$

If the dimensionless parameter

$$\gamma_G = -\frac{V}{\theta}\frac{d\theta}{dV}, \tag{6.84}$$

is introduced, the equation of state takes the form:

$$P = \frac{3k_0TN}{V}\gamma_G = \frac{E(T)}{V}\gamma_G; \quad T \gg \theta, \tag{6.85}$$

where γ_G is called *the Grüneisen parameter*. Using the definition of the Debye temperature $\theta = \hbar\omega_{max}/k_0$, the Grüneisen parameter can be presented in the different forms

$$\gamma_G = -\frac{V}{\theta}\frac{d\theta}{dV} = -\frac{d\ln\theta}{d\ln V} = -\frac{d\ln\omega_{max}}{d\ln V} = -\frac{\Delta\omega_{max}/\omega_{max}}{\Delta V/V}. \tag{6.86}$$

Then the parameter γ_G has the following physical sense: *The Grüneisen parameter characterizes the relative change in the Debye temperature or the relative change in the maximum frequency, accounting for the unit of relative change in the volume of a solid.*

It can be shown that the parameter γ_G is associated with anharmonicity of vibrations of atoms in a crystal. Let us demonstrate this with the example of a simple one-dimensional crystal.

The maximum value of the wave number q_{max} for a one-dimensional crystal of the length L with the number of atoms N is determined from the condition analogous to (6.74):

$$\frac{L}{2\pi}\int_0^{q_{max}} dq = N. \tag{6.87}$$

Then from the relationship $\omega_{max} = \upsilon_0 q_{max}$ and (6.87) we get

$$\omega_{max} = 2\pi\upsilon_0\frac{N}{L} = \frac{2\pi\upsilon_0}{R_0}. \tag{6.88}$$

Here $R_0 \equiv a$ is the distance between two atoms in the equilibrium state (see Fig. 6.1). If we take into account the speed of propagation of sound (6.21), in expression (6.88), we get: $\omega_{max} = 2\pi\sqrt{\beta/M}$ or

$$\omega_{max}^2 = (2\pi)^2\frac{\beta}{M}. \tag{6.89}$$

If we take into account that the constant of elasticity of a linear crystal, as seen from (6.2), is $\beta = U''(R_0)$, expression (6.89) takes the form:

$$\omega_{max}^2 = \frac{(2\pi)^2}{M}U''(R_0). \tag{6.90}$$

Because of the vibrations, the distance between atoms changes by ΔR_0, and, consequently, the frequency changes by $\Delta\omega_{\max}$. Then (6.90) can be presented in the form:

$$(\omega_{\max} + \Delta\omega_{\max})^2 = \frac{(2\pi)^2}{M} U''(R_0 + \Delta R_0). \tag{6.91}$$

Inasmuch as $\Delta R_0 \ll R_0$ and $\Delta\omega_{\max} \ll \omega_{\max}$, on the left-hand side of (6.91) we can neglect $(\Delta\omega_{\max})^2$, and expand the right-hand side into a series in ΔR_0, restricting it to the linear approximation. As a result, (6.91) takes the form:

$$\frac{\Delta\omega_{\max}}{\omega_{\max}} = -\gamma \frac{\Delta R_0}{\beta}, \tag{6.92}$$

where $U'''(R_0) = -2\gamma$ [see (6.2)]. As a result, according to (6.86) and (6.92), for the Grüneisen parameter of a one-dimensional crystal we get:

$$\gamma_{\mathrm{G}} = -\frac{\Delta\omega_{\max}/\omega_{\max}}{\Delta R_0/R_0} = \gamma \frac{R_0}{\beta}, \tag{6.93}$$

where γ is the coefficient of anharmonicity of vibrations [see (6.2)]. Thus, for the pre-assigned linear crystal $\gamma_{\mathrm{G}} \sim \gamma$. Naturally, in the expression for the potential (6.2) the ratio of the anharmonic item to the harmonic one ought to equal the relation of the displacement x to R_0

$$\frac{\gamma x}{\beta} \approx \frac{x}{R_0}. \tag{6.94}$$

Hence it is seen that $\beta \approx R_0\gamma$ and from (6.93) it follows that the Grüneisen parameter is a constant quantity on the order of unity $\gamma_{\mathrm{G}} \approx 1$, not depending on temperature.

From the equations of state (6.95) and (6.93) it follows that in the harmonic approximation ($\gamma = 0$) a crystal does not exert pressure on the surrounding bodies $P = 0$.

6. *Heat capacity.* With regard to definitions of the heat capacity $C_V = (\partial E/\partial T)$ or $C_V = T\,(\partial S/\partial T)$, from (6.82) or (6.81) it follows that

$$C_V = 3k_0 N; T \gg \theta, \tag{6.95}$$

i.e. at high temperatures $T \gg \theta$ the isochoric heat capacity is a constant quantity, depending on neither the nature of the crystal nor the temperature. This theoretical result corresponds to the experimental *Dulong–Petit law* [see Sect. 6.4, Fig. 6.7].

The difference of isobaric and isochoric heat capacities has the appearance

$$C_P - C_V = -T\left(\frac{\partial P}{\partial T}\right)_V^2 \left(\frac{\partial P}{\partial V}\right)_T^{-1}. \tag{6.96}$$

Hence and from equation of state (6.85) we get

$$C_P - C_V = 3k_0 N\gamma_G; T \gg \theta. \tag{6.97}$$

It is seen that in the harmonic approximation ($\gamma_{\rm G} = 0$) isobaric and isochoric heat capacities coincide $C_P = C_V$ and the difference between capacities is determined only by the anharmonicity of vibrations.

Note the following interesting fact. For a three-dimensional simple crystal from the expression of the Debye temperature (6.79) and from the definition of the Grüneisen parameter (6.84) it follows that $\gamma_{\rm G} = 1/3$. If we take this into account in (6.85) and (6.97), we get known results for an ideal gas.

7. *Thermal expansion.* One of the thermodynamic properties of solids is thermal expansion. It can occur at the isobaric process (P = const) and is determined by the coefficient:

$$\alpha_P = \frac{1}{V}\left(\frac{\partial V}{\partial T}\right)_P. \tag{6.98}$$

If the basic thermodynamic relationship is used (2.104), (6.98) takes the form:

$$\alpha_P = \gamma_T \left(\frac{\partial P}{\partial T}\right)_V, \tag{6.99}$$

where $\gamma_T = -\frac{1}{V}\left(\frac{\partial V}{\partial P}\right)_T$ is the coefficient of the isothermal compressibility. From the equations of state (6.85), (6.97) and (6.99) we get

$$\alpha_P = \gamma_T \frac{3k_0 N}{V}\gamma_{\rm G} = \gamma_T \frac{C_V}{V}\gamma_{\rm G}. \tag{6.100}$$

Hence stems the *Grüneisen relationship*

$$V\alpha_P = \gamma_T C_V \gamma_{\rm G}. \tag{6.101}$$

This relation between thermodynamic coefficients and the parameter $\gamma_{\rm G}$ was experimentally corroborated by Grüneisen in 1908.

From the expression (6.100) it is seen that the isobaric coefficient of the thermal expansion α_P as well as the thermal equation of state and the difference $(C_P - C_V)$ is proportional to the Grüneisen parameter $\gamma_{\rm G}$. Consequently, all the three indicated quantities are associated with the anharmonicity of vibrations.

6.4 Quantum Theory of Heat Capacity of Solids: Einstein and Debye Models

Among the thermodynamic coefficients of solids, the temperature dependence of heat capacity occupies a special place, inasmuch as it is very sensitive to the internal structure of the substance, to the kinds of interaction between

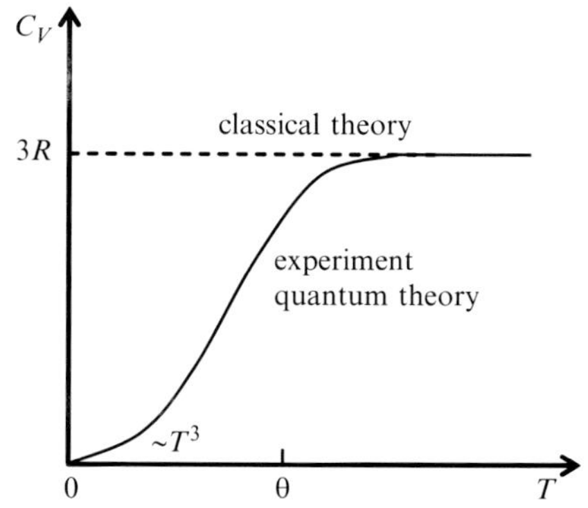

Fig. 6.7. The temperature dependence of heat capacity of solids

particles forming it, and to the nature (classical or quantum) of their motion. Because of this, at all times, interest in the theoretical and experimental investigation of the heat capacity of solids is significant. It is no mere chance that in 1907 Einstein got interested in the heat capacity of solids and for the first time created its quantum theory.

As far back as the beginning of the nineteenth century (1819), having generalized available experimental facts on the heat capacity of solids, P.-L. Dulong and A.-T. Petit determined the following appropriateness: *At room temperature and higher, the heat capacity of a solid depends on neither the kind of solid nor temperature; the specific heat capacity of one mole equals* $C_V \approx 3R \approx 6\,\mathrm{cal/mol \cdot K}$.

Note that this experimental result, bearing the name *Dulong–Petit law*, is very easily explained on the basis of classical theory. Indeed, if the vibrational motion of the atoms at lattice points is regarded as classical, and the law about equal distribution of energy over degrees of freedom is also taken into account, stemming from the Boltzmann statistics [see also result (6.95)], then in the region $T \gg \theta$ the classical theory completely coincides with experiment (Fig. 6.7).

This conformity between theory and experiment was observed as long as the heat capacity was measured only in the range of high temperatures ($T > \theta$). At the end of the nineteenth century with the development of cryogenic techniques (the production of low temperatures) the measurement of heat capacity at much lower than room temperature was begun. It was determined that the Dulong–Petit law is not fulfilled: with the lowering of temperature the heat capacity drops, and as temperature approaches absolute zero it tends to zero, which is in conformity with the Nernst principle. Thus, at low temperatures ($T < \theta$) there arose the nonconformity between experiment and the only theory existing at that time, classical theory (Fig. 6.7). This phenomenon was observed right up to the beginning of the twentieth century, till the time when Einstein formulated the quantum theory of the heat capacity of a solid.

6.4.1 Einstein's Theory

The quantum theory of the heat capacity of a solid was proposed by Einstein in 1907 for the first time. At that time, quantum mechanics did not exist. However, as far back as in 1900, Planck introduced for the first time the concept of the energetic portion – *quantum*. According to the Planck postulate, an oscillator vibrating with the frequency ω irradiates energy by whole portions $h\omega$, i.e. the radiation occurs by quanta. On the basis of this postulate Planck found the function of the spectral density of energy of the black radiation (the Planck known formula) and, in this way, explained all available experimental facts on the radiation in wide temperature and frequency regions. Note that in particular cases the Wien and Rayleigh–Jeans laws stem from the Planck formula for the spectral density of energy of radiation.

As a result Planck, for the first time proposed[1] the quantum expression for the mean energy of a linear harmonic oscillator with the frequency ω:

$$\varepsilon = \frac{\hbar\omega}{e^{\hbar\omega/k_0T} - 1}. \tag{6.102}$$

It is seen that at high temperatures, when $k_0T \gg \hbar\omega$, from (6.102) stems the known classical expression for a linear harmonic oscillator $\varepsilon = k_0T$.

Formula (6.102) was obtained by Planck for energy of a linear inter-atom oscillator, providing absorption and emission of light. In 1907 Einstein supposed that expression (6.102) can be applied to the thermal vibration of atoms at crystalline lattice points. This supposition made it possible to calculate the mean energy of a vibrating crystal. To do this Einstein proposed a very simple model, according to which *a solid consists of N atoms at crystalline lattice points that vibrate independent of each other with an identical frequency* $\omega = \omega_0$. The Einstein model allows one to replace a crystalline solid, consisting of N number of atoms, with a system of N non-interacting three-dimensional harmonic oscillators, vibrating with an identical frequency ω_0, i.e. with an ideal gas of harmonic oscillators. Schematically this model is depicted in Fig. 6.8.

If (6.102) is used, then according this model, the total energy of a crystal, consisting of N atoms, can be presented as the total energy of $3N$ non-interacting harmonic linear oscillators with an identical frequency ω_0:

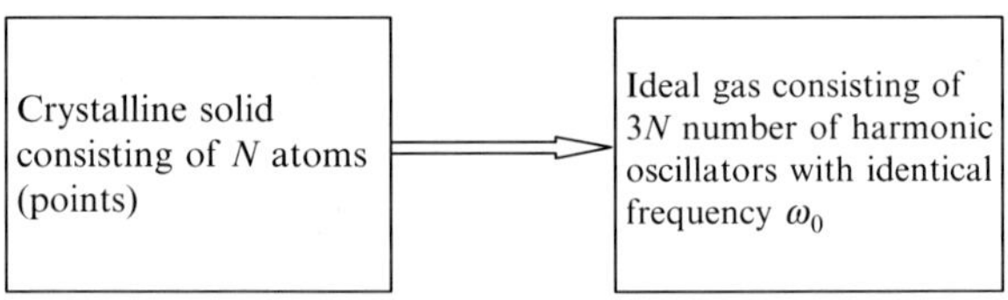

Fig. 6.8. The illustration of the Einstein model

[1] Notice that at that time the concept of zero vibrations did not exist.

$$E = 3N\varepsilon(\omega_0) = \frac{3N\hbar\omega_0}{\mathrm{e}^{\hbar\omega_0/k_0T} - 1}. \tag{6.103}$$

Hence the heat capacity of a solid $C_V = (\partial E/\partial T)_V$ equals

$$C_V = 3k_0N\left(\frac{\hbar\omega_0}{k_0T}\right)^2 \frac{\mathrm{e}^{\hbar\omega_0/k_0T}}{(\mathrm{e}^{\hbar\omega_0/k_0T} - 1)^2}. \tag{6.104}$$

Introduce the concept of the Einstein characteristic temperature $\theta_0 = \hbar\omega_0/k_0$. Then expression (6.104) takes the form:

$$C_V = 3k_0N\left(\frac{\theta_0}{T}\right)^2 \frac{\exp(\theta_0/T)}{[\exp(\theta_0/T) - 1]^2}. \tag{6.105}$$

If we accept the value of the frequency equal to $\omega_0 \approx 3 \times 10^{13}\,\mathrm{s}^{-1}$, the Einstein temperature is on the order of room temperature: $\theta_0 \approx 3 \times 10^2\,\mathrm{K}$.

In the region of *high temperatures* ($T \gg \theta_0$) from (6.105) the known classical result for the heat capacity of a solid $C_V = 3k_0N = 3R$ follows.

In the region of *low temperatures* ($T \ll \theta_0$), if in the denominator of (6.105), we neglect unity, for the heat capacity we get

$$C_V \approx 3k_0N\,(\theta_0/T)^2 \exp(-\theta_0/T)\,; \quad T \ll \theta_0 \tag{6.106}$$

As is seen from (6.106), in the region of low temperatures ($T \ll \theta_0$) the heat capacity very strongly depends on temperature and in the limiting case ($T \to 0\,\mathrm{K}$) C_V *exponentially* tends to zero. Though such dependence satisfies the known Nernst principle, with experiment it is consistent only qualitatively. Numerous experiments show that in the region of absolute zero the function $C_V(T)$ behaves not as an exponent, but has the appearance $C_V(T) \sim T^3$.

Such a nonconformity of the theory with experiment shows that in spite of the fact that Einstein's idea about mean energy of the quantum oscillator (6.102) was right, the model proposed by him (Fig. 6.8) in the region of low temperatures did not reflect reality. In this situation perfection of the Einstein model was required. This was done by Debye.

6.4.2 Debye's Theory

In 1912 the Dutch physicist Debye, having accepted Einstein's idea as the basis, proposed a new model, more exactly reflecting reality when calculating energy of a crystal, thereby creating a theory, consistent with experiment.

How is the Debye model distinct from the Einstein model?

According to the Einstein model, *atoms (ions) at crystalline lattice points perform harmonic vibration, independent of each other, with the identical frequency* $\omega = \omega_0$. Therefore, the total energy of a crystal can be presented in the form (6.103). According to the Debye model, however, between atoms (ions) at crystalline lattice points an elastic bond exists. Therefore, at some point

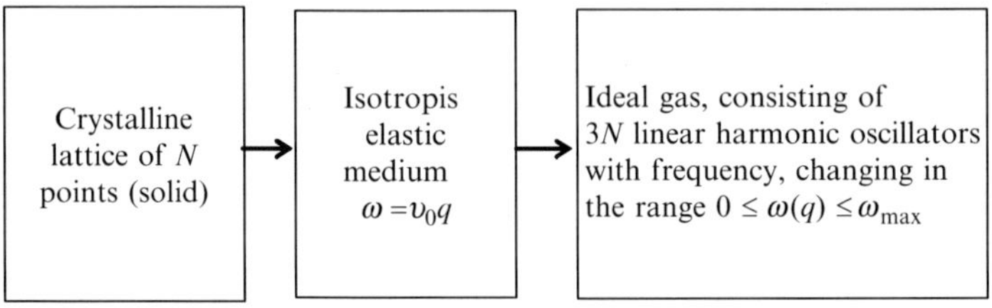

Fig. 6.9. The illustration of the Debye model

vibrations arise with the frequency ω propagating over the entire crystal in the form of an elastic wave of corresponding length. For each crystal there exists the specified relationship $\omega = \omega(q)$ between the frequency ω and the wave number $q = 2\pi/\lambda$ or the frequency ω and the wavelength λ. Debye proposed *that in the specified approximation, a crystal possessing a discrete structure can be replaced with an isotropic continuum medium (the continuum approximation). Then the relation between ω and q can be presented in the form $\omega(q) = \upsilon_0 q$, where υ_0 is the velocity of the elastic wave* (see Sect. 6.1, Fig. 6.5).

Note that for continuum media the wave number q and frequency ω change in the infinite range $0 \le q < \infty$, $0 \le \omega < \infty$ and take on the infinite number of values.

Taking into account that a crystal has a discrete, but periodic structure, and that the wave vector and frequency change in the finite range $0 \le q < q_{\max}$ и $0 \le \omega < \omega_{\max}$ and have a finite number of values in this range, Debye proposed a model of a crystal, depicted in Fig. 6.9.

1. *Mean energy.* On the basis of the Debye model and the Einstein idea, the total energy of a solid with the simple crystalline lattice can be presented in the form:

$$E = 3\sum_{\boldsymbol{q}} \varepsilon\left(\omega(\boldsymbol{q})\right) = 3\sum_{\boldsymbol{q}} \frac{\hbar\omega(\boldsymbol{q})}{\mathrm{e}^{\hbar\omega(\boldsymbol{q})/k_0T} - 1}. \tag{6.107}$$

Here the factor 3 is associated with the fact that to each value of the wave vector $\boldsymbol{q}$, three coinciding frequencies correspond.

Note that the sum (6.107) consists of N number of items. Each of them corresponds to one value of $\boldsymbol{q}$. Inasmuch as the number of points N is a very large number, it can be regarded that the wave vector $\boldsymbol{q}$ changes continuously (quasi-discretely). Therefore, taking into account (6.36) in (6.107), it can be passed from the sum to the integral

$$E = \frac{3V}{(2\pi)^3}\int \frac{\hbar\omega(\boldsymbol{q})}{\mathrm{e}^{\hbar\omega(\boldsymbol{q})/k_0T} - 1}\mathrm{d}\boldsymbol{q}. \tag{6.108}$$

If we take into account that in the considered case the $\boldsymbol{q}$-space has a spherical symmetry and the frequency depends only on the magnitude of the wave vector ($\boldsymbol{q}$), i.e. $\omega(\boldsymbol{q}) = \omega(q)$, (6.108) takes the form:

$$E = \frac{3V}{2\pi^2} \int\limits_0^{q_{\max}} \frac{\hbar\omega(q)}{e^{\hbar\omega(q)/k_0T} - 1} q^2 dq. \tag{6.109}$$

Here in the $\boldsymbol{q}$-space we pass to the spherical coordinate system, where an element of volume $d\boldsymbol{q} = q^2 dq \sin\theta d\theta d\phi$ and consider that the integration over angles gives 4π. If we restrict ourselves to the continuum approximation $\omega(q) = \upsilon_0 q$, in (6.109) from the integration over dq we can pass to the integration over $d\omega$. As a result, we get

$$E = \frac{3V\hbar}{2\pi^2\upsilon_0^3} \int\limits_0^{\omega_{\max}} \frac{\omega^3 d\omega}{e^{\hbar\omega/k_0T} - 1}. \tag{6.110}$$

Here $\omega_{\max} = \upsilon_0 q_{\max}$ is the maximum possible frequency, which on the basis of the Debye model is found from condition (6.75) and given by expression (6.78).

If it is passed to the dimensionless variable $x = \hbar\omega/k_0T$, the expression for energy (6.110) takes the form:

$$E = \frac{3Vk_0T}{2\pi^2} \left(\frac{k_0T}{\hbar\upsilon_0}\right)^3 \int\limits_0^{x_{\max}} \frac{x^3 dx}{e^x - 1}, \tag{6.111}$$

where $x_{\max} = \frac{\hbar\omega_{\max}}{k_0T}$.

If we use the definition of the Debye characteristic temperature $\theta = \hbar\omega_{\max}/k_0$ in the expression (6.111), we get

$$E = 3k_0TND\left(\frac{\theta}{T}\right), \tag{6.112}$$

where

$$D\left(\frac{\theta}{T}\right) = 3\left(\frac{T}{\theta}\right)^3 \int\limits_0^{\theta/T} \frac{x^3 dx}{e^x - 1} \tag{6.113}$$

is the Debye function. As is seen, mean energy of a crystal (6.112) at an arbitrary temperature is distinct from energy (6.82) for the classical case ($T \gg \theta$) by the factor, and namely by the Debye function. Since the Debye function cannot be calculated analytically, we consider its different asymptotes.

High temperatures: $T \gg \theta$. In this case, inasmuch as $x \ll 1$, in (6.113) the factor $(e^x - 1)^{-1}$ can be expanded into a series

$$\frac{1}{e^x - 1} \approx \frac{1}{x + x^2/2 + x^3/6 + \ldots} \approx \frac{1}{x}\left(1 - \frac{x}{2} + \frac{x^2}{12}\right); x \ll 1. \tag{6.114}$$

Taking this into account in (6.113) and carrying out the integration, for the asymptote of the Debye function in the high-temperature region we get:

$$D\left(\frac{\theta}{T}\right) = 1 - \frac{3}{8}\left(\frac{\theta}{T}\right) + \frac{1}{20}\left(\frac{\theta}{T}\right)^2 ; T \gg \theta. \tag{6.115}$$

It is seen that in the limiting case

$$\lim_{T\to\infty} D\left(\frac{\theta}{T}\right) = 1. \tag{6.113a}$$

Substituting (6.115) into (6.112), energy in the high-temperature region can be written as:

$$E = 3k_0NT - \frac{9}{8}k_0N\theta + \frac{3}{20}k_0NT\left(\frac{\theta}{T}\right)^2 ; T \gg \theta. \tag{6.116}$$

Low temperatures: $T \ll \theta$. In this case, the upper boundary of the integration in the function can be accepted as equal to $\theta/T \to \infty$. Then we get:

$$D\left(\frac{\theta}{T}\right) = 3\left(\frac{T}{\theta}\right)^3 \int_0^\infty \frac{x^3 \mathrm{d}x}{\mathrm{e}^x - 1}; T \ll \theta. \tag{6.117}$$

If from Appendix A we take into account that

$$\int_0^\infty \frac{x^3 \mathrm{d}x}{\mathrm{e}^x - 1} = \Gamma(4)\xi(4) = 6 \cdot \frac{\pi^4}{90} = \frac{\pi^4}{15}, \tag{6.118}$$

then at low temperatures for the Debye function we have the asymptote

$$D\left(\frac{\theta}{T}\right) = \frac{\pi^4}{5}\left(\frac{T}{\theta}\right)^3 ; T \ll \theta. \tag{6.119}$$

Hence it follows that

$$\lim_{T\to 0} D\left(\theta/T\right) = 0. \tag{6.113b}$$

From asymptotic values (6.113a) and (6.113b) it follows that the Debye function changes in the limits 0–1. A plot of this function is schematically presented in Fig. 6.10.

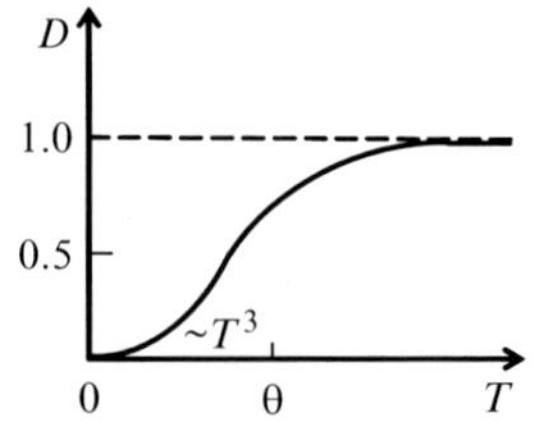

Fig. 6.10. A plot of the Debye function

According to (6.112) and (6.119), the mean energy of a crystal in the region of low temperatures has the appearance:

$$E = \frac{3\pi^4}{5} k_0 N\theta \left(\frac{T}{\theta}\right)^4 ; T \ll \theta \tag{6.120}$$

2. *Heat capacity.* In order to find the expression of the heat capacity of a solid at an arbitrary temperature from the general expression of energy (6.112), take the derivative with respect to temperature. Then we get

$$C_V = 3k_0 N \left[D\left(\theta/T\right) + T\frac{\partial}{\partial T} D\left(\theta/T\right) \right]. \tag{6.121}$$

If we take into account that the derivative of the Debye function with respect to temperature equals

$$\frac{\partial}{\partial T} D\left(\frac{\theta}{T}\right) = \frac{3}{T} D\left(\frac{\theta}{T}\right) - \frac{3\theta}{T^2}\left(e^{\theta/T} - 1\right)^{-1}, \tag{6.122}$$

for the heat capacity at an arbitrary temperature we get:

$$C_V = 3k_0 N L_V\left(\frac{\theta}{T}\right), \tag{6.123}$$

where the function

$$L_V\left(\theta/T\right) = 4D\left(\frac{\theta}{T}\right) - 3\left(\frac{\theta}{T}\right)\left(e^{\theta/T} - 1\right)^{-1} \tag{6.124}$$

In the general form $C_V(T)$ determines the temperature dependence of the heat capacity.

If we integrate the integral entering into the Debye function (6.113) once in parts and substitute the result into (6.124), then the function $L_V(\theta/T)$ can be presented in the form:

$$L_V(\theta/T) = 3\left(\frac{T}{\theta}\right)^3 \int\limits_0^{\theta/T} \frac{x^4 e^x}{(e^x - 1)^2} dx. \tag{6.125}$$

Inasmuch as the function $L_V(\theta/T)$ at any temperature does not have an analytical form, consider its asymptotic behavior.

High temperatures: $T \gg \theta$. In this case, in (6.125) the integration variable is $x \ll 1$ and the integrand expression can be simplified. To do this on the basis of expansion (6.114) it can be written as:

$$\frac{1}{(e^x - 1)^2} \approx \frac{1}{x^2}\left(1 - x + \frac{5x^2}{12}\right); x \ll 1 \tag{6.126}$$

and, conformably,

$$\frac{e^x}{(e^x-1)^2} \approx \frac{1}{x^2}\left(1-x+\frac{5x^2}{12}\right)\left(1+x+\frac{x^2}{2}\right) \approx \frac{1}{x^2}\left(1-\frac{x^2}{12}\right); \quad x << 1. \tag{6.127}$$

Now we can easily calculate the integral (6.125) with accuracy to $(\theta/T)^2$. As a result, we get

$$L_V(\theta/T) = 1 - \frac{1}{20}\left(\frac{\theta}{T}\right)^2 ; T \gg \theta. \tag{6.128}$$

Substituting this expression into (6.123), for the heat capacity in the region of high temperatures we have

$$C_V = 3k_0N\left[1 - \frac{1}{20}\left(\frac{\theta}{T}\right)^2\right] ; T \gg \theta. \tag{6.129}$$

Note that quantization of vibrations of a crystal decreases the heat capacity and, as is seen from (6.129), contribution of the quantization to the heat capacity is negative

$$\Delta C_V = C_V^q - C_V^c = -\frac{3}{20}\left(\frac{\theta}{T}\right)^2 k_0N. \tag{6.130}$$

Low temperatures: $T \ll \theta$. In this case, the upper limit of the integral (6.125) can be replaced with infinity. Then, the integral obtained, according to Appendix A, gives

$$\int_0^\infty \frac{x^4 e^x}{(e^x-1)^2}dx = \frac{4\pi^4}{15} \tag{6.131}$$

and as a result we get

$$L_V(\theta/T) = \frac{4\pi^4}{5}\left(\frac{T}{\theta}\right)^3 ; T \ll \theta. \tag{6.132}$$

Consideration of this asymptote of the $L_V(\theta/T)$-function in (6.123) for the heat capacity in the region of low temperatures gives:

$$C_V = \frac{12\pi^4}{5}k_0N\left(\frac{T}{\theta}\right)^3 ; T \ll \theta. \tag{6.133}$$

The dependence $C_V \sim T^3$ bears the name *the Debye law*. As is seen, according to the Debye theory, in contradistinction to exponential dependence (6.106) obtained by Einstein, at very low temperatures ($T << \theta$) the heat capacity tends to zero as $C_V \sim T^3$, which is well consistent with experiment. According to (6.128) and (6.132), the function $L_V(\theta/T)$ determining the heat capacity in all regions of temperature, as well as the Debye function, changes in the limits

0–1; therewith in the region of absolute zero both these functions behave as T^3 (see Fig. 6.10).

From the expression of the heat capacity (6.123) it is seen that only one parameter – the Debye temperature θ – enters into the function $C_V(T)$. This parameter characterizes the nature of a solid, inasmuch as is determined by the speed of sound v_0 in a solid and the lattice constant $V/N = a$ [see (6.21) and (6.79)].

The microscopic sense of the Debye temperature follows from the formula (6.70). Simultaneously note that θ, as a macroscopic parameter, determines boundaries of the classical and the quantum theory: as $T \gg \theta$ the vibrational motion of atoms at crystalline lattice points is classical, and as $T \leq \theta$ it bears the quantum character.

Measuring the temperature dependence of the heat capacity $C_V(\theta/T)$ and using (6.123), the Debye temperature can be experimentally determined. In Table 6.1 theoretically and experimentally found values of θ for some solids are adduced.

As is seen from the table, theoretical and experimental values are close to each other and are on the order of $(0.7\text{–}4.1)10^2$ K. Of all known crystals, the crystal of the diamond is an exception, for which $\theta = 1{,}850$ K.

We will briefly dwell upon the comparison of the Debye theory with experiment. Suppose that the Debye temperature in all temperature regions $\theta = \hbar\omega_{\max}/k_0 = \text{const}$, and on the basis of the Debye theory (6.123) calculate the temperature dependence $C_V(T)$, and thereupon compare it with experiment; we see that they do not always coincide. In particular, the law $C_V(T) \sim T^3$ is fulfilled only at temperature $T < 5$ K. At temperature $T > 10$ K results of the Debye theory and experimental data are distinct from each other. In order to circumvent this contradistinction they suppose that the Debye parameter θ itself depends on temperature: $\theta = \theta(T)$. This means that for the function $\theta(T)$, entering into $L_V(\theta/T)$, it is necessary to pre-assign such a dependence in order that the Debye theory and experiment would coincide. The weak dependence of $\theta(T)$ is schematically presented in Fig. 6.11.

The true cause of the contradistinction of the Debye theory from experiment resides in the fact that the Debye theory has used the simplest dispersion law $\omega(q) = v_0 q$ (Fig. 6.5) and hence stems the function of the density of frequencies $g(\omega) \sim \omega^2$ (Fig. 6.6). For real crystals, however, functions $\omega(q)$ and $g(\omega)$ have a very complex form.

Table 6.1. The theoretical and experimental values of the Debye temperature for some solids

Substance	Al	Cu	Ag	Au	Pb	NaCl
θ (K) experiment	410	310	220	185	88	275
θ (K) theory	394	342	212	158	73	302

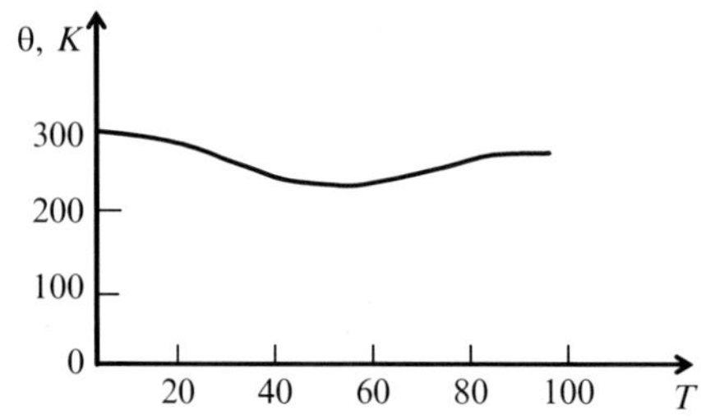

Fig. 6.11. The temperature dependence of the Debye temperature

In conclusion, note that having determined the Debye temperature experimentally for the given crystal, according to (6.79), we can calculate the speed of sound v_0, and thereupon from (6.21) determine the coefficient of elasticity β.

6.5 Quantum Theory of Thermodynamic Properties of Solids

In Sect. 6.4 the quantum theory of the heat capacity of a solid proposed by Einstein and perfected by Debye was expounded. It was constructed on the basis of Einstein's supposition, which made it possible to calculate energy of a solid and, consequently, the heat capacity C_V in an arbitrary temperature region. However, this theory did not allow the consideration of other thermodynamic properties such as the thermal equation of state, difference of heat capacities $(C_P - C_V)$, coefficient of thermal expansion, etc. To solve these questions it is necessary to construct a consecutive theory, using quantum mechanics and the Gibbs method. This method enables the immediate calculation of the free energy, finding the equation of state and considering the thermodynamic properties of a solid.

In Sect. 6.3 on the basis of the Gibbs method the classical theory of the thermodynamic properties of a solid in the high-temperature region $(T \gg \theta)$, in which thermal vibrational motions of a lattice are classical, was developed.

In this section we dwell upon the quantum theory of thermodynamic properties, based on the Gibbs method.

On the basis of the quantum theory there lies the spectrum of frequencies of the vibrational motion of a crystalline lattice (Fig. 6.4). Suppose that a simple crystalline lattice consists of N elementary points. In Sect. 6.2 we showed that energy (the Hamilton function) of the vibrational motion of such a crystal equals the sum of energies of $3N$ harmonic oscillators (6.78).

Knowing the Hamilton function and using the correspondence principle, the Hamilton operator of the given system can be found. To do this, we need to replace the impulse entering into the Hamilton function with the corresponding operator. If in the Hamilton function for a crystalline lattice (6.37) the generalized impulse is replaced with the operator

$$\hat{\boldsymbol{P}}_{\boldsymbol{q}j} \rightarrow \frac{\hbar}{i} \frac{\partial}{\partial X_{\boldsymbol{q}j}}, \tag{6.134}$$

the Hamilton operator takes the form:

$$\hat{\mathscr{H}} = \sum_{\boldsymbol{q}} \sum_{j=1}^{3} \left(-\frac{\hbar^2}{2M} \frac{\partial^2}{\partial X_{\boldsymbol{q}j}^2} + \frac{M}{2} \omega_{\boldsymbol{q}j}^2 X_{\boldsymbol{q}j}^2 \right), \tag{6.135}$$

where $\hbar = h/2\pi$, and h is the Planck constant.

Hence it is seen that the Hamilton operator (6.135) can also be presented in the form

$$\hat{\mathscr{H}} = \sum_{\boldsymbol{q}} \sum_{j=1}^{3} \hat{\mathscr{H}}_{\boldsymbol{q}j}, \tag{6.136}$$

where

$$\hat{\mathscr{H}}_{qj} = -\frac{\hbar^2}{2M} \frac{\partial^2}{\partial X_{\boldsymbol{q}j}^2} + \frac{M}{2} \omega_{\boldsymbol{q}j}^2 X_{\boldsymbol{q}j}^2 \tag{6.137}$$

is the Hamilton operator of a harmonic oscillator of the $(\boldsymbol{q}, j)th$ type with frequency $\omega_{\boldsymbol{q}j}$. It is known that the eigenvalues of the operator (6.136), i.e. the energy spectrum of a crystal, are

$$E_{(n_{qj})} = \sum_{\boldsymbol{q}} \sum_{j=1}^{3} \varepsilon_{n_{\boldsymbol{q}j}}, \tag{6.138}$$

where

$$\varepsilon_{n_{qj}} = (n_{\boldsymbol{q}j} + 1/2)\hbar\omega_{\boldsymbol{q}j} \tag{6.139}$$

is the energy spectrum of a harmonic oscillator with the frequency $\omega_{\boldsymbol{q}j}$, i.e. eigenvalues of the Hamilton operator (6.137).

The symbol $(n_{\boldsymbol{q}j})$ shows the totality of quantum numbers of $3N$ number of oscillators

$$(n_{\boldsymbol{q}j}) \equiv (n_{11}, n_{12}, n_{13}; n_{21}, n_{22}, n_{23}, \ldots; n_{N1}, n_{N2}, n_{N3}), \tag{6.140}$$

where $n_{\boldsymbol{q}j} = 0, 1, 2, \ldots$ is the oscillatory quantum number of the pre-assigned oscillator of the $(\boldsymbol{q}, j)th$ type.

As is seen from (6.138) and (6.140), each quantum state of a solid is determined by the totality of quantum numbers of $3N$ oscillators.

1. *Free energy.* Knowing the energy spectrum (6.138), the free energy of a crystal can be calculated:

$$F = -k_0 T \ln Z, \tag{6.141}$$

where

$$Z = \sum_{(n_{\boldsymbol{q}j})} \exp\left(-\frac{1}{k_0 T} \sum_{\boldsymbol{q}} \sum_{j=1}^{3} \varepsilon_{n_{\boldsymbol{q}j}} \right) \tag{6.142}$$

is the statistical sum. If (6.139) is substituted into (6.142), the statistical sum takes the form:

$$Z = \sum_{(n_{\boldsymbol{q}j})} \exp\Big(-\frac{1}{k_0 T} \sum_{\boldsymbol{q}} \sum_{j=1}^{3} (n_{\boldsymbol{q}j} + 1/2)\hbar\omega_{\boldsymbol{q}j} \Big). \tag{6.143}$$

If we add up with respect to oscillatory quantum number $n_{\boldsymbol{q}j} = 0, 1, 2, \ldots$ and take into account that

$$\sum_{n_{qj}=0}^{\infty} \exp\left(-\frac{1}{k_0 T}(n_{\boldsymbol{q}j} + 1/2)\hbar\omega_{\boldsymbol{q}j} \right) = \frac{\mathrm{e}^{-\hbar\omega_{qj}/2k_0 T}}{1 - \mathrm{e}^{-\hbar\omega_{qj}/k_0 T}}, \tag{6.144}$$

for the statistical sum we get

$$Z = \prod_{\boldsymbol{q}} \prod_{j=1}^{3} \frac{\mathrm{e}^{-\hbar\omega_{qj}/2k_0 T}}{1 - \mathrm{e}^{-\hbar\omega_{qj}/k_0 T}}. \tag{6.145}$$

Thus, from (6.145) and (6.141) we get the general expression for the free energy as:

$$F = E_0 + k_0 T \sum_{q} \sum_{j=1}^{3} \ln(1 - \mathrm{e}^{-\hbar\omega_{qj}/k_0 T}), \tag{6.146}$$

where

$$E_0 = \sum_{\boldsymbol{q}} \sum_{j=1}^{3} \frac{\hbar\omega_{\boldsymbol{q}j}}{2} \tag{6.147}$$

is the energy of zero vibrations.

For further calculations of free energy suppose that a crystal is isotropic and all three vibrational branches, shown in Fig. 6.4, coincide: $\omega_1(\boldsymbol{q}) = \omega_2(\boldsymbol{q}) = \omega_3(\boldsymbol{q}) = \omega(\boldsymbol{q})$. Then the expression for free energy (6.146) takes the form:

$$F = E_0 + 3k_0 T \sum_{\boldsymbol{q}} \ln(1 - \mathrm{e}^{-\hbar\omega_q/k_0 T}), \tag{6.148}$$

where the zero energy in conformity with (6.147) equals

$$E_0 = \frac{3}{2} \sum_{\boldsymbol{q}} \hbar\omega(\boldsymbol{q}). \tag{6.149}$$

If, according to (6.36), we pass from the summation with respect to the quasi-discrete variable $\boldsymbol{q}$ to the integral, we get

$$F = E_0 + \frac{3k_0 T V}{(2\pi)^3} \int \ln(1 - \mathrm{e}^{-\hbar\omega_q/k_0 T}) \mathrm{d}\boldsymbol{q}. \tag{6.150}$$

In order to bring computations to the end, we use the continuum approximation $\omega(\boldsymbol{q}) = \omega(q) = v_0 q$ and the Debye model (Fig. 6.5, $0 \le q \le q_{\max}$ $0 \le \omega(q) \le \omega_{\max}$). As a result, we have

$$F = E_0 + \frac{3k_0TV}{2\pi^2 v_0^3} \int_0^{\omega_{\max}} \ln(1 - e^{-\hbar\omega_q/k_0T})\omega^2 d\omega \tag{6.151}$$

and

$$E_0 = \frac{3V\hbar}{4\pi^2 v_0^3} \int_0^{\omega_{\max}} \omega^3 d\omega. \tag{6.152}$$

Introduce the dimensionless parameter $x = \hbar\omega/k_0T$ and make use of the definition of the Debye temperature $\theta = \hbar\omega_{\max}/k_0$. Then for the free energy we get:

$$F = E_0 + 9Nk_0T \left(\frac{T}{\theta}\right)^3 \int_0^{\theta/T} \ln(1 - e^{-x})x^2 dx, \tag{6.153}$$

and for the zero energy it is:

$$E_0 = \frac{9}{8} Nk_0\theta. \tag{6.154}$$

Here we take into account expressions (6.78) and (6.79) for $\omega_{\max}$ and the Debye temperature θ, conformably.

The integral entering into expression of the free energy (6.153) integrates up by parts. As a result, we have

$$F = E_0 + 3Nk_0T \ln(1 - e^{-\theta/T}) - 3Nk_0T \left(\frac{T}{\theta}\right)^3 \int_0^{\theta/T} \frac{x^3 dx}{e^x - 1}. \tag{6.155}$$

If we use the definition of the Debye function (6.113), (6.155) takes the form:

$$F = E_0 + 3Nk_0T \ln(1 - e^{-\theta/T}) - Nk_0TD\left(\frac{\theta}{T}\right). \tag{6.156}$$

In the particular case, if we take into account the asymptote of the Debye function (6.115) in the region of high temperatures ($T \gg \theta$), for the free energy we get a classical result (6.80).

Knowing the explicit form of the expression of free energy (6.156), in the general form basic thermodynamic parameters of a solid with regard to quantization of the vibrational motion of atoms at lattice points can be calculated.

2. *Entropy.* From the definition $S = -(\partial F/\partial T)_V$ and (6.156) for entropy we get the expression

$$S = -3Nk_0 \ln(1 - e^{-\theta/T}) + 4Nk_0D\left(\frac{\theta}{T}\right), \tag{6.157}$$

from which at the high temperature ($T \gg \theta$) it follows the known classical result (6.81). A new result is obtained in the low-temperature region $T \ll \theta$. If in this case we take into account the asymptote of the Debye function (6.119), entropy takes the form

$$S = \frac{4\pi^4}{5} k_0 N \left(\frac{T}{\theta}\right)^3 ; T \ll \theta. \tag{6.158}$$

3. *Mean energy.* From the relationship $E = F + TS$, and also from the expressions (6.156) and (6.157) in the Debye approximation for the mean energy we get
$$E = \frac{9}{8} N k_0 \theta + 3k_0 NTD\left(\frac{\theta}{T}\right). \tag{6.159}$$
4. *Isochoric heat capacity.* Using the definition $C_V = (\partial E/\partial T)_V$ and the expression (6.159) for the heat capacity we have
$$C_V = 3k_0 N\left[D\left(\frac{\theta}{T}\right) + T\frac{\partial}{\partial T}D\left(\frac{\theta}{T}\right)\right] = 3k_0 N L_V\left(\frac{\theta}{T}\right), \tag{6.160}$$
where the function $L_V(\theta/T)$ is determined by expressions (6.124) or (6.125).
5. *Thermal equation of state.* Taking into account the dependence $\theta = \theta(V)$ and using the expression of free energy (6.156), for the thermal equation of state $P = -(\partial F/\partial V)_T$ we get
$$P = P_0 + \frac{3k_0 NT}{V} D\left(\frac{\theta}{T}\right)\gamma_G. \tag{6.161}$$

Here
$$\left(\frac{\partial D}{\partial \theta}\right)_T = -\frac{3}{\theta}D\left(\frac{\theta}{T}\right) + \frac{3}{T}(e^{\theta/T-1})^{-1} \tag{6.162}$$
and
$$P_0 = -\frac{\partial E_0}{\partial V} = \frac{9}{8}\frac{Nk_0\theta}{V}\gamma_G, \tag{6.163}$$
were taken into account where P_0 is the zero pressure, associated with zero vibrations of the lattice and not depending on temperature. This is purely a quantum effect; in the classical case, such a concept does not exist.

Finally the equation of state (6.161) in an arbitrary temperature region takes the form

$$P(V,T) = \frac{9}{8}\frac{Nk_0\theta}{V}\gamma_G\left[1 + \frac{8}{3}\frac{T}{\theta}D\left(\frac{\theta}{T}\right)\right]. \tag{6.164}$$

Using the asymptotes of the Debye function, the thermal equation of the state of a solid (6.164) can be written for different temperature regions as:
at high temperatures ($T \gg \theta$)

$$P(V,T) = \frac{3k_0NT}{V}\gamma_G \tag{6.165}$$

and at *low temperatures* $(T \ll \theta)$

$$P(V,T) = \frac{9}{8}\frac{k_0N\theta}{V}\gamma_G\left[1 + \frac{8\pi^4}{15}\left(\frac{T}{\theta}\right)^4\right]. \tag{6.166}$$

6. *Isobaric heat capacity.* Knowing the thermal equation of the state of a solid (6.161), from relationship (6.96) for the isobaric heat capacity we get

$$C_P = C_V - T\left(\frac{\partial P}{\partial T}\right)^2\left(\frac{\partial P}{\partial V}\right)^{-1}, \tag{6.167}$$

where C_V is determined by formula (6.160). If we use relationship (6.122) for the derivatives of the Debye function with respect to temperature $(\partial/\partial T)D(\theta/T)$, and also (6.162), the isobaric heat capacity can be written down as

$$C_P = 3k_0NL_P\left(\frac{\theta}{T}\right), \tag{6.168}$$

where the functions

$$L_P(\theta/T) = L_V(\theta/T) + \gamma_G M_1(\theta/T, \gamma_G) \tag{6.169}$$

and

$$M_1\left(\frac{\theta}{T}, \gamma_G\right) = \frac{L_V^2\left(\frac{\theta}{T}\right)}{\left[\frac{3\theta}{8T} + D\left(\frac{\theta}{T}\right)\right](1+\gamma_G) - \gamma_G L_V\left(\frac{\theta}{T}\right)}. \tag{6.170}$$

are introduced.

As a result, for the difference of heat capacities $(C_P - C_V)$ and their ratio C_P/C_V we get, conformably,

$$C_P - C_V = 3k_0N\gamma_G M_1\left(\frac{\theta}{T}, \gamma_G\right) \tag{6.171}$$

and

$$\frac{C_P}{C_V} = 1 + \gamma_G M_2\left(\frac{\theta}{T}, \gamma_G\right), \tag{6.172}$$

where

$$M_2\left(\frac{\theta}{T}, \gamma_G\right) = \frac{L_V\left(\frac{\theta}{T}\right)}{\left[\frac{3\theta}{8T} + D\left(\frac{\theta}{T}\right)\right](1+\gamma_G) - \gamma_G L_V\left(\frac{\theta}{T}\right)}. \tag{6.173}$$

We adduce asymptotes of functions entering into expressions of heat capacities. Asymptotes of the $L_V(\theta/T)$- function, determining the isochoric heat capacity C_V, are given by formulae (6.84) and (6.88).

For other functions we have:
in *the region of high temperatures* $T >> \theta$

$$M_1\left(\frac{\theta}{T},\gamma_{\mathrm{G}}\right)=1-\frac{1}{20}(3+2\gamma_{\mathrm{G}})\left(\frac{\theta}{T}\right)^2, \tag{6.174}$$

$$M_2\left(\frac{\theta}{T},\gamma_{\mathrm{G}}\right)=1-\frac{1}{10}(1+\gamma_{\mathrm{G}})\left(\frac{\theta}{T}\right)^2. \tag{6.175}$$

and in *the region of low temperatures* $T \ll \theta$

$$M_1\left(\frac{\theta}{T},\gamma_{\mathrm{G}}\right)=\frac{8}{3(1+\gamma_{\mathrm{G}})}\left(\frac{4\pi^4}{5}\right)^2\left(\frac{T}{\theta}\right)^7, \tag{6.176}$$

$$M_2\left(\frac{\theta}{T},\gamma_{\mathrm{G}}\right)=\frac{32\pi^4}{15(1+\gamma_{\mathrm{G}})}\left(\frac{T}{\theta}\right)^4. \tag{6.177}$$

Inasmuch as in the low-temperature region $M_1 \sim T^7$ and $M_2 \sim T^4$, from (6.171) and (6.172) it is seen that in the limiting case $T \to 0$ isobaric and isochoric heat capacities of a solid coincide: $C_P \approx C_V$.

From the asymptotes adduced above, it follows that as $T \to 0$ isobaric and isochoric heat capacities equally tend to zero as $\sim T^3$, and their difference, according to the Nernst principle, tends to zero more strongly: $(C_P - C_V) \sim T^7$.

Temperature dependences of isochoric C_V and isobaric C_P heat capacities are adduced in Fig. 6.12 for clarity. And temperature dependences of their difference $(C_P - C_V)$ and relation C_P/C_V are presented in Figs. 6.13 and 6.14, conformably.

7. *Isobaric thermal expansion.* This effect in the classical case in the region of high temperatures ($T \gg \theta$)) was considered in Sect. 6.3, where the known Grüneisen relationship (6.101) was obtained. It can be shown that the Grüneisen relationship (6.101) takes place in the general case, too. Using expressions (6.99) and (6.161), for α_P we get

$$\alpha_P=\frac{3k_0N\gamma_T}{V}\gamma_{\mathrm{G}}L_V\left(\frac{\theta}{T}\right)=\frac{\gamma_T}{V}\gamma_{\mathrm{G}}C_V. \tag{6.178}$$

This expression is distinct from (6.100) by the fact that in the classical case $C_V = 3k_0N = 3R$, and in the general case C_V is determined by formula (6.160). From (6.178) the Grüneisen relationship follows

$$V\frac{\alpha_P}{C_V}=\gamma_T\gamma_{\mathrm{G}}. \tag{6.179}$$

Hence it is seen that, inasmuch as the isothermal compressibility γ_T and the Grüneisen constant γ_{G} do not depend on temperature, the ratio $\alpha_P/C_V =$ const.

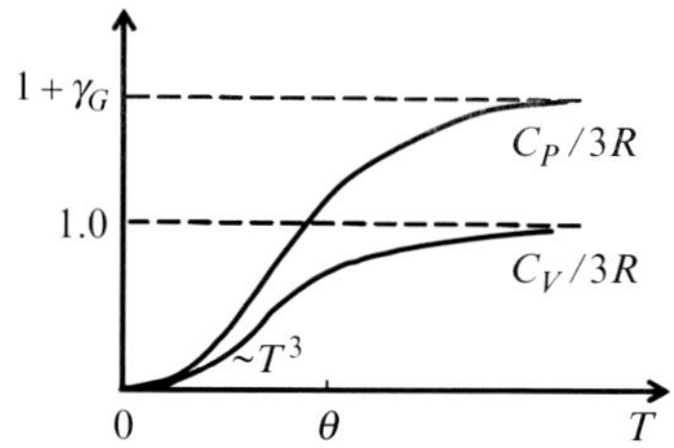

Fig. 6.12. The temperature dependences of isochoric C_V and isobaric C_P heat capacities

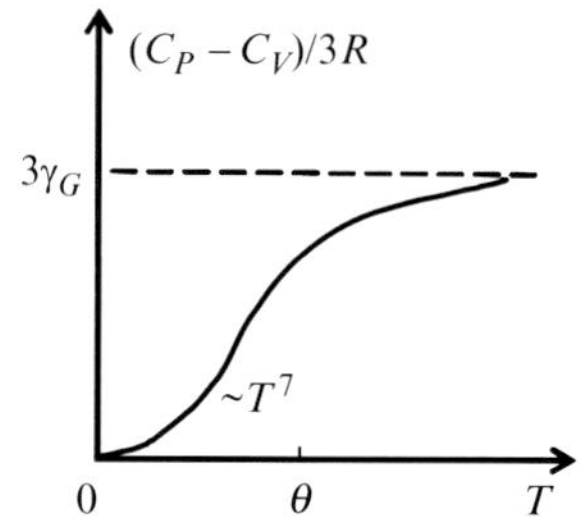

Fig. 6.13. The temperature dependences of difference $(C_P - C_V)$

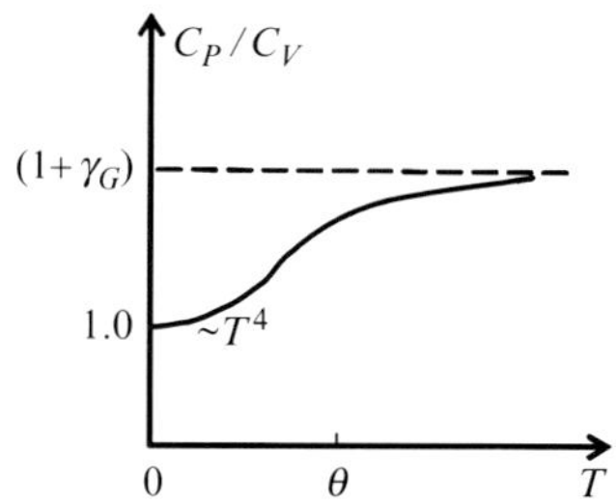

Fig. 6.14. The temperature dependences of relation C_P/C_V

The relationship between thermodynamic coefficients (6.179) has the general character and is true for any temperature. As noted in Sect. 6.3, this was experimentally corroborated by Grüneisen.

From the relationships adduced above, it follows that pressure (the thermal equation of state), the isobaric heat capacity and the coefficient of thermal expansion are proportional to the Grüneisen constant γ_G and, consequently, P, C_P, α_P in a solid are distinct from zero only with consideration of anharmonism of vibrations of atoms at lattice points.

In conclusion, note that in terms of the Debye theory, thermodynamic properties of a solid are determined by the four functions adduced above:

$D(\theta/T)$ – the Debye function determining entropy (6.157), mean energy (6.159) and thermal equation of state (6.161) of a solid;

$L_V(\theta/T)$ – the function determining isochoric heat capacity (6.160) and isobaric coefficient of thermal expansion (6.178);

$M_1(\theta/T, \gamma_G)$ – the function determining the difference $(C_P - C_V)$ of isobaric and isochoric heat capacities (6.171);

$M_2(\theta/T, \gamma_G)$ – the function determining the ratio C_P/C_V of isobaric and isochoric heat capacities (6.172).

All these four functions have one common property. As temperature changes in the range $0 < T < \infty$ each of them changes in the limits from zero to unity:

$$D\left(\frac{\theta}{T}\right) = L_V\left(\frac{\theta}{T}\right) = M_1\left(\frac{\theta}{T}, \gamma_G\right) = M_2\left(\frac{\theta}{T}, \gamma_G\right) = \begin{cases} 0; T \to 0 \\ 1; T \to \infty. \end{cases} \tag{6.180}$$

Notice that the isobaric heat capacity C_P is determined by the function $L_P(\theta/T, \gamma_G)$ [see (6.168)], which changes in the limits from 0 to $(1 + \gamma_G)$:

$$L_P\left(\frac{\theta}{T}, \gamma_G\right) = \begin{cases} 0; \quad T \to 0 \\ 1 + \gamma_G; T \to \infty. \end{cases} \tag{6.181}$$

7

Quantum Statistics: Equilibrium Electron Gas

Summary. It is known that statistical physics studies two kinds of macroscopic systems: *classical and quantum.* Systems in which the motion of particles that constitute them is described by the classical (newtonian) equation of motion and whose state as a whole is described by the Hamilton function are called *classical systems.* The condition of classicity of the motion has the appearance

$$s \gg h, \quad \text{(condition 1)}$$

where h is the planck constant, $s = mvL$ is the action of motion, m is the mass of a particle, v is the velocity of motion, and L are linear dimensions of the space in which the motion occurs. The condition of classicity of the motion (see Sect. 1.2) can also be presented in the form

$$L \gg \lambda \text{ or } L \gg h / \sqrt{mk_0T}, \qquad \text{(condition 2)}$$

where $\lambda = h / mv$ is the de Broglie wavelength of a particle, and k_0 is the Boltzmann constant: $mv^2 \approx k_0T$.

If the above conditions are not fulfilled, the motion is described by the Schrödinger equation and the energy spectrum of a particle becomes discrete, i.e. the motion is quantized. Systems consisting of such particles are called *quantum systems.*

In the preceding chapters, on the basis of the Gibbs canonical distribution, we considered both the classical and the quantum state of some systems and constructed the classical and quantum theories of thermodynamic properties, conformably. For instance, we became acquainted with the classical and quantum theories of the heat capacity of a diatomic ideal gas and a solid. Note that in both cases the statistics were classical; only in one case it was needed to calculate the statistical integral, and in the other it was the statistical sum.

This chapter is devoted to quantum statistics (Fermi–Dirac and Bose–Einstein distribution functions) and the theory of thermodynamic properties of quantum ideal gases. What does the quantum statistics represent?

In order to answer this question, consider the following problem. Imagine that in a vessel of volume V, N particles with identical masses and spins are found: for instance, a free electron gas in a metal or a photon gas in a closed vessel.

To such systems we apply *the quantum statistics*, taking into account *the principle of indistinguishability of particles* (see Sect. 7.2).

Classical statistics is applicable only in the case in which *the distance d between the particles of a system is much more than the de Broglie wavelength* λ (see Sect. 7.1):

$$d \gg \lambda \text{ or } n^{-1/3} \gg h / \sqrt{2mk_0T}, \quad \text{(condition 3)}$$

where m and n are the mass and concentration of particles in a system, respectively, and T is the absolute temperature.

Note that quantum theory (the theory of quantum systems) should not be confused with quantum statistics. From a comparison of conditions of classicity of the motion (condition 2) and classicity of statistics (condition 3), it is seen that cases where the former is satisfied, the latter can be violated because $L \gg d$. *Thus, the motion of particles in a system can be classical, whereas their statistics can be quantum and vice versa.*

Quantum statistics can be applied both to classical and quantum systems. Quantum statistics is more general. In the particular case where condition (3) is satisfied, results of quantum statistics pass onto the corresponding results of classical statistics.

Before moving on to an exposition of quantum statistics, we consider the Boltzmann distribution which forms the basis of classical statistics and expound the basic difficulties of classical statistics.

7.1 Boltzmann Distribution: Difficulties of Classical Statistics

The distribution of velocities of the molecules of an ideal gas was given for the first time by the famous Scottish physicist Maxwell in 1859. In 1871, the outstanding Austrian physicist Boltzmann generalized the Maxwell distribution to the case in which a gas is found in an external potential field (the gravitational field of the earth) and obtained the well-known barometric formula.

The Boltzmann distribution function was found to satisfy classical systems, i.e. for the case when the energy of a particle is determined by its impulse and coordinates, and is equal to the sum of kinetic and potential energies. Here, we assume that the state of each particle forming an ideal gas, i.e. its energy, is determined by a set of quantum numbers, but the statistics is classical.

The statement of the problem is as follows: assume that in volume V there are N particles. Let k be the totality of quantum numbers determining the state of one particle, ε_k be its energy and n_k be the number of particles in this state. It is required to find the mean number of particles $\bar{n}_k$ in the state k. In order that the exchange interaction be absent, the gas ought to be rarefied:

$$\bar{n}_k \ll 1. \tag{7.1}$$

When solving the stated problem, we use the Gibbs grand canonical distribution for open systems. If we take the quantum state k with energy ε_k as an open subsystem, and the remaining part of the gas to function as a thermostat, the Gibbs grand canonical distribution can be presented in the form

$$W_{n_k} = \mathrm{e}^{\frac{\Omega_k + \mu n_k - \varepsilon_k n_k}{k_0 T}}, \tag{7.2}$$

where Ω_k is the grand canonical potential of the considered subsystem, and W_{n_k} is the probability that in the quantum state k, i.e. in the subsystem, there are n_k particles. Then, $W_0 = \exp(\Omega_k / k_0 T) \approx 1$ is the probability that the state k is empty, because according to condition (7.1) the gas is very rarefied and the number of quantum states is far more than the number of particles. It is clear that the probability of one particle being found in the k -state ought to be

$$W_1 = \mathrm{e}^{\frac{\Omega_k + (\mu - \varepsilon_k)}{k_0 T}} \approx \mathrm{e}^{\frac{\mu - \varepsilon_k}{k_0 T}} \ll 1, \tag{7.3}$$

where it is taken into account that $\exp(\Omega_k / k_0 T) = 1$. It is known that Ω_k entering into distribution (7.2) is determined from the normalization condition $\sum\limits_{n_k} W_{n_k} = 1$. Thus,

$$\Omega_k = -k_0 T \ln \sum_{n_k} \mathrm{e}^{\frac{(\mu - \varepsilon_k) n_k}{k_0 T}}. \tag{7.4}$$

If we take into account condition (7.3) and in the sum (7.4) keep only two terms ($n_k = 0; 1$), we get

$$\Omega_k = -k_0 T \ln \left(1 + \mathrm{e}^{\frac{\mu - \varepsilon_k}{k_0 T}}\right). \tag{7.5}$$

Because the second term in (7.5) is very small, at small x we can use the approximation $\ln(1 + x) \approx x$. Then

$$\Omega_k = -k_0 T \cdot \mathrm{e}^{\frac{\mu - \varepsilon_k}{k_0 T}}. \tag{7.6}$$

As a result, the mean number of particles in the k -state equals

$$\bar{n}_k = -\left(\frac{\partial \Omega_k}{\partial \mu}\right)_T = \mathrm{e}^{\frac{\mu - \varepsilon_k}{k_0 T}}. \tag{7.7}$$

Denote the mean number of particles in the k -state by $\bar{n}_k \equiv f_k$. Then we get

$$f_k = \mathrm{e}^{\frac{\mu - \varepsilon_k}{k_0 T}}. \tag{7.8}$$

This distribution function in classical statistics bears the name *Boltzmann distribution*, and a gas obeying such a distribution is called *the classical ideal gas or the Boltzmann gas.*

The grand thermodynamic potential of the complete system has the appearance

$$\Omega_{\text{cl}} = \sum_k \Omega_k = -k_0 T \sum_k \mathrm{e}^{\frac{\mu-\varepsilon_k}{k_0 T}}. \tag{7.9}$$

In order to find the explicit form of the thermodynamic potential, it is necessary to know the dependence of energy ε_k on the wave vector $\boldsymbol{k}$. For simplicity, the energy of a free particle is presented in the form

$$\varepsilon_k = \frac{\hbar^2 k^2}{2m}, \tag{7.10}$$

where m is the mass of the particle.

In the quasi-classical case, in order to pass from the summation with respect to $\boldsymbol{k}$ to the integral, according to (6.36), we use the rule

$$\sum_k \varphi(k) \Rightarrow \frac{V g_0}{(2\pi)^3} \int \varphi(\boldsymbol{k}) \mathrm{d}\boldsymbol{k}, \tag{7.11}$$

where $\varphi(k)$ is an arbitrary smooth function, $g_0 = 2s+1$-fold degeneracy over the spin and s is the spin quantum number. Then we get

$$\Omega_{\text{cl}} = -\frac{k_0 T V g_0}{(2\pi)^3} \int \mathrm{e}^{\frac{\mu-\varepsilon_k}{k_0 T}} \mathrm{d}\boldsymbol{k}. \tag{7.12}$$

If we take into account that the energy of the particle depends only on the magnitude of the wave vector, in (7.12), changing the $\boldsymbol{k}$-space to the spherical coordinate system $\mathrm{d}\boldsymbol{k} = k^2 \mathrm{d}k \, \sin\theta \, \mathrm{d}\theta \, \mathrm{d}\varphi$ and integrating, we get

$$\Omega_{\text{cl}} = -\frac{k_0 T V g_0}{2\pi^2} \int \mathrm{e}^{\frac{\mu-\varepsilon_k}{k_0 T}} k^2 \mathrm{d}k. \tag{7.13}$$

Here, from the integral over $\mathrm{d}k$, according to (7.10), we can pass to the integral over $\mathrm{d}\varepsilon$:

$$\Omega_{\text{cl}} = -\frac{k_0 T V (2m)^{3/2} g_0}{4\pi^2 \hbar^3} \mathrm{e}^{\mu/k_0 T} \int_0^\infty \mathrm{e}^{-\varepsilon/k_0 T} \varepsilon^{1/2} \mathrm{d}\varepsilon, \tag{7.14}$$

where we have taken into account

$$k^2 \mathrm{d}k = \frac{(2m)^{3/2}}{2\hbar^3} \varepsilon^{1/2} \mathrm{d}\varepsilon. \tag{7.15}$$

Using the dimensionless variable of integration $x = \varepsilon / k_0 T$, and according to Appendix A, we get

$$\Omega_{\rm cl} = -\frac{g_0 V (k_0 T)^{5/2} (2m)^{3/2}}{8\pi^{3/2}\hbar^3} e^{\mu/\,k_0 T}. \tag{7.16}$$

Hence, we can find the total number of particles N and the pressure P of the system [see (2.206)]:

$$\begin{cases} P = -\left(\dfrac{\partial \Omega}{\partial V}\right)_{T,\mu} = \dfrac{g_0 (k_0 T)^{5/2} (2m)^{3/2}}{8\pi^{3/2}\hbar^3} e^{\mu/\,k_0 T}, \\ N = -\left(\dfrac{\partial \Omega}{\partial \mu}\right)_{V,T} = \dfrac{g_0 V (2m k_0 T)^{3/2}}{8\pi^{3/2}\hbar^3} e^{\mu/\,k_0 T}. \end{cases} \tag{7.17}$$

The solution obtained by combining (eliminating μ) these equations gives the equation of state of a classical ideal gas: $P = k_0 NT\,/\,V$.

Using the condition of classicity of statistics $\bar{n}_k \equiv f_k \ll 1$ (7.8), we can write

$$e^{\frac{\mu - \varepsilon_k}{k_0 T}} \ll 1. \tag{7.18}$$

In order that this condition be fulfilled at all values of ε_k, including for $\varepsilon_k = 0$, the chemical potential ought to satisfy the inequality

$$e^{\mu/\,k_0 T} \ll 1. \tag{7.19}$$

To determine the explicit form of condition of classicity of statistics (7.234) from the second equation of system (7.17), we find the chemical potential. Then, the condition of classicity of statistics takes the form

$$A_0 \equiv e^{\frac{\mu}{k_0 T}} = \frac{N}{V g_0} \frac{(2\pi\hbar^2)^{3/2}}{(m k_0 T)^{3/2}} \approx \left(\frac{\lambda}{d}\right)^3 \ll 1, \tag{7.20}$$

where $d = (N\,/\,V)^{-1/3} = n^{-1/3}$ is the mean distance between particles, and $\lambda = h\,/\,(2mk_0 T)^{1/2}$ is the de Broglie wavelength. This condition of classicity coincides with the condition (3) introduced at the beginning of this chapter.

A gas whose chemical potential satisfies (7.20) is called *a classical ideal gas*, *Boltzmann gas* or *non-degenerate gas*. To satisfy this condition, it is necessary that the chemical potential take on large negative values, i.e. $(-\mu\,/\,k_0 T) \gg 1$, and to do this the concentration of particles $n = N/V$ ought to be small, the mass of particles m large and the temperature T high.

With the aid of the Boltzmann distribution (7.8), we find the mean energy of a non-degenerate gas. To do this, we calculate the sum

$$\overline{E} = \sum_k \varepsilon_k f_k = \sum_k \varepsilon_k e^{\frac{\mu - \varepsilon_k}{k_0 T}}. \tag{7.21}$$

In order to pass from the summation to the integration, we use (7.11) and (7.15) and compare the obtained result with the first equation of system (7.17). As a result, we get the very simple relationship between energy E and pressure P:

$$P = \frac{2}{3}\frac{E}{V}. \tag{7.22}$$

Using expression (7.20) for the chemical potential μ from (7.21), we get the well-known expression for the mean energy of a classical ideal gas:

$$E = \frac{3}{2}k_0 TN. \tag{7.23}$$

Hence, it is seen that for each degree of freedom of a free particle, there accounts energy equal to

$$\frac{E}{3N} = \frac{1}{2}k_0 T. \tag{7.24}$$

Entropy of a classical ideal gas can be found using (7.16) and the relationship

$$S_{\rm cl}(V,T,\mu) = -\left(\frac{\partial \Omega_{kl}(V,T,\mu)}{\partial T}\right)_{V,\mu}. \tag{7.25}$$

As a result, we get

$$S_{\rm cl}(V,T,\mu) = \frac{5}{2}\frac{k_0 g_0 V(2mk_0T)^{3/2}}{8\pi^{3/2}\hbar^3}\left(1-\frac{2}{5}\frac{\mu}{k_0T}\right)\mathrm{e}^{\mu/\,k_0T}. \tag{7.26}$$

If we substitute expression (7.20) into (7.26), the dependence of entropy on V, T, N takes the form

$$S_{\rm cl}(V,T,N) = \frac{5}{2}k_0 N - k_0 N\ \ln\left[\frac{N}{V g_0}\left(\frac{2\pi\hbar^2}{mk_0T}\right)^{3/2}\right] \tag{7.27}$$

[compare with expression (4.10)]. Hence follows the well-known expression of the isochoric heat capacity: $C_V = 3k_0N\,/\,2$.

It can be shown that all atomic and molecular gases satisfy the condition of classicity (7.20). Indeed, at $N\,/\,V \approx 10^{19}\,\mathrm{cm}^{-3}$, $m \approx 10^{-24}\,\mathrm{g}$ and $T = 300\,\mathrm{K}$, from expression (7.20) it follows that $\exp(\mu\,/\,k_0T) \approx 10^{-4} \ll 1$. Thus, the Boltzmann statistics (the distribution) can be applied to all atomic and molecular gases. With the obtained theoretical results, the known experimental data are explained well.

However, there exist gases for which, when applying the Boltzmann distribution, non-explainable principal difficulties arise between theory and experiment. These difficulties mainly arise when applying the Boltzmann distribution to *a free electron gas in metals* and to *a photon gas.*

We recall these difficulties.

1. *The difficulty associated with the theory of the heat capacity of an electron gas in metals.*

To explain electrical and thermal conductivity of metals, the following model is assumed: A conducting metal consists of a crystal lattice formed by ions and electrons (an electron gas) moving freely in the metal. The electron gas can weakly interact with the vibrating lattice. The Drude theory of metals, which is based on this model, well explains the Wiedemann–Franz law and the temperature dependence of the specific resistivity of metals. However, the application of Boltzmann statistics when calculating the heat capacity of a free electron gas leads to nonconformity between the theory and the experiment. Indeed, according to this model, the energy of a metal is equal to the sum of energy of the vibratory motion of the crystalline lattice and of energy of the translational motion of free electrons:

$$E_{\text{met}} = E_{\text{lat}} + E_{\text{el}}, \tag{7.28}$$

and the heat capacity, consequently, is

$$C_V^{\text{met}} = C_V^{\text{lat}} + C_V^{\text{el}}. \tag{7.29}$$

For simplicity, consider a uni-valent metal. In this case, the number of ions and, consequently, free electrons is identical and equals N.

It is known that at temperatures higher than the Debye temperature ($T \gg \theta$), the vibratory motion of ions at the crystalline lattice points is classical (see Chap. 6) and the mean energy of the lattice $E_{\text{lat}} = 3k_0NT$. In order to calculate energy of the free translational motion of electrons, we apply the Boltzmann distribution and take into account that each degree of freedom corresponds to $k_0T\,/\,2$ energy (7.24). Then we get $E_{\text{el}} = 3k_0NT\,/\,2$. Thus, the total mean energy of a conducting crystal (a metal) equals

$$E_{\text{met}} = 3k_0NT + \frac{3}{2}k_0NT = \frac{9}{2}k_0\,TN, \tag{7.30}$$

and the heat capacity is

$$C_V^{\text{met}} = \frac{9}{2}k_0N = \frac{9}{2}R = 9\,\text{cal/ mol K}. \tag{7.31}$$

However, according to experimental data, at high temperatures in metals, as well as in dielectrics, the heat capacity $C_V^{\text{met}} \approx 3R \approx 6\,\text{cal/ mol K}$. Consequently, in the formation of the heat capacity of metals an electron gas almost does not play a role. Why? The answer to this question can be found in Sect. 7.9.

2. *The second difficulty of classical statistics is associated with the Pauli paramagnetism*, i.e. the contribution of an electron gas to the magnetic properties of a metal. It is known that a free electron possesses an intrinsic magnetic moment, equal to the Bohr magneton $\mu_{\text{B}} = e\hbar/2mc = 0.93 \times 10^{-20}\,\text{erg/G}$, which is associated with its spin $s = 1/2$. Therefore, in a magnetic field $\mathbf{H}$, at the expense of free electrons in the metal the paramagnetic property ought to arise. If classical statistics apply to an electron gas, when calculating the paramagnetic susceptibility χ we get the expression [see Sect. 7.10, (7.190)]:

$$\chi = \frac{M}{VH} = \frac{n\mu_{\mathrm{B}}^2}{k_0 T} \approx 10^{-4}\,\mathrm{erg/G^2\,cm^3}, \tag{7.32}$$

where, to evaluate it, we have taken $T = 300\,K$, and $n = 5 \times 10^{22}\,\mathrm{cm}^{-3}$. However, by experiment, the value of χ obtained was seen to be lower by two orders than that deduced in (7.32). At the same time, it was experimentally established that χ is almost independent of temperature, whereas according to classical statistics $\chi \sim 1/T$.

The question arises as to what causes such a strong divergence of results of classical statistics and experiment. The comprehensive answer to this question can be found in Sect. 7.10. Going ahead, suppose that the cause of the indicated divergence is because the electron gas does not behave like a classical gas. In order to be convinced of this, it is necessary to check the criterion of classicity (7.20) for an electron gas in metals. If we take the temperature of an electron gas as $T = 300\,\mathrm{K}$ and the concentration $n = 5 \times 10^{22}\,\mathrm{cm}^{-3}$, then according to (7.20) $\exp(\mu / k_0 T) \approx 10^2$, which contradicts the condition of classicity of the statistics, i.e. $\exp(\mu / k_0 T) \ll 1$. Consequently, *a free electron gas in metals is not classical and therefore the Boltzmann statistics cannot be applied to it.*

Such a situation in the 1920s led to the development of the new *quantum statistics.* During this time, quantum mechanics emerged (*the principle of indistinguishability of particles*) and based on that *the Fermi–Dirac quantum statistics* was formulated (see Sect. 7.3). In 1927, on the basis of quantum statistics Pauli developed the theory of the electron paramagnetism, and in 1928 Sommerfeld circumvented the nonconformity associated with the heat capacity of an electron gas in metals. They showed that an electron gas in metals is not classical but *a statistically degenerate quantum gas* (see Sects. 7.9 and 7.10).

3. *The difficulty associated with the application of the Boltzmann classical statistics to a photon gas.* It is well-known that in the rise of quantum physics the experimental study of the black-body radiation played a big role. In this field, many experimental results were accumulated and appropriate formulae were derived, such as the Wien, Rayleigh–Jeans, Stefan–Boltzmann laws, which in 1900 were explained by the well-known German physicist Max Planck. It is founded on the hypothesis that atoms on the walls of a hollow vessel act as oscillators with a specified frequency ω and irradiate electromagnetic energy inside, not continually but by portions. As a result, for the spectral density of energy of the black-body radiation Planck obtained the well-known formula

$$\rho(\omega, T) = \frac{\mathrm{d}E(\omega, T)}{V\,\mathrm{d}\omega} = \frac{\hbar\omega^3}{\pi^2 c^3} \frac{1}{\mathrm{e}^{\hbar\omega/k_0 T} - 1}, \tag{7.33}$$

where $E(\omega, T)$ is the total energy of radiation inside the vessel of volume V, ω is the frequency of the radiation and T is temperature of the walls of the vessel.

The newly introduced Planck hypothesis, in conformity with (7.33), successfully explained all experimental characteristics of black-body radiation from formula (7.33), particular cases of which are the Rayleigh–Jeans and Wien laws.

In 1905, Einstein put forward the following very bold hypothesis: light, i.e. the electromagnetic field, of a particular frequency ω consists of particles – *photons* – with energy and impulse:

$$\begin{aligned} \varepsilon &= \hbar\omega, \\ p &= \hbar\omega/c, \end{aligned} \tag{7.34}$$

where $\hbar = h/2\pi$; h is the Planck constant, and ω and c are the frequency and speed of light, respectively.

Consequently, an electromagnetic field filling up the hollow region of a vessel can be presented as an ideal gas – *a photon gas* – consisting of photons with different energies and impulses. Hence, naturally, the idea arises that by applying Boltzmann statistics to a photon gas, we can calculate the energy of photons with frequencies in the range $\mathrm{d}\omega$ and, thereby, derive the Planck formula (7.33). However, as a result of such a computation, what was obtained was not the general Planck formula but the Wien formula, just in the particular case of $k_0T \gg \hbar\omega$. Indeed, according the Boltzmann statistics, the number of photons with frequencies in the range $\mathrm{d}\omega$ equals

$$\mathrm{d}N(\omega) = \mathrm{const} \cdot \mathrm{e}^{-\varepsilon/k_0T} p^2 \mathrm{d}p = \mathrm{const} \cdot \mathrm{e}^{-\hbar\omega/k_0T} \omega^2 \mathrm{d}\omega, \tag{7.35}$$

and, consequently, the energy of photons with frequencies in the range $\mathrm{d}\omega$ is

$$\mathrm{d}E(\omega) = \hbar\omega\, \mathrm{d}N(\omega) = \mathrm{const} \cdot \mathrm{e}^{-\hbar\omega/k_0T} \omega^3 \mathrm{d}\omega. \tag{7.36}$$

With regard to this expression, for the spectral density we get

$$\rho(\omega) = \frac{\mathrm{d}E(\omega)}{\mathrm{d}\omega} = \mathrm{const} \cdot \mathrm{e}^{-\hbar\omega/k_0T} \omega^3. \tag{7.37}$$

The obtained expression (7.37) is not the Planck formula, but its particular case - Wien's law. Hence, it follows that the Boltzmann distribution cannot be applied to a photon gas, i.e. a photon gas is not a Boltzmann gas.

The difficulty associated with the photon gas was circumvented by the Indian physicist Satyendra Nath Bose on the basis of a new statistics worked out especially for *a photon gas* (see Sect. 7.14). Slightly later, Einstein generalized the Bose statistics to the case of a gas consisting of particles with the mass of rest different from zero. Therefore, Bose statistics got the name *Bose–Einstein statistics* (see Sect. 7.3) in 1924.

7.2 Principle of Indistinguishability of Particles: Fermions and Bosons

In Sect. 7.1 we noted that difficulties associated with the application of classical statistics to photon and free electron gases in metals were circumvented only after the introduction of quantum statistics. Before going into Fermi–Dirac and Bose–Einstein distributions, which constitute the basis of quantum statistics, we introduce *the principle of indistinguishability of particles.*

Consider a system consisting of N identical particles with identical physical parameters (mass, charge, spin, and so on).

According to classical mechanics, in spite of the fact that all the particles are identical, they can be distinguished. Thus, for instance, if in the initial position we enumerate particles and at subsequent moments trace their motion in the trajectory (according to the solution for the Hamilton equation), it can be exactly said at which point of space the particle is found. In other words, in classical mechanics, in spite of the fact that the characteristics of all particles are identical, they do not lose their individuality and are distinct from each other.

In quantum mechanics, according to the uncertainty principle, the concept of the trajectory is absent, and, therefore, even if the initial position of a particle is known, in subsequent moments its location cannot be completely determined, i.e. if we enumerate particles in the initial state, we cannot exactly determine their positions in space at subsequent moments of time. Consequently, in quantum mechanics, in contrast to classical mechanics, particles of a system consisting of identical particles when changing the state of the system do not conserve their individuality and are not distinct from each other. Therefore, in quantum mechanics of a system consisting of identical particles, we can only speak of the system as a whole, and not of the state of individual particles in it. In order to show this, we write the Hamilton operator for the considered system. Assume that a system consisting of N interacting particles is found in an external potential field that does not depend on time. Then, for such a stationary state ($\partial\hat{\mathbf{H}} / \partial t = 0$), the Hamilton has the appearance

$$\hat{\mathbf{H}}(1,2,\ldots,N) = \sum_{i=1}^{N}\left[-\frac{\hbar^2}{2m}\nabla_i^2 + W(i)\right] + \sum_{i\neq k=1}^{N} U(i,k), \tag{7.38}$$

where i and k are the totality of all coordinates of a particle of the system, including the spin, with the corresponding number, e.g. $(i) \Rightarrow (x_i, y_i, z_i, s_i)$; ∇_i is the Laplace operator, $W(i)$ is the potential energy of the ith particle in the external potential field, m is the particle mass and $U(i,k)$ is the energy of interaction between particles.

Since energy of the interaction depends only on the distance between particles, $U(i,k) = U(k,i)$. It we take this into account, from (7.38) it is seen that at the mutual transposition of particles ($i \leftrightarrows k$) the Hamilton operator does not change, because this transposition leads only to the transposition of the items in sum (7.38). Consequently, the property of the Hamilton operator

$$\hat{\mathbf{H}}(1,2,\ldots,i,\ldots,k,\ldots,N)=\hat{\mathbf{H}}(1,2,\ldots,k,\ldots,i,\ldots,N) \tag{7.39}$$

is the condition of identity of the particles. Indeed, if out of N particles only one is distinct from the others, then on the transposition of this particle with an arbitrary other one, condition (7.39) is not fulfilled.

Now we introduce the concept of the operator of the mutual transposition of particles and denote it by $\hat{P}_{ik}$. The action of this operator leads to the fact that particles i and k in the system exchange places, i.e. coordinates (i) and (k) on which the wave function of the system depends mutually exchange:

$$\hat{P}_{ik}\psi(1,2,\ldots,i,\ldots,k,\ldots,N)=\psi(1,2,\ldots,k,\ldots,i,\ldots,N) \tag{7.40}$$

or in the shortened form

$$\hat{P}_{ik}\psi(i,k)=\psi(k,i). \tag{7.41}$$

From the Hamiltonian property (7.39) and the definition of the operator of the mutual transposition of particles (7.40) it is seen that $\hat{\mathbf{H}}$ and $\hat{P}_{ik}$ commute:

$$\hat{P}_{ik}\hat{\mathbf{H}}-\hat{\mathbf{H}}\hat{P}_{ik}=0. \tag{7.42}$$

Note that this commutation, which is also a property of the Hamilton operator (7.39), is the mathematical expression of the identity of particles of the system.

The wave function of the system $\psi(1,2,\ldots,i,\ldots,k,\ldots,N)\equiv\psi(i,k)$ satisfies the Schrödinger stationary equation

$$\hat{\mathbf{H}}\psi(i,k)=E\psi(i,k) \tag{7.43}$$

where E is the total energy of the system.

If we act by the operator $\hat{P}_{ik}$ on each side of (7.43), we get

$$\hat{P}_{ik}\hat{\mathbf{H}}\psi(i,k)=E\hat{P}_{ik}\psi(i,k). \tag{7.44}$$

If we take into account the property of commutativity of operators of Hamilton and mutual transposition (7.42), then (7.44) takes the form

$$\hat{\mathbf{H}}(\hat{P}_{ik}\psi(i,k))=E(\hat{P}_{ik}\psi(i,k)). \tag{7.45}$$

Hence, it is seen that both the function $\psi(i,k)$ and the function

$$\psi^{(1)}(i,k)=\hat{P}_{ik}\psi(i,k)=\psi(1,2,\ldots,k,\ldots,i,\ldots,N) \tag{7.46}$$

satisfy the Schrödinger equation; in other words, the function $\psi^{(1)}(i,k)$ is also the wave function characterising the state of the system. The functions $\psi^{(1)}$ and ψ are distinguished only by the fact that in the state $\psi^{(1)}$ particles with coordinates i and k exchange places with reference to the state ψ. Continuing the process of transposition (the exchange), we get $\psi^{(2)}, \psi^{(3)}, \ldots, N!$ wave functions. It is clear that each of these functions is a wave function of the state with the total energy E. Consequently, to one value of energy E of the system of identical particles, $N!$ wave functions correspond ($N!$-fold degeneracy). Such a degeneracy is called *the exchange degeneracy.*

Thus, we have shown that separate particles of a system comprising identical particles do not conserve their individuality, i.e. independently of a particle being found at some point or other of the space, the system is found in the same state. In other words, we cannot speak about the state of separate particles, but only about the system as a whole, i.e. it is impossible to distinguish the particles. This conclusion in quantum mechanics is called *the principle of indistinguishability of particles* and formulated as follows: *A system consisting of identical particles can have only such states that do not change at the transposition of particles.*

From the linearity of the Hamilton operator and the principle of superposition, it follows that the linear combination

$$\psi(1, 2, \ldots, N) = \sum_n C_n \psi^{(n)}(1, 2, \ldots, N) \tag{7.47}$$

of wave functions $\psi^{(1)}, \psi^{(2)}, \ldots$ also satisfies the Schrödinger equation; however, it is necessary to choose the coefficients C_n so that they describe one of the possible states of the system.

From commutativity (7.42), it follows that eigenfunctions of operators of transposition $\hat{P}_{ik}$ and Hamilton $\hat{\mathbf{H}}$ of system (7.38) are identical and eigenvalues of the operator $\hat{P}_{ik}$ are real and conservation quantities, i.e. they are an integral of motion. Then we can write

$$\hat{P}_{ik}\psi(1, 2, \ldots, i, \ldots, k, \ldots, N) = \lambda\, \psi(1, 2, \ldots, i, \ldots, k, \ldots, N), \tag{7.48}$$

where λ is the eigenvalue of the operator $\hat{P}_{ik}$, which is a real and conservation quantity. To determine this quantity, act by the operator $\hat{P}_{ik}$ on both sides of (7.48):

$$\hat{P}_{ik}^2\psi(1, 2, \ldots, i, \ldots, k, \ldots, N) = \lambda \hat{P}_{ik}\psi(1, 2, \ldots, i, \ldots, k, \ldots, N). \tag{7.49}$$

Because on the left-hand side of the equation the operators $\hat{P}_{ik}$ acts twice, and then ψ does not change, and on the right-hand side if we take into account (7.48), (7.49) takes the form

$$\psi(1, 2, \ldots, i, \ldots, k, \ldots, N) = \lambda^2 \psi(1, 2, \ldots, i, \ldots, k, \ldots, N). \tag{7.50}$$

As a result, we get

$$\lambda^2 = 1 \text{ or } \lambda \pm 1. \tag{7.51}$$

Consequently, eigenfunctions of the operator of the transposition $\hat{P}_{ik}$ and, correspondingly, the Hamiltons on transposition $i \rightleftarrows k$ of identical particles either do not change or only change their sign:

$$\hat{P}_{ik}\psi(1,2,\ldots,i,\ldots,k,\ldots,N) = \pm\psi(1,2,\ldots,i,\ldots,k,\ldots,N). \tag{7.52}$$

In the first case, the wave function is called *a symmetric wave function*

$$\hat{P}_{ik}\psi_{\mathrm{s}}(1,2,\ldots,i,\ldots,k,\ldots,N) = +\psi_{\mathrm{s}}(1,2,\ldots,i,\ldots,k,\ldots,N), \tag{7.53}$$

and in the second case it is *an antisymmetric wave function.*

$$\hat{P}_{ik}\psi_{\mathrm{a}}(1,2,\ldots,i,\ldots k,\ldots,N) = -\psi_{\mathrm{a}}(1,2,\ldots,i,\ldots,k,\ldots,N), \tag{7.54}$$

where i and k take on any integer value from 1 to N.

Thus, according to the principle of indistinguishability, the state of a system of identical particles can be described by a symmetric or an antisymmetric wave function. Because the eigenvalues $\lambda = \pm 1$ of the operator of the transposition $\hat{P}_{ik}$ are an integral of motion (a conservation quantity), the symmetry of the wave function (symmetric or antisymmetric) is absolute. This means that if the wave function of a system possesses a specified symmetry, it can never have the other symmetry.

Furthermore, the wave function of systems of identical particles cannot possess a mixed symmetry, i.e. if for one transposition the wave function is symmetric, for other transpositions it cannot be antisymmetric. Such a state is impossible. In order to show these, suppose the reverse: assume that ψ function in reference to the transposition $i\rightleftarrows k$ is antisymmetric, and in reference to the transpositions $i\rightleftarrows j$ and $j\rightleftarrows k$ it is symmetric. Then we can write

$$\begin{aligned}
&\psi(\ldots,i,\ldots,k,\ldots,j,\ldots) = -\psi(\ldots,k,\ldots,i,\ldots,j,\ldots)\\
&= -\psi(\ldots,k,\ldots,j,\ldots,i,\ldots) = -\psi(\ldots,j,\ldots,k,\ldots,i,\ldots)\\
&= -\psi(\ldots,i,\ldots,k,\ldots,j,\ldots).
\end{aligned} \tag{7.55}$$

Hence it follows that

$$2\psi(\ldots,i,\ldots,k,\ldots,j,\ldots) = 0;\ \psi(\ldots,i,\ldots,k,\ldots,j,\ldots) = 0. \tag{7.56}$$

Consequently, the state with the mixed symmetry is impossible.

Experimentally, it has been shown that in nature there exist systems described by both symmetric and antisymmetric wave functions, and the following rules have been established:

1. If a system consists of identical particles with a spin as a multiple of the Planck constant, i.e.

$$s = 0;\ h;\ 2h, \tag{7.57}$$

it is described by the symmetric wave function and the particles themselves are called *Bose particles* or briefly *bosons.* π- and K-mesons with the spin $s = 0$ and photons with the spin $s = h$ are examples of such particles.

2. If a system consists of identical particles with a spin as a multiple of the odd semi-integer Planck constant, i.e.

$$s = \frac{h}{2};\ 3\frac{h}{2};\ 5\frac{h}{2}, \tag{7.58}$$

it is described by the antisymmetric wave function, and the particles themselves are called *Fermi particles* or briefly *fermions*. An electron, proton, neutron and their antiparticles with the spin $s = h/2$ are examples of fermions.

Note that these rules obtained as a result of generalization of experimental facts were subsequently theoretically substantiated by Pauli.

If particles forming a system are not elementary, i.e. the system consists of several elementary particles, whether they belong to fermions or bosons is determined by the number of elementary fermions entering into their composition: if the number of fermions is even, the complex particle is a boson, and if it is odd, the particle is concerned with fermions. According to this rule, an atom of hydrogen and an α particle are bosons, and He^3 is a fermion.

It is evident that in the general form the Schrödinger equation for interacting particles cannot be solved. Therefore, we consider the particular case – an ideal gas, consisting of N particles. In this case, because the energy of interaction $U(i,k) = 0$, the operator (7.38) takes the form

$$\hat{\mathbf{H}}_0(1,2,\ldots,N) = \sum_{i=1}^{N} \hat{\mathbf{H}}_i, \tag{7.59}$$

where

$$\hat{\mathbf{H}}_i = -\frac{\hbar^2}{2m}\nabla_i^2 + W(i) \tag{7.60}$$

is the Hamilton operator of one particle.

It is known that the eigenfunction $\psi_0(1,2,\ldots,N)$ of operator (7.59) can be presented in the form

$$\psi_0(1,2,\ldots,N) = \varphi_{\alpha_1}(1)\varphi_{\alpha_2}(2)\ldots\varphi_{\alpha_N}(N) = \prod_{i=1}^{N}\varphi_{\alpha_i}(i), \tag{7.61}$$

and its eigenvalue

$$E_0 = \varepsilon_{\alpha_1} + \varepsilon_{\alpha_2} + \cdots + \varepsilon_{\alpha_N} = \sum_{i=1}^{N}\varepsilon_{\alpha_i}. \tag{7.62}$$

Here $\varphi_{\alpha_i}(i)$ and ε_{α_i} are, respectively, the wave function and the energy of one particle obtained as a solution of the equation

$$\hat{\mathbf{H}}_i\varphi_{\alpha_i} = \varepsilon_{\alpha_i}\varphi_{\alpha_i}, \tag{7.63}$$

where α_i is a totality of quantum numbers determining the state of the ith particle.

It is evident that wave function (7.61) does not satisfy the condition of symmetry, which follows from the principle of indistinguishability (7.52). Indeed, at transposition $i \rightleftarrows k$, the particle i ought to be found in the state α_k, and particle k in the state α_i; thereby, the wave function ψ_0 changes. Therefore, for the wave function (7.61) it is necessary to compose linear combinations such that the obtained functions would satisfy the condition of symmetry (7.52) and, thereby, the principle of indistinguishability would be fulfilled. In order words, the solution (7.61) should comprise of symmetric or antisymmetric wave functions.

In order to obtain from (7.61) *the symmetric* wave function describing the state of an ideal gas consisting of *bosons*, it is sufficient to add up all functions (the number of such function is $N!$) obtained as a result of the pair transposition of particles:

$$\psi_{0\mathrm{s}}(1, 2, \ldots, N) = \sum_{\nu} P_{ik}^{(\nu)} \varphi_{\alpha_1}(1)\varphi_{\alpha_2}(2) \ldots \varphi_{\alpha_N}(N), \tag{7.64}$$

where ν is the number of the transposition and the summation is carried out with respect to the $N!$ number of transpositions.

In order to obtain from (7.61) the *antisymmetric* wave function describing the state of an ideal gas consisting of *fermions*, when adding up with respect to ν, in contrast to (7.64), we need to take items for even variables with the sign "+" and for odd ones with the sign "−":

$$\psi_{0\mathrm{a}}(1, 2, \ldots, N) = \sum_{\nu} (-1)^{\nu} \hat{P}_{ik}^{(\nu)} \varphi_{\alpha_1}(1)\varphi_{\alpha_2}(2) \ldots \varphi_{\alpha_N}(N). \tag{7.65}$$

This antisymmetric wave function can also be presented in the form of the Slater determinant

$$\psi_{0\mathrm{a}}(1, 2, \ldots, N) = \begin{vmatrix} \varphi_{\alpha_1}(1) & \varphi_{\alpha_1}(2) & \ldots & \varphi_{\alpha_1}(N) \\ \varphi_{\alpha_2}(1) & \varphi_{\alpha_2}(2) & \ldots & \varphi_{\alpha_2}(N) \\ \ldots & \ldots & \ldots & \ldots \\ \varphi_{\alpha_N}(1) & \varphi_{\alpha_N}(2) & \ldots & \varphi_{\alpha_N}(N) \end{vmatrix}. \tag{7.66}$$

From such a form of the entry of the antisymmetric wave function (in the form of a determinant), it follows that the wave function $\psi_{0\mathrm{a}}(1, 2, \ldots, N)$ satisfies the condition (7.54). Indeed, in determinant (7.66), on mutual transposition $i \rightleftarrows k$ two corresponding columns exchange places, which leads to the change in the sign of the determinant to the opposite one.

From the entry of the antisymmetric wave function (7.66) in the form of the determinant follows one more important result: if the quantum states of two arbitrarily taken particles are identical ($\alpha_i = \alpha_k$), then in (7.66) two rows are identical, and this is the condition for the entire determinant to be zero, i.e. the wave function $\psi_{0\mathrm{a}}(1, 2, \ldots, N)$ is zero. This is precisely *the Pauli exclusion principle* for fermions: *In one quantum state, two and more fermions cannot be found.*[1]

From the symmetric wave function (7.64), it is seen that quantum states of two arbitrary bosons can be identical ($\alpha_i = \alpha_k$); even all of them can be found in one quantum state ($\alpha_1 = \alpha_2 = \cdots = \alpha_N$) and therefore $\psi_{0s}(1, 2, \ldots, N) \neq 0$.

General conclusion: From the principle of indistinguishability of particles, it follows that in nature two kinds of particles exist:

1. *Bosons.* The spin of these particles (a photon, π -and k -mesons) equals a multiple of the Planck constant $h : s = 0;\ h;\ 2h; \ldots$. In one quantum state one can find an arbitrary number n_k of bosons:

$$n_k = 0;\ 1;\ 2;\ 3; \ldots \tag{7.67}$$

A system consisting of bosons (a Bose system) is described by the symmetric wave function.

2. *Fermions.* The spin of these particles (an electron, proton, neutron and their anti-particles) equals an odd multiple semi-integer of the Planck constant h: $s = \frac{h}{2};\ 3\frac{h}{2};\ 5\frac{h}{2};\ \ldots$. In one quantum state one can find only one fermion

$$n_k = 0;\ 1. \tag{7.68}$$

A system consisting of fermions (a Fermi system) is described by the antisymmetric wave function.

Statistics for Fermi and Bose systems formulated on the basis of quantum mechanics (with regard to the principle of indistinguishability of particles) is called *quantum statistics.* If interaction between particles is neglected, Fermi and Bose systems are called *ideal Fermi* and *Bose gases* or *ideal quantum gases.*

The principal propositions of quantum statistics are explained in subsequent sections.

In conclusion, we note the following. From the indistinguishability of particles, it follows that to one value of energy of the system $N!$-fold exchange degeneracies correspond. The principle of indistinguishability of particles circumvents this degeneracy, since the state of the system is described by only the symmetric or antisymmetric wave function. On the other hand, it is known that the degeneracy is removed by external forces or by interactions inside the system. In the quantum case, such an interaction can be "*the exchange interaction*", which does not have an analogue in classical physics. Later we will show that this interaction manifests itself when the de Broglie wavelength of a particle λ is of the order of the mean distance d between particles ($\lambda \approx d$). Only in this case, particles of a system comprising identical particles lose their individuality and one cannot assert which particle is found in which state.

[1] Note that the Pauli principle is true not only for an ideal gas, but also for any system of fermions with the antisymmetric wave function.

It is interesting also to note that, because in the particular quantum state there can be any number of bosons but only one fermion, the exchange interaction between bosons can be treated as "attraction" and that between fermions as "repulsion".

7.3 Distribution Functions of Quantum Statistics

In this section, on the basis of the principle of indistinguishability of particles we will find the distribution function of an ideal gas consisting of fermions or bosons. Assume that a gas of volume V contains N non-interacting fermions or bosons. It is required to find the mean number $\bar{n}_k$ of fermions or bosons in the quantum state $\boldsymbol{k}$ with energy ε_k at temperature T. This problem can be solved in different ways. We make use of the method proposed by Landau.

Assume that N fermions or bosons are distributed over quantum states $\boldsymbol{k}$ with energy ε_k. The number of particles in these states is denoted by n_k. Particles can come into or move out of this state; i.e. n_k can change. If we take a multitude of particles in the k state as a subsystem and the rest as a thermostat, we can apply the Gibbs grand canonical distribution for an open system [see (3.68)]. Because in our case the particles do not interact, in (3.68) we can replace the energy of the subsystem by

$$E_{nN} \Rightarrow n_k \varepsilon_k \tag{7.69}$$

and the number of particles by

$$N \Rightarrow n_k . \tag{7.70}$$

As a result, (3.68) takes the form

$$W_{n_k} = \mathrm{e}^{\frac{\Omega_k + \mu n_k - \varepsilon_k n_k}{k_0 T}}, \tag{7.71}$$

where W_{n_k} is the probability of n_k particles found in the quantum state $\boldsymbol{k}$ with energy ε_k, and Ω_k is the grand thermodynamic potential of a subsystem, which is found from the normalization condition

$$\sum_{n_k} W_{n_k} = 1. \tag{7.72}$$

Thus,

$$\Omega_k = -k_0 T \ln \sum_{n_k} \mathrm{e}^{\frac{(\mu - \varepsilon_k) n_k}{k_0 T}} . \tag{7.73}$$

Hence we can find the mean value of the number of particles $\bar{n}_k$ in the quantum state $\boldsymbol{k}$, i.e. the distribution sought [see (2.206)], by the formula

$$\bar{n}_k = -\left(\frac{\partial \Omega_k}{\partial \mu}\right)_{T,V} . \tag{7.74}$$

Consequently, to determine the explicit form of the distribution function $\bar{n}_k$ for fermions and bosons, it is necessary to calculate the sum in the expression (7.73).

Because systems of fermions and bosons are described by wave functions of different symmetries, it is necessary to consider them separately.

Fermi–Dirac distribution. This distribution, which is applicable to any system of fermions, was proposed by Fermi in 1926 for an electron gas the same year Dirac determined its relation with quantum mechanics. In 1927, Pauli, on the basis of the Fermi–Dirac statistics, resolved the difficulties associated with paramagnetism of free electrons in metals (see Sect. 7.10). In 1928, Sommerfeld, on applying this distribution to the free electron gas, circumvented the contradiction associated with the heat capacity of metals (see Sect. 7.9).

In Sect. 7.2, we showed that a system of fermions, in the particular case of a quantum ideal gas, is described by the antisymmetric wave function and that fermions are governed by the Pauli exclusion principle, according to which the number of fermions in the quantum state k cannot be more than one:

$$n_k = 0;\ 1. \tag{7.75}$$

In this case, the sum in (7.73) ought to consist of only two terms:

$$\Omega_k = -k_0 T \ln(1 + \mathrm{e}^{(\mu-\varepsilon_k)/k_0 T}). \tag{7.76}$$

Note that the expression of the grand thermodynamic potential (7.76) is different from (7.5); as a matter of fact, in expression (7.5) the second term under the logarithm sign is much smaller than unity, while in (7.76) this term can take on arbitrary values. This means that in (7.5), in the case of the Boltzmann distribution, the chemical potential ought to be negative ($\mu < 0$) in order that $\exp(\mu/k_0 T) \ll 1$ is satisfied. And in the case of Fermi distribution, the chemical potential can change in the range $-\infty < \mu < +\infty$, i.e. it is an arbitrary quantity.

On substituting the expression of the grand thermodynamic potential (7.76) into (7.74), we get the final expression for the Fermi–Dirac distribution

$$\bar{n}_k \equiv f(\varepsilon_k) = \frac{1}{\mathrm{e}^{(\varepsilon_k-\mu)/k_0 T} + 1};\quad -\infty < \mu < +\infty. \tag{7.77}$$

We denote by $\bar{n}_k \equiv f(\varepsilon_k) \equiv f(k) \equiv f(\varepsilon)$ the distribution function for fermions and henceforth will use these notations. This distribution can be interpreted in two ways: Distribution function (7.77) shows the mean number of fermions in the quantum state $\boldsymbol{k}$ with energy ε_k at temperature T, or it is the probability of fermions found in the quantum state $\boldsymbol{k}$ with energy ε_k at temperature T.

We analyse the Fermi–Dirac distribution function (7.77) for different values of the chemical potential and temperature entering into the distribution as parameters. We first consider the case where the chemical potential is positive ($\mu = \mu_0 > 0$). The positive quantity μ_0 is called *the Fermi boundary energy* or, briefly, *the Fermi boundary*. If $\varepsilon_k < \mu_0$, then $(\varepsilon_k - \mu_0) < 0$. Then in the

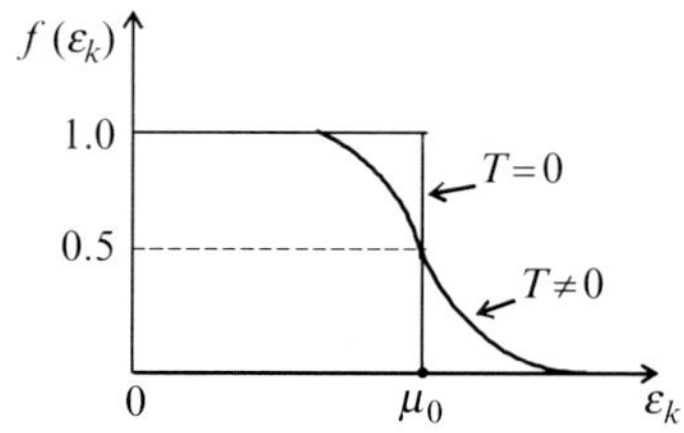

Fig. 7.1. The Fermi-Dirac distributions function

limit $T \to 0$, the exponent in (7.33) tends to zero and $f(\varepsilon_k) = 1$. In other words, all states with energy $\varepsilon_k < \mu_0$ are filled up. If, however, $\varepsilon_k > \mu_0$, then $(\varepsilon_k - \mu_0) > 0$ and as $T \to 0$ the index of the exponent in (7.77) becomes a very large positive quantity and as a result $f(\varepsilon_k > \mu) = 0$.

At finite temperatures $T \neq 0$, the function $f(\varepsilon) = 1$ only in the case where the energy is much smaller than the Fermi boundary energy. It is seen that at $\varepsilon_k = \mu_0$ the function $f(\varepsilon_k) = 1/2$. If energy ε_k is less than the Fermi boundary energy μ_0, at finite temperatures $T \neq 0$ the Fermi distribution function is less than unity: $f(\varepsilon_k < \mu_0) < 1$, and in the limiting case of large energies

$$\lim_{\varepsilon_k \to \infty} f(\varepsilon_k) = 0. \tag{7.78}$$

The Fermi–Dirac distributions for positive values of the chemical potential $\mu = \mu_0 > 0$ at $T = 0$ and $T \neq 0$ are schematically presented in Fig. 7.1.

With the aid of the Fermi–Dirac distribution (7.77), we can find the total number of fermions

$$N = \sum_k \bar{n}_k = \sum_k f(\varepsilon_k) = \sum_k \frac{1}{\mathrm{e}^{(\varepsilon_k - \mu)/k_0 T} + 1} \tag{7.79}$$

and the mean value of the total energy of fermions

$$E = \sum_k \varepsilon_k \bar{n}_k = \sum_k \varepsilon_k f(\varepsilon_k) = \sum_k \frac{\varepsilon_k}{\mathrm{e}^{(\varepsilon_k - \mu)/k_0 T} + 1}. \tag{7.80}$$

Using (7.76), we can calculate the grand thermodynamic potential of a Fermi gas:

$$\Omega = \sum_k \Omega_k = -k_0 T \sum_k \ln\left(1 + \mathrm{e}^{(\mu - \varepsilon_k)/k_0 T}\right). \tag{7.81}$$

Bose–Einstein distribution. This distribution of bosons was proposed by the Indian physicist Satyendra Nath Bose for a photon gas. Thereby, he statistically substantiated the Planck formula. Subsequently, Einstein generalized this distribution to bosons with the mass at rest, distinct from zero.

In Sect. 7.2, we showed that a Bose gas is described by the symmetric wave function and for its particles the Pauli exclusion principle does not exist, i.e. in one quantum state there can be an arbitrary number of bosons:

$$n_k = 0;\ 1;\ 2;\ 3;\dots \tag{7.82}$$

If we take into account this fact in (7.73), the grand thermodynamic potential for a Bose gas Ω_k takes the form

$$\Omega_k = -k_0 T \ln\left(1 + \mathrm{e}^{(\mu-\varepsilon_k)/k_0 T} + \mathrm{e}^{2(\mu-\varepsilon_k)/k_0 T} + \cdots\right). \tag{7.83}$$

Because the energy is always positive, i.e. $\varepsilon_k \geq 0$, in order that the series in (7.83) would converge also at $\varepsilon_k = 0$, the chemical potential ought to be negative, i.e. $\mu < 0$. Under this condition, the series under the logarithm becomes an infinitely decreasing geometric progression, the sum of which is easily determined:

$$\left(1 + \mathrm{e}^{(\mu-\varepsilon_k)/k_0 T} + \mathrm{e}^{2(\mu-\varepsilon_k)/k_0 T} + \cdots\right) = \left(1 - \mathrm{e}^{(\mu-\varepsilon_k)/k_0 T}\right)^{-1}. \tag{7.84}$$

Thus, for the grand thermodynamic potential of a Bose gas we finally get

$$\Omega_k = k_0 T \ln\left(1 - \mathrm{e}^{(\mu-\varepsilon_k)/k_0 T}\right), \tag{7.85}$$

and the Bose–Einstein distribution sought, i.e. $\bar{n}_k = -\left(\partial\Omega_k/\partial\mu\right)_{T,V}$, takes the form

$$\bar{n}_k = \frac{1}{\mathrm{e}^{(\varepsilon_k-\mu)/k_0 T} - 1}; \quad \mu < 0. \tag{7.86}$$

This distribution can be also treated in two ways: Distribution function (7.86) gives the mean number of bosons in the quantum state $\boldsymbol{k}$ with energy ε_k at temperature T, or it is the probability of bosons being found in the quantum state $\boldsymbol{k}$ with energy ε_k at temperature T.

Temperature T and the chemical potential μ enter into the distribution function (7.86) as parameters. We now analyse the dependence of the Bose–Einstein distribution on energy ε_k for different limiting values of these parameters.

In the limiting case when the chemical potential $\mu \to -0$ and temperature $T \to 0$ at small values of energy $\varepsilon_k \to 0$, the exponent in (7.86) tends to unity and the distribution function becomes infinite: $\bar{n}(\varepsilon_k) \to \infty$.

If the chemical potential takes on finite negative values ($\mu < 0$), even for the value of energy $\varepsilon_k = 0$ the distribution function takes on finite values $n_k(0) \neq 0$, and with increasing energy ε_k, it decreases. At very large negative values of the chemical potential ($\mu \to -\infty$), the distribution function (7.86) passes into the Boltzmann distribution (Fig. 7.2).

For comparison, plots of the Bose–Einstein and the Fermi–Dirac distribution function at $T = 0$ are shown in Fig. 7.3.

From the figure it is seen that in the case $\mu \to -0$ at $T = 0$, the majority of bosons (it can be said, all) pass to the lowermost energy level with $\varepsilon_k = 0$: $\bar{n}_k(0) \to N$. This phenomenon is called *Bose–Einstein condensation* (see Sect. 7.13).

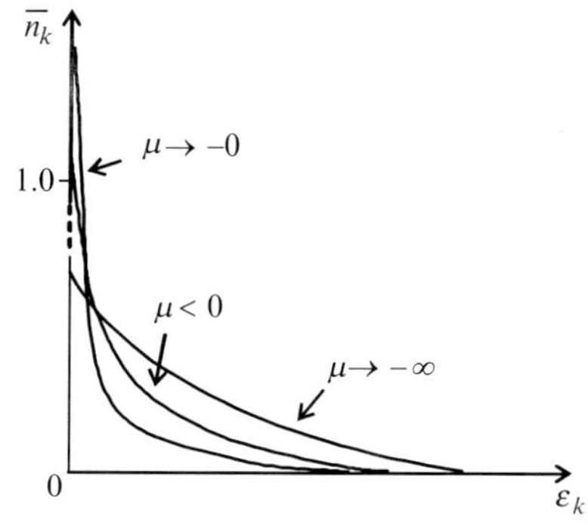

Fig. 7.2. The comparison of the Boltzmann, the Bose-Einstein and the Fermi-Dirac distribution functions

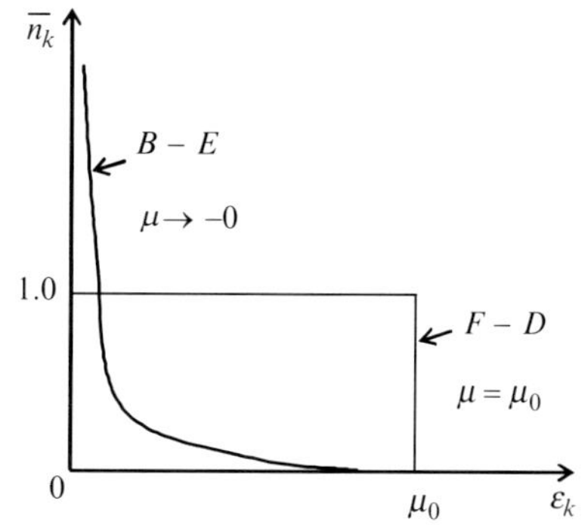

Fig. 7.3. The comparison of the Bose-Einstein and the Fermi-Dirac distribution function at $T = 0$

With the aid of the Bose–Einstein distribution (7.86), we can find the total number of bosons

$$N = \sum_k \bar{n}_k = \sum_k \frac{1}{\mathrm{e}^{(\varepsilon_k - \mu)/k_0 T} - 1} \tag{7.87}$$

and the mean value of the total energy of bosons

$$E = \sum_k \varepsilon_k \bar{n}_k = \sum_k \frac{\varepsilon_k}{\mathrm{e}^{(\varepsilon_k - \mu)/k_0 T} - 1}. \tag{7.88}$$

And using (7.85), we can calculate the grand thermodynamic potential of a Bose gas:

$$\Omega = \sum_k \Omega_k = +k_0 T \sum_k \ln\left(1 - \mathrm{e}^{(\mu - \varepsilon_k)/k_0 T}\right). \tag{7.89}$$

In conclusion, note that just as symmetric and antisymmetric wave functions of quantum ideal gases cannot pass into each other, so also the distribution functions do not pass into each other. However, under some conditions (such as $\mu \to -\infty$), both quantum distributions pass into the classical Boltzmann distribution (7.8). Indeed, the passage

$$\lim_{\mu \to -\infty} \left[e^{(\varepsilon_k - \mu)/k_0 T} \pm 1 \right]^{-1} = e^{(\mu - \varepsilon_k)/k_0 T} \tag{7.90}$$

does not depend on which particles (fermions or bosons) the gas consists of.

7.4 Equations of States of Fermi and Bose Gases

The thermal equation of the state of Fermi and Bose gases[2] can be found from the system of equations [see (2.206)]

$$\begin{cases} P = -\left(\dfrac{\partial \Omega(V, T, \mu)}{\partial V} \right)_{T,\mu}, \\ N = -\left(\dfrac{\partial \Omega(V, T, \mu)}{\partial \mu} \right)_{V,T}, \end{cases} \tag{7.91}$$

where $\Omega(V, T, \mu)$ is the grand thermodynamic potential of the system.

Solving these equations together, by eliminating the chemical potential μ, i.e. finding $\mu = \mu(T, V, N)$ from the second equation of (7.91), and substituting it into the first, we get the thermal equation of state

$$P = P(V, T, N). \tag{7.92}$$

Thus, from (7.91) it is seen that the determination of the thermal equation of state is reduced to finding the explicit form of the grand thermodynamic potential $\Omega = \Omega(V, T, \mu)$.

Assume that an ideal quantum gas of volume V consists of N fermions or bosons. The expression of the grand thermodynamic potential of such gases has the form [see (7.81) and (7.89)]

$$\Omega = \mp k_0 T \sum_{\boldsymbol{k}} \ln \left(1 \pm e^{(\mu - \varepsilon_k)/ k_0 T} \right). \tag{7.93}$$

Here and henceforth, in formulae the upper sign refers to fermions and the lower sign refers to bosons. Then, according to this rule, for the upper sign the chemical potential changes in the range $-\infty < \mu < +\infty$, and for the lower one it is $\mu < 0$.

To determine the explicit form of the grand thermodynamic potential $\Omega = \Omega(V, T, \mu)$, it is necessary to know the dependence of ε on $\boldsymbol{k}$, which can be quite complex. Therefore, consider a simple case when the energy of a particle ε_k is pre-assigned by the quadratic dependence on the wave vector $\boldsymbol{k}$ (7.10).

[2] Note that here and further all results received for Fermi-gas are true for electron gas

This model is well suited not only for free particles with the mass m but also can be applied for free electrons in metals as well as conduction electrons and holes in semiconductors; however, in these cases it is necessary to replace the mass m with the effective mass m^*.

To calculate the sum (7.93), it is necessary, according to rule (7.11), to pass from the summation with respect to $\boldsymbol{k}$ to the integration

$$\Omega = \mp \frac{k_0 TV g_0}{(2\pi)^3} \int \ln(1 \pm \mathrm{e}^{(\mu-\varepsilon_k)/\,k_0 T}) \mathrm{d}\boldsymbol{k}, \tag{7.94}$$

where $g_0 = (2s+1)$ is the multiplicity of degeneracy with respect to the spin s, which is the spin quantum number of a particle. Because the energy ε_k of a particle depends only on the magnitude of the wave vector $\boldsymbol{k}$, in (7.94) we move to the spherical coordinate system $\mathrm{d}\boldsymbol{k} = \sin\theta\, \mathrm{d}\theta\, \mathrm{d}\varphi\, k^2\, \mathrm{d}k$ and take into account that the integral over angles equals 4π. Then we get

$$\Omega = \mp \frac{k_0 TV g_0}{2\pi^2} \int\limits_0^\infty \ln\left(1 \pm \mathrm{e}^{(\mu-\varepsilon_k)/\,k_0 T}\right) k^2 \mathrm{d}k. \tag{7.95}$$

On the strength of the model (7.10), it is more favourable to pass from the integral over $\mathrm{d}k$ to the integral over $\mathrm{d}\varepsilon$. Then, the expression (7.95) takes the following final form:

$$\Omega = \mp \frac{k_0 TV g_0}{(2\pi)^2} \frac{(2m)^{3/2}}{\hbar^3} \int\limits_0^\infty \ln\left(1 \pm \mathrm{e}^{(\mu-\varepsilon)/\,k_0 T}\right) \varepsilon^{1/2} \mathrm{d}\varepsilon. \tag{7.96}$$

Here and henceforth, we will follow the notation $\varepsilon_k \equiv \varepsilon$.

If we take into account the expression of the grand thermodynamic potential (7.96) in (7.91), we get the following expression for pressure:

$$P = \pm \frac{k_0 T g_0}{(2\pi)^2} \frac{(2m)^{3/2}}{\hbar^3} \int\limits_0^\infty \ln\left(1 \pm \mathrm{e}^{(\mu-\varepsilon)/\,k_0 T}\right) \varepsilon^{1/2} \mathrm{d}\varepsilon. \tag{7.97}$$

This expression can be integrated by parts once; thereupon take the derivatives from (7.96) with respect to μ and substitute it in the second equation of (7.91). As a result, the system of equations (7.91) acquires the form

$$\begin{cases} P = \dfrac{2}{3} \dfrac{g_0 (2m)^{3/2}}{(2\pi)^2 \hbar^3} \int\limits_0^\infty \dfrac{\varepsilon^{3/2} \mathrm{d}\varepsilon}{\mathrm{e}^{(\varepsilon-\mu)/\,k_0 T} \pm 1}, \\ N = \dfrac{V g_0 (2m)^{3/2}}{(2\pi)^2 \hbar^3} \int\limits_0^\infty \dfrac{\varepsilon^{1/2} \mathrm{d}\varepsilon}{\mathrm{e}^{(\varepsilon-\mu)/\,k_0 T} \pm 1}. \end{cases} \tag{7.98}$$

This system of equations in the general case is the parametric form of the thermal equation of state of Fermi and Bose gases, in which μ is a parameter.

Solving this system of equations and eliminating from it the parameter μ, we get the thermal equation of state $P = P(N, V, T)$.

Now, we find the caloric equation of state of an ideal quantum gas, i.e. the dependence of energy on V, T, N. To do this, we combine expressions (7.80) and (7.88), pass from the summation with respect to $\boldsymbol{k}$ to the integration [see (7.11)] and take into account (7.10). As a result, the mean value of the total internal energy takes the form

$$E = \frac{Vg_0(2m)^{3/2}}{(2\pi)^2\hbar^3}\int\limits_0^\infty \frac{\varepsilon^{3/2}\mathrm{d}\varepsilon}{\mathrm{e}^{(\varepsilon-\mu)/k_0T} \pm 1}. \tag{7.99}$$

Hence, it is seen that energy is a function of temperature, volume and chemical potential: $E = E(T, V, \mu)$. If from the second equation of (7.98) we find $\mu = \mu(T, V, N)$ and substitute it into (7.99), we get the dependence of energy on temperature, volume and number of particles: $E = E(T, V, N)$.

Comparing (7.98) with (7.99), find the simple relationship between the energy density and pressure

$$P = \frac{2}{3}\frac{E}{V}, \tag{7.100}$$

which under the accepted model is general and is independent of the kind of particles (fermions or bosons) and the degree of degeneracy of the gas.

On the other hand, the energy density can be presented in the form

$$\frac{E}{V} = \int\limits_0^\infty g(\varepsilon)\varepsilon f(\varepsilon)\mathrm{d}\varepsilon. \tag{7.101}$$

From a comparison of this equation with (7.99), we can find *the function of the density of states* $g(\varepsilon)$, i.e. the number of quantum states accounting for the unit range of energy close to ε:

$$g(\varepsilon) = \frac{g_0(2m)^{3/2}}{(2\pi)^2\hbar^3}\varepsilon^{1/2} \sim \varepsilon^{1/2}. \tag{7.102}$$

This dependence is schematically presented in Fig. 7.4.

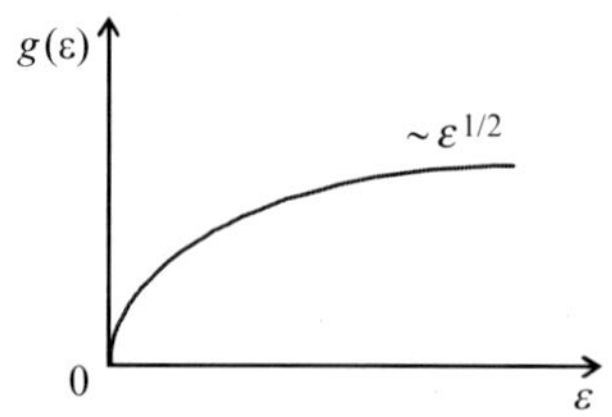

Fig. 7.4. The function of the density of states $g(\varepsilon)$

7.5 Thermodynamic Properties of Weakly Degenerate Fermi and Bose Gases

In order to construct a statistical theory of thermodynamic properties of quantum gases, i.e. to find the explicit form of thermal and caloric equations of state, as well as the temperature dependences of the heat capacity and other thermodynamic quantities, it is necessary to solve the system of equations (7.98) and (7.99). It is evident that it is impossible to solve these equations in the general form (for any degree of degeneracy). Therefore, we consider the particular limiting cases of weak and strong degeneracy. In this section we analyse the influence of weak degeneracy on the thermodynamic properties of Bose and Fermi gases. Thermodynamics of a strongly degenerate quantum gas is considered in subsequent sections.

A gas whose statistical properties are described by the Boltzmann distribution (7.8), i.e. they satisfy conditions (7.18) or (7.234), is called *non-degenerate*. If these conditions are not fulfilled, the Boltzmann distribution cannot be applied. In such a case, the statistical properties of the gas are determined by Fermi–Dirac or Bose–Einstein distributions. Such gases are called *quantum* or *degenerate gases*.

Under some conditions, the behaviour of a gas does not obey the Boltzmann distribution (the classical statistics), but is only slightly different from the classical one, i.e. degeneracy begins to manifest itself. Such gases are called *weakly degenerate*. For them, criteria (7.18) and (7.234) are not very strongly satisfied:

$$\mathrm{e}^{(\mu-\varepsilon_k)/\,k_0T} < 1 \,\text{or}\, \mathrm{e}^{\mu/\,k_0T} < 1. \tag{7.103}$$

In the non-degenerate case, when conditions (7.18) and (7.234) are fulfilled, the disparity between Bose and Fermi gases disappears [see (7.90)] and the gas is called *classical*.

We will consider the statistics of the thermodynamic properties of Bose and Fermi gases satisfying conditions (7.103). Because in this approximation Bose and Fermi gases are not different from each other, from classical theory they can be analysed simultaneously on the basis of the general equation of state (7.98).

The system of equations determining the thermal equation of state can be presented in the following form:

$$\begin{cases} P = \dfrac{2}{3}\dfrac{g_0(2m)^{3/2}}{(2\pi)^2\hbar^3} A \int\limits_0^\infty \dfrac{\varepsilon^{3/2}\mathrm{e}^{-\varepsilon/k_0T}\mathrm{d}\varepsilon}{1 \pm A\,\mathrm{e}^{-\varepsilon/\,k_0T}}, \\[2ex] N = \dfrac{Vg_0(2m)^{3/2}}{(2\pi)^2\hbar^3} A \int\limits_0^\infty \dfrac{\varepsilon^{1/2}\mathrm{e}^{-\varepsilon/k_0T}\mathrm{d}\varepsilon}{1 \pm A\,\mathrm{e}^{-\varepsilon/\,k_0T}}, \end{cases} \tag{7.104}$$

where

$$A = \mathrm{e}^{\mu/k_0T} < 1. \tag{7.105}$$

We solve the system of equations (7.104) in consecutive approximations with respect to the parameter $A < 1$.

In the zeroth approximation, in the denominator of the integrand expression the second item can be neglected compared to unity. Then, the system of equations (7.104) passes into (7.7) and we get the equation of state of an ideal classical gas, i.e. $P = k_0 NT / V$. The parameter A, which in the zeroth approximation we denote by A_0, is

$$A_0 = \frac{N}{Vg_0}\left(\frac{2\pi\hbar^2}{mk^0T}\right)^{3/2} \ll 1. \tag{7.106}$$

We now analyse the influence of weak degeneracy on the thermodynamic properties of a gas. To do this, we take into account weak degeneracy (7.103), expand into a series the fraction under the integral sign (7.104) and restrict ourselves to the first approximation. Then, (7.104) takes the form

$$\begin{cases} P = \dfrac{2g_0(2m)^{3/2}(k_0T)^{5/2}}{3(2\pi)^2\hbar^3} A \int\limits_0^\infty x^{3/2}(1 \mp A_0\,\mathrm{e}^{-x})\mathrm{e}^{-x}\mathrm{d}x, \\ N = \dfrac{Vg_0(2m)^{3/2}(k_0T)^{3/2}}{(2\pi)^2\hbar^3} A \int\limits_0^\infty x^{1/2}(1 \mp A_0\,\mathrm{e}^{-x})\mathrm{e}^{-x}\mathrm{d}x. \end{cases} \tag{7.107}$$

Here, in the second item we replaced the small parameter A with its value in the zeroth approximation A_0, as in (7.106), and introduced the dimensionless integration variable $x = \varepsilon / k_0T$.

Fulfilling the elementary integration (see Appendix A), from the second equation of system (7.107) in the first approximation for the parameter A we have

$$A = A_0(1 \pm 2^{-3/2}A_0). \tag{7.108}$$

On substituting this expression into the first equation of (7.107) and taking into account (7.106), for the explicit form of the thermal equation of state in the first approximation we get

$$P = \frac{k_0TN}{V}\left[1 \pm \frac{\pi^{3/2}\hbar^3}{2g_0(mk_0T)^{3/2}}\frac{N}{V}\right]. \tag{7.109}$$

As can be seen, $\Delta P = (P - P_{\mathrm{cl}}) \sim \pm T^{-1/2}$. If we remember that in (7.109) the upper sign refers to a Fermi gas and the lower one to a Bose gas, it can be said that even by considering weak degeneracy, the pressure of a Fermi gas compared to a Boltzmann gas increases, and the pressure of a Bose gas decreases (Fig. 7.5). This shows that the exchange interaction between bosons bears the character of "attraction", and that between fermions is a "repulsion". An interaction of such a kind has no classical analogue and is a purely quantum effect, i.e. it is associated with the principle of indistinguishability of particles.

The caloric equation of state of Fermi and Bose gases can be determined by solving the system of equations consisting of (7.111) and the second equation of (7.98) in the first approximation, or by simply using the relationships (7.100) and (7.109):

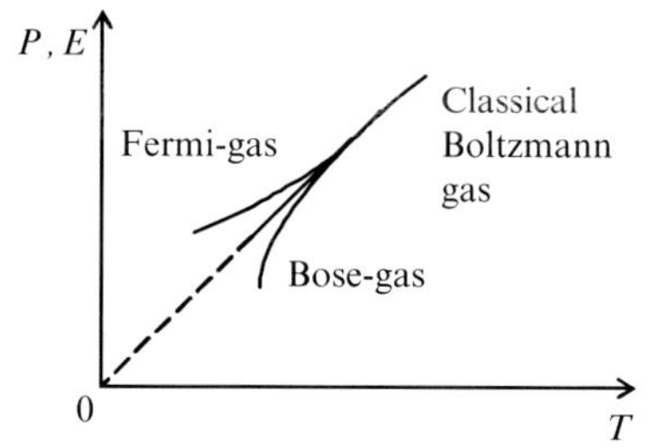

Fig. 7.5. The temperature dependence of the pressure and energy of the weakly degenerate Fermi and Bose gases

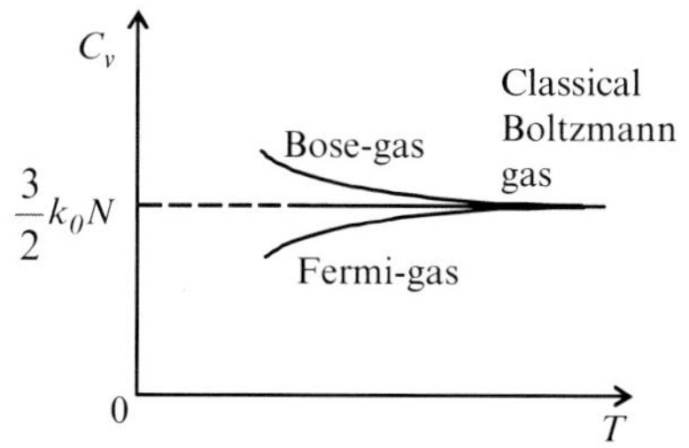

Fig. 7.6. The temperature dependence of the heat capacity of the weakly degenerate Fermi and Bose gases

$$E = \frac{3}{2}PV = \frac{3k_0 TN}{2}\left[1 \pm \frac{\pi^{3/2}\hbar^3}{2g_0(mk_0T)^{3/2}}\frac{N}{V}\right]. \tag{7.110}$$

Notice that the temperature dependence of energy is the same as for pressure (Fig. 7.5).

If from (7.110) we take the derivative with respect to T, we get the expression for the heat capacity of a weakly degenerate ideal quantum gas:

$$C_V = \frac{3k_0N}{2}\left[1 \mp \frac{\pi^{3/2}\hbar^3}{4g_0(mk_0T)^{3/2}}\frac{N}{V}\right]. \tag{7.111}$$

It is seen that $\Delta C_V = (C_V - C_V^{cl}) \sim \mp T^{-3/2}$.

The expression of the heat capacity also shows that the exchange interaction between bosons bears the character of "attraction", and that between fermions is "repulsion" (see Fig. 7.6).

Note that small corrections to pressure (7.109), energy (7.110) and heat capacity (7.111), naturally, coincide with conditions of classicity of the statistics (7.106), and as $T \to \infty$ they also tend to zero and, thereby, the obtained results coincide with those for a classical gas (see Figs. 7.5 and 7.6).

In order to determine the contribution of weak degeneracy to the entropy of Bose and Fermi gases, it is necessary to use the expression of the grand thermodynamic potential (7.96) in the approximation $e^{\mu/k_0T} < 1$. Then, in this approximation we get $\Omega = \Omega(V, T, \mu)$ and find entropy $S = -(\partial\Omega / \partial T)_{V,\mu} = S(V, T, \mu)$. Thereupon, substituting (7.108) into the obtained expression, we find entropy as a function of volume, temperature and the number of particles: $S = S(V, T, N)$.

7.6 Completely Degenerate Fermi Gas: Electron Gas: Temperature of Degeneracy

While the thermodynamic properties of weakly degenerate Bose and Fermi gases can be analysed altogether (Sect. 7.5), it is necessary to consider strongly degenerate gases separately, because the distribution function and statistical properties of degenerate, especially completely degenerate, Bose and Fermi gases sharply differ from each other (Fig. 7.3).

In this section we will analyse the statistical properties of a completely degenerate Fermi gas – electron gas. Assume that an ideal gas consisting of N fermions (electrons) occupies a volume V. It can be both classical and quantum. If the gas satisfies condition (7.20), it behaves as a Boltzmann gas (Sect. 7.1). In the general case, assume that the gas does not satisfy condition (7.20) and is described by the Fermi distribution function. Simultaneously, assume that energy of fermions is given by a simple parabolic dispersion law, i.e. by expression (7.10). Then the equation of state of quantum gases is given by the system of equations (7.98). This system for a Fermi gas can be rewritten in the form

$$\begin{cases} P = \dfrac{2}{3}\dfrac{g_0(2m)^{3/2}}{(2\pi)^2\hbar^3}\displaystyle\int_0^\infty \varepsilon^{3/2} f(\varepsilon)\mathrm{d}\varepsilon, \\ N = \dfrac{V g_0(2m)^{3/2}}{(2\pi)^2\hbar^3}\displaystyle\int_0^\infty \varepsilon^{3/2} f(\varepsilon)\mathrm{d}\varepsilon, \end{cases} \tag{7.112}$$

where $f(\varepsilon)$ is the Fermi distribution function, determined by (7.77).

Here, we consider the case of the absolute zero of temperature ($T = 0$).[3] To do this, we have to evaluate the integrals entering into (7.112) by integrating by parts once. Then, this system of equations takes the form

$$\begin{cases} P = \dfrac{4}{15}\dfrac{g_0(2m)^{3/2}}{(2\pi)^2\hbar^3}\displaystyle\int_0^\infty \left(-\frac{\partial f}{\partial \varepsilon}\right) \varepsilon^{5/2}\mathrm{d}\varepsilon, \\ N = \dfrac{2}{3}\dfrac{V g_0(2m)^{3/2}}{(2\pi)^2\hbar^3}\displaystyle\int_0^\infty \left(-\frac{\partial f}{\partial \varepsilon}\right) \varepsilon^{3/2}\mathrm{d}\varepsilon. \end{cases} \tag{7.113}$$

Because it is assumed that the total number of fermions in this case is constant, i.e. $N = \text{const}$, at $T = 0$ the filling of the energy spectrum by fermions has the appearance shown in Fig. 7.7. In other words, the energy spectrum is completely filled up to the Fermi boundary μ_0, and the levels above are empty.

[3] The case of finite temperatures ($T \neq 0$) will be considered in Sect. 7.7.

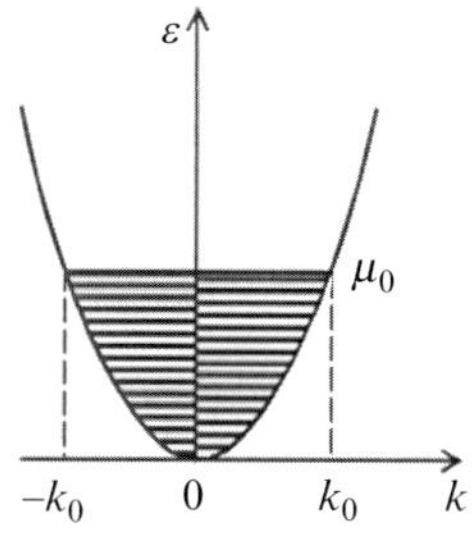

Fig. 7.7. The filling of the energy spectrum by fermions at $T = 0$

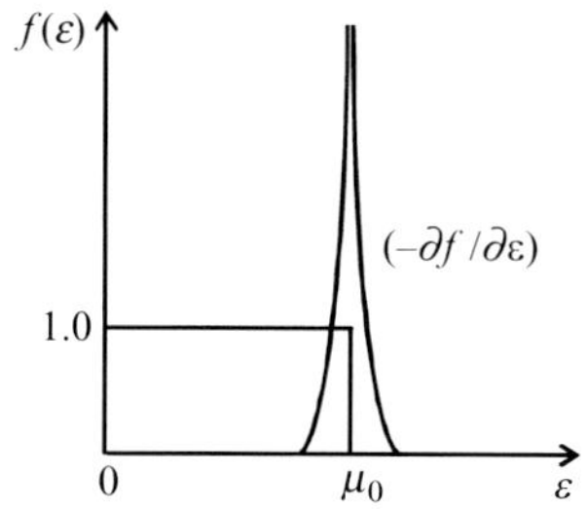

Fig. 7.8. The plots of the distribution function Fermi and its derivative at $T = 0$

The distribution function in this case has a step-like form:

$$\lim_{T\to 0} f(\varepsilon) = \begin{cases} 1, & \varepsilon \le \mu_0 \\ 0, & \varepsilon > \mu_0. \end{cases} \tag{7.114}$$

Plots of the distribution function and its derivative are schematically presented in Fig. 7.8.

As seen from Fig. 7.8, at $T = 0$ the derivative of the distribution function with respect to ε behaves like the δ-function:

$$\lim_{T\to 0}\left(-\frac{\partial f}{\partial \varepsilon}\right) = \delta(\varepsilon - \mu_0). \tag{7.114a}$$

A gas described by such a distribution is called a *completely degenerate* Fermi gas.

If we take into account the property of the distribution function (7.114a), then the integration in the system of equations (7.113) is easily fulfilled. As a result, we get

$$\begin{cases} P_0 = \dfrac{4}{15}\dfrac{g_0(2m)^{3/2}}{(2\pi)^2\hbar^3}\mu_0^{5/2}, \\[2ex] N = \dfrac{2}{3}\dfrac{Vg_0(2m)^{3/2}}{(2\pi)^2\hbar^3}\mu_0^{3/2}. \end{cases} \tag{7.115}$$

Hence, for the considered gas we can find the following:

Fermi boundary energy

$$\mu_0 = \frac{\hbar^2}{2m}\left(\frac{6\pi^2}{g_0}\frac{N}{V}\right)^{2/3}, \tag{7.116}$$

Fermi impulse

$$P_0 = \hbar k_0 = (2m\mu_0)^{1/2} = \hbar\left(\frac{6\pi^2}{g_0}\frac{N}{V}\right)^{1/3}, \tag{7.117}$$

Zero pressure of a Fermi gas

$$P_0 = \frac{2}{5}\frac{\hbar^2}{2m}\left(\frac{6\pi^2}{g_0}\right)^{2/3}\left(\frac{N}{V}\right)^{5/3} \tag{7.118}$$

Zero energy

$$E_0 = \frac{3}{5}\frac{\hbar^2}{2m}\left(\frac{6\pi^2}{g_0}\frac{N}{V}\right)^{2/3} N. \tag{7.119}$$

Here, we have used the relationship (7.100), i.e. $E_0 = 3\,VP_0\,/\,2$.

Applying the general expression (7.96) or using the known expression $\Omega_0 = -P_0 V$ for the grand thermodynamic potential at $T = 0$, we get

$$\Omega_0 = -\frac{4}{15}\frac{Vg_0(2m)^{3/2}}{(2\pi)^2\hbar^3}\mu_0^{5/2}, \tag{7.120}$$

where we introduced the notation $\Omega_0 = \Omega(0, V, \mu_0)$.

The fact that, in contrast to a classical gas, at the absolute zero temperature, a Fermi gas possesses the finite energy E_0, pressure P_0 and the limiting impulse fermion p_0 is a purely quantum effect. Naturally, this effect is associated with the exchange interaction of the type of "repulsion" between fermions, which follows from the Pauli exclusion principle. Also note that as a result of the quantum effect the zero pressure P_0 depends on the concentration $n = N\,/\,V$ not linearly but more strongly: $P_0 \sim (N\,/\,V)^{5/3}$.

From the expression of the total zero energy (7.119) and Fermi boundary energy (7.116), it follows that at the absolute zero temperature, energy accounting for each fermion comprises 60% of the Fermi boundary energy:

$$\varepsilon_0 = \frac{E_0}{N} = \frac{3}{5}\mu_0. \tag{7.121}$$

From (7.121) follows the simple relationship between the total energy and the boundary energy μ_0:

$$E_0 = \frac{3}{5}\mu_0 N. \tag{7.122}$$

Note one more peculiarity of an ideal Fermi gas. According to the classical theory, the small concentration of the gas is one of the conditions for its ideality, i.e. the gas ought to be sufficiently rarefied. In a Fermi gas, on the contrary, when increasing the concentration, a gas still further approaches the ideal one. We show this with an example of an electron gas in metals. Indeed, according to the condition of ideality, the interaction energy between particles ought to be much less than the mean energy: $\varepsilon_{\text{int}} \ll \varepsilon_0$. Interaction between free electrons in metals bears the Coulomb character: $\varepsilon_{\text{int}} \sim e^2/d$, where $d = (V/N)^{1/3}$ is the mean distance between electrons, and e is the electric charge of an electron. Therefore, the condition of ideality of the gas takes the form

$$\frac{e^2}{d} \ll \varepsilon_0. \tag{7.123}$$

If we use the expression for ε_0 (7.121) and boundary energy (7.116), the condition of ideality takes the following explicit form:

$$\frac{3\hbar^2}{10e^2 m}\left(\frac{6\pi^2}{g_0}\right)^{2/3}\left(\frac{N}{V}\right)^{1/3} \gg 1. \tag{7.124}$$

As seen from (7.124), as the concentration increases, an electron gas satisfies the condition of ideality better.

Temperature of degeneracy. Consider one more important characteristic parameter of the Fermi gas, called *the temperature of degeneracy* T_0, which is defined as

$$k_0 T_0 = \mu_0 \quad \text{or} \quad T_0 = \mu_0 \,/\, k_0. \tag{7.125}$$

With regard to the expression of the Fermi boundary energy μ_0 (7.116), the temperature of degeneracy takes the following form:

$$T_0 = \frac{\hbar^2}{2mk_0}\left(\frac{6\pi^2}{g_0}\frac{N}{V}\right)^{2/3}. \tag{7.126}$$

Note that *temperature of degeneracy is that temperature at which all fermions (even fermions at the zero level $\varepsilon = 0$) filling up the energy spectrum participate in the thermal motion.* However, at temperature $T = T_0$ not all fermions possess the identical energy $3k_0T\,/\,2$.

A Fermi gas with a constant number of fermions ($N = \text{const}$) in different temperature regions (in reference to temperature of degeneracy) is found in different statistical states:

1. As $T \to 0$, a Fermi gas is *completely degenerate* (Sect. 7.6, Figs. 7.7 and 7.8);
2. As $T \ll T_0$, the gas is found in a state of *strong degeneracy* (Sect. 7.7, Figs. 7.9 and 7.10);
3. As $T \approx T_0$, the gas is found in a state of *an arbitrary degree of degeneracy* (Sect. 7.8);
4. As $T \gg T_0$, the gas is *weakly degenerate* (Sect. 7.5);
5. As $T \to \infty$, the gas is found in a *non-degenerate* (classical) state (Sect. 7.1).

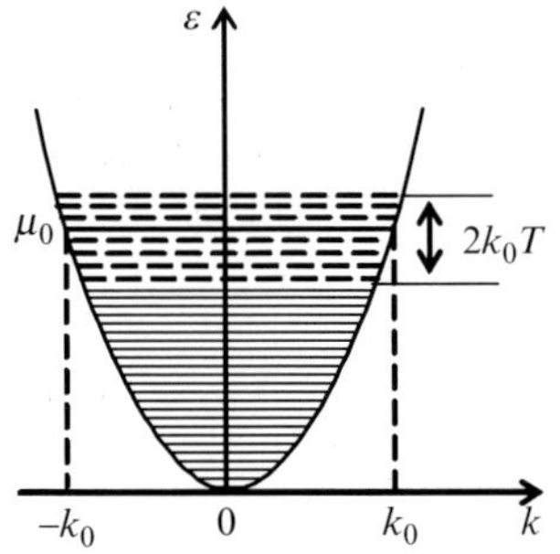

Fig. 7.9. The filling of the energy spectrum by fermions at temperature distinct from zero

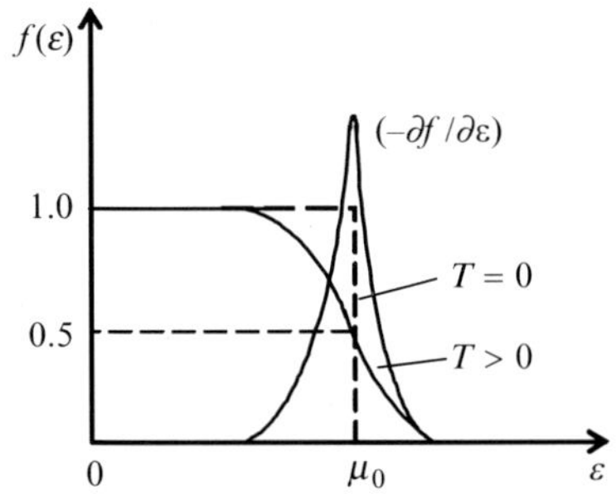

Fig. 7.10. The plots of the distribution function Fermi and its derivative at temperature distinct from zero

7.7 Thermodynamic Properties of Strongly Degenerate Fermi Gas: Electron Gas

When we speak of a strongly degenerate Fermi gas – electron gas – we imply a gas that is found in the statistical state schematically presented in Figs. 7.9 and 7.10. This state is characterized by the fact that under the action of finite but small ($T \ll T_0$) temperatures, a small part of fermions – electrons – found in quantum states below the Fermi boundary pass to higher levels. As a result, the Fermi boundary acquires a finite width on the order of $2k_0T$. However, note that this width is much smaller than the boundary energy: $2k_0T \ll \mu_0$.

We analyse the thermodynamic properties of a Fermi gas – electron gas – found in such a statistical state. To do this, we take the system of equations (7.113) as the basis. In the zeroth approximation (at $T = 0$), this system, as a result of the property of the Fermi distribution function (7.114a), passes into the system (7.115), which we solved in Sect. 7.6.

Here, we solve the system of equations (7.113) in the first approximation with respect to the dimensionless small parameter $k_0T \,/\, \mu_0 = T \,/\, T_0 \ll 1$ and consider the influence of strong degeneracy on the thermodynamic properties of the gas. Here, $T_0 = \mu_0 \,/\, k_0$ is temperature of degeneracy.

From the system of equations (7.113), it is seen that to find the equation of state it is necessary to calculate integrals of the type

$$I = \int_0^\infty \left(-\frac{\partial f}{\partial \varepsilon}\right)\varphi(\varepsilon)\mathrm{d}\varepsilon \tag{7.127}$$

in the first approximation with respect to the parameter $k_0T/\mu_0 \ll 1$. In our case, $\varphi(\varepsilon) = \varepsilon^{5/2}$ and $\varphi(\varepsilon) = \varepsilon^{3/2}$. The distribution function of a strongly degenerate gas and its derivative are presented in Fig. 7.10. From the figure, it is seen that at finite temperature the derivative $(-\partial f \,/\, \partial \varepsilon)$ for the value of energy $\varepsilon = \mu(T)$ is maximum, where μ is the Fermi level at temperature T. Therefore, to calculate the integral (7.127) the function $\varphi(\varepsilon)$ can be expanded into a series around μ in powers of $(\varepsilon - \mu)$. Then, integral (7.127) takes the form

$$I = \varphi(\mu) + I_1\left(\frac{\mathrm{d}\varphi}{\mathrm{d}\varepsilon}\right)_{\varepsilon=\mu} + \frac{1}{2}I_2\left(\frac{\mathrm{d}^2\varphi}{\mathrm{d}\varepsilon^2}\right)_{\varepsilon=\mu} + \cdots \tag{7.128}$$

Here, we used the property of the distribution function (7.114a) and employed the following notations:

$$I_1 = \int_0^\infty (\varepsilon-\mu)\left(-\frac{\partial f}{\partial \varepsilon}\right)\mathrm{d}\varepsilon; \quad I_2 = \int_0^\infty (\varepsilon-\mu)^2\left(-\frac{\partial f}{\partial \varepsilon}\right)\mathrm{d}\varepsilon. \tag{7.129}$$

We introduce the new dimensionless variable $x = (\varepsilon - \mu) \,/\, k_0T$. At temperature $T \ll T_0$, when $\mu \,/\, k_0T \gg 1$, the lower boundary of integrals (7.129) with respect to x can be replaced with $-\infty$. Simultaneously, we take into account that the function $(-\partial f \,/\, \partial x) = \mathrm{e}^x \,/\, (\mathrm{e}^x + 1)^2$ is even: $\mathrm{e}^x \,/\, (\mathrm{e}^x+1)^2 = \mathrm{e}^{-x} \,/\, (\mathrm{e}^{-x}+1)^2$. Then, because the function $x(-\partial f \,/\, \partial x)$ under the first integral sign is odd, the first integral is equal to zero, i.e.

$$I_1 = k_0T\int_{-\infty}^\infty x\left(-\frac{\partial f}{\partial x}\right)\mathrm{d}x = 0, \tag{7.130}$$

and the second one

$$I_2 = (k_0T)^2\int_{-\infty}^\infty x^2\left(-\frac{\partial f}{\partial x}\right)\mathrm{d}x = 2(k_0T)^2\int_0^\infty x^2\left(-\frac{\partial f}{\partial x}\right)\mathrm{d}x. \tag{7.131}$$

On integrating (7.131) once by parts, we get

$$I_2 = 4(k_0T)^2\int_0^\infty \frac{x}{\mathrm{e}^x+1}\mathrm{d}x. \tag{7.132}$$

If we take into account that (according to Appendix A)

$$\int_0^\infty \frac{x\,\mathrm{d}x}{\mathrm{e}^x+1} = \frac{\pi^2}{12}, \tag{7.133}$$

then integral (7.127) takes the form

$$\int_0^\infty \left(-\frac{\partial f}{\partial \varepsilon}\right)\varphi(\varepsilon)\mathrm{d}\varepsilon = \varphi(\mu) + \frac{\pi^2}{6}(k_0T)^2\left(\frac{\mathrm{d}^2\varphi}{\mathrm{d}\varepsilon^2}\right)_{\varepsilon=\mu} + \cdots \tag{7.134}$$

On applying this formula for the approximate integration to the system of equations (7.113) and in the first approximation, distinct from zero, with respect to the parameter $(k_0T\,/\,\mu)$ we get

$$\begin{cases} P = \dfrac{4}{15}\dfrac{g_0(2m)^{3/2}}{(2\pi)^2\hbar^3}\mu^{5/2}\left[1+\dfrac{5\pi^2}{24}\left(\dfrac{k_0T}{\mu_0}\right)^2\right]; & T \ll T_0, \\ N = \dfrac{2}{3}\dfrac{Vg_0(2m)^{3/2}}{(2\pi)^2\hbar^3}\mu^{3/2}\left[1+\dfrac{\pi^2}{8}\left(\dfrac{k_0T}{\mu_0}\right)^2\right]; & T \ll T_0, \end{cases} \tag{7.135}$$

where in the terms proportional to $(\sim T^2)$ we replaced $\mu = \mu_0$.

From the second equation of this system in the first approximation with respect to degeneracy, we can find the temperature dependence of the Fermi level (boundary):

$$\mu(T) = \mu_0\left[1-\frac{\pi^2}{12}\left(\frac{k_0T}{\mu_0}\right)^2\right]; \quad T \ll T_0. \tag{7.136}$$

Here μ_0 is the Fermi boundary energy at $T=0$, derived in (7.116). It is seen that at finite but small temperatures the Fermi boundary decreases. The reason for this lies in the fact that the density of quantum states increases as $g(\varepsilon) \sim \sqrt{\varepsilon}$ (Fig. 7.4), i.e. the density of quantum states above the Fermi boundary μ_0 is somewhat more than that below the Fermi boundary. Therefore, a specified amount of fermions – electrons – passing the Fermi boundary upwards occupies a narrower strip of energy. As a result, on average the Fermi level decreases.

If we substitute the expression for the chemical potential (7.136) into the first equation of the system (7.135), at very low temperatures ($T \ll T_0$) the equation of the state of a Fermi gas takes the form

$$P(T) = P_0\left[1+\frac{5\pi^2}{12}\left(\frac{k_0T}{\mu_0}\right)^2\right]; \quad T \ll T_0, \tag{7.137}$$

where $P_0 = \frac{2}{5}\frac{N}{V}\mu_0$ is the zero pressure of a Fermi gas [see (7.115)].

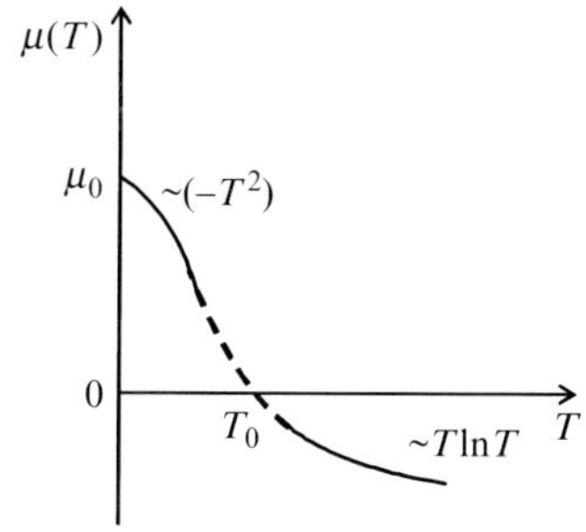

Fig. 7.11. The temperature dependence of the chemical potential

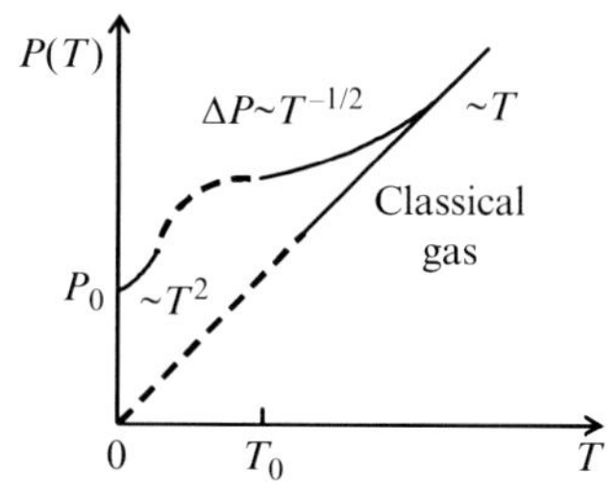

Fig. 7.12. The temperature dependence of pressure of a Fermi gas

We derive one more expression for the chemical potential (7.20) of a classical non-degenerate electron gas:

$$\mu(T) = -k_0 T \ln\left[\frac{Vg_0}{N}\left(\frac{mk_0T}{2\pi\hbar^2}\right)^{3/2}\right]; \quad T \gg T_0. \qquad (7.138)$$

The temperature dependence of the chemical potential $\mu(T)$ based on the expressions (7.136) and (7.138) is schematically presented in Fig. 7.11. It is seen that at the temperature of degeneracy T_0 the chemical potential passes through zero: $\mu(T_0) = 0$.

In order to gravarphically present the dependence of pressure of a Fermi gas in all temperature regions of degeneracy, we turn our attention to (7.109), (7.137) and Fig. 7.5. Then we get the dependence schematically shown in Fig. 7.12.

The mean energy of a degenerate Fermi gas – electron gas – in the first approximation can be found using the relationship $E = 3PV2/$ and also expressions (7.135) and (7.136):

$$E = E_0 + \frac{\pi^2}{4}\mu_0 N\left(\frac{k_0T}{\mu_0}\right)^2; \quad T \ll T_0. \qquad (7.139)$$

In the same approximation, the heat capacity of a Fermi gas – electron gas – equals

$$C_V = \frac{\pi^2}{2} k_0 N \left(\frac{k_0 T}{\mu_0} \right); \quad T \ll T_0 \tag{7.140}$$

In order to calculate the entropy of a Fermi gas – electron gas – in the considered approximation, we use the general expression of the grand thermodynamic potential (7.96): $S = -(\partial \Omega(V,T,\mu) \,/\, \partial T)_{V,\mu}$. If we preliminarily integrate the expression (7.96) twice by parts, we get

$$\Omega = -\frac{4}{15} \frac{V g_0}{(2\pi)^2} \frac{(2m)^{3/2}}{\hbar^3} \int_0^\infty \varepsilon^{5/2} \left(-\frac{\partial f}{\partial \varepsilon} \right) \mathrm{d}\varepsilon. \tag{7.141}$$

Note that from comparison of expressions (7.113) and (7.141) follows the known thermodynamic relationship $\Omega = -PV$.

On applying the approximation (7.134) to the expression of the grand thermodynamic potential (7.141), with the necessary accuracy we get

$$\Omega = -\frac{4}{15} \frac{V g_0}{(2\pi)^2} \frac{(2m)^{3/2}}{\hbar^3} \mu^{5/2} \left[1 + \frac{5\pi^2}{8} \left(\frac{k_0 T}{\mu} \right)^2 \right]; \; T \ll T_0. \tag{7.142}$$

If in (7.142) we take the derivative with respect to T at constant volume V and the chemical potential $\mu (\mu = \mu_0)$, the entropy of a Fermi gas – electron gas – takes the very simple form

$$S = \frac{\pi^2}{2} k_0 N \left(\frac{k_0 T}{\mu_0} \right); \quad T \ll T_0. \tag{7.143}$$

Note that the expression of the heat capacity $C_V = T (\partial S \,/\, \partial T)_V$ taken from (7.143) coincides with (7.140) obtained with the aid of the relationship $C_V = (\partial E \,/\, \partial T)_V$.

If we combine the result for a Fermi-gas – electron gas – coming from (7.111) with the temperature dependence (7.140), we get the schematic dependence $C_V(T)$ depicted in Fig. 7.13.

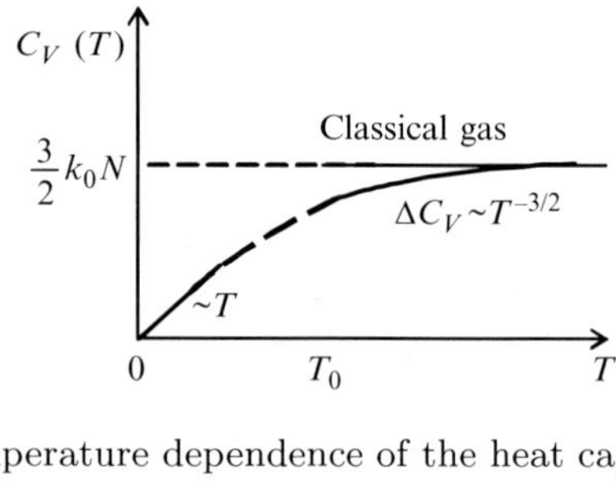

Fig. 7.13. The temperature dependence of the heat capacity of a Fermi gas

7.8 General Case: Criteria of Classicity and Degeneracy of Fermi Gas: Electron Gas

If we introduce the concept of the reduced chemical potential of an ideal gas

$$\eta = \frac{\mu}{k_0 T}, \tag{7.144}$$

the Fermi–Dirac distribution function can be presented as follows:

$$f(\varepsilon) = (\mathrm{e}^{\varepsilon/\,k_0 T - \eta} + 1)^{-1}. \tag{7.145}$$

It is seen that if the reduced chemical potential satisfies the condition

$$A \equiv \exp(\eta) \ll 1, \tag{7.146}$$

the distribution function (7.145) passes into the Boltzmann distribution:

$$\lim_{\eta \to -\infty} (\mathrm{e}^{\varepsilon/\,k_0 T - \eta} + 1)^{-1} = \mathrm{e}^{\eta - \varepsilon/\,k_0 T} = \mathrm{e}^{(\mu - \varepsilon)/\,k_0 T}. \tag{7.147}$$

Therefore, if the chemical potential of a gas satisfies condition (7.146), such a gas is called *non-degenerate* or *classical*; in other words, it is a Boltzmann gas. Note that a classical gas, non-degenerate gas and Boltzmann gas are equivalent concepts.

Thus, in order that an ideal gas be non-degenerate, its reduced chemical potential η ought to satisfy condition (7.146), i.e. η ought to be a very large negative quantity:

$$(-\eta) \gg 1. \tag{7.148}$$

The basic criterion of classicity or non-degeneracy of a Fermi gas – electron gas (7.146) can be rewritten also in another form [see (7.20)]:

$$\frac{n}{2}\left(\frac{2\pi\hbar^2}{mk_0 T}\right)^{3/2} \ll 1, \tag{7.149}$$

where $n = N\,/\,V$ is the concentration, and $g_0 = 2$.

From the condition of classicity presented in form (7.146), it follows that in order that a gas be classical (non-degenerate), it is necessary that the concentration be small, temperature high and the mass of particles large.

If we take into account the de Broglie wavelength $\lambda = h/\sqrt{2mk_0 T}$ of particles and the mean distance between them $d = n^{-1/3}$, condition (7.149) takes the form

$$\frac{1}{2\pi^{3/2}}\left(\frac{\lambda}{d}\right)^3 \ll 1. \tag{7.150}$$

Hence it follows that in order that a gas be non-degenerate, it is necessary that the de Broglie wavelength λ of a particle be much less than the mean distance d between them; in other words, the gas ought to be sufficiently rarefied. Note that only when fulfilling the condition $\lambda \ll d$, the particles conserve their individuality; therefore the principle of indistinguishability of particles does not play any role, in the system the exchange interaction is absent and the gas behaves as a Boltzmann gas.

Note that the above-presented inequalities (7.148)–(7.150) are different forms of the criterion of non-degeneracy (classicity) of an ideal gas (7.146) and they are all equivalent.

A gas not satisfying the indicated conditions, especially (7.146), is called a *degenerate, quantum* or *Fermi gas.* Consequently, degenerate gases are those whose chemical potential satisfies the condition

$$\exp(\eta) \geq 1 \quad \text{or} \quad \eta \geq 0. \tag{7.151}$$

A gas can exist in different degrees of degeneracy: *completely degenerate, strongly degenerate, moderately degenerate and weakly degenerate.*

A concretely pre-assigned gas (with the known concentration n and mass m), depending on which temperature region it is found in, can possess different degrees of degeneracy. These temperature regions are determined in reference to the *temperature of degeneracy* T_0, the expression of which is given in (7.125) or (7.126).

The temperature of degeneracy T_0 characterises the statistical state of a Fermi gas. At temperature $T \ll T_0$, i.e. in the approximation $T \to 0$, a Fermi gas is called *completely degenerate.* This case was considered in Sect. 7.6.

Finite temperatures with reference to the temperature of degeneracy T_0 can be divided into three regions:

7.8.1 Low Temperatures

$$T \ll T_0 \quad \text{or} \quad T \ll \mu_0 \,/\, k_0. \tag{7.152}$$

A Fermi gas satisfying this condition is called *a strongly degenerate gas.* Because in this case there exists the small parameter $\eta^{-1} = k_0 T \,/\, \mu_0 \ll 1$, we can construct the approximate analytical statistics of this gas (see Sect. 7.7).

If we use the temperature of degeneracy (7.126), the condition for strong degeneracy (7.152) takes the form

$$\frac{\hbar}{2mk_0T}(3\pi^2 n)^{2/3} \gg 1. \tag{7.153}$$

Thus, in order that a gas be found in the strongly degenerate state, it is necessary that its concentration n be large, the mass of fermions small and the temperature low.

The condition for strong degeneracy (7.153) can be rewritten in the form

$$\frac{1}{4}\left(\frac{3}{\pi}\right)^{2/3}\left(\frac{\lambda}{d}\right)^2 \gg 1, \tag{7.154}$$

where

$$d = n^{-1/3}, \ \lambda = h \,/\, \sqrt{2mk_0T}.$$

Consequently, in order that a gas be found in the strongly degenerate state, the de Broglie wavelength λ ought to be much larger than the mean distance d between fermions. When fulfilling this condition, particles lose their individuality, and therefore it is necessary to take into account the principle of indistinguishability.

7.8.2 High Temperatures

$$T > T_0 \quad \text{or} \quad T > \mu_0 \,/\, k_0. \tag{7.155}$$

A gas satisfying this condition is called *a weakly degenerate gas.* Because in this case there exists the small parameter $\eta = \mu_0 \,/\, k_0T < 1$, we can construct the approximate analytical theory of this gas (see Sect. 7.5).

The condition of weak degeneracy (7.155) can be rewritten in the form

$$\frac{\hbar^2}{2mk_0T}(3\pi^2 n)^{2/3} > 1 \tag{7.156}$$

or

$$\frac{1}{4}\left(\frac{3}{\pi}\right)^{2/3}\left(\frac{\lambda}{d}\right)^2 < 1. \tag{7.157}$$

If temperature satisfies the condition $T \gg T_0$, (7.157) passes into a strong inequality

$$\frac{1}{4}\left(\frac{3}{\pi}\right)^{2/3}\left(\frac{\lambda}{d}\right)^2 \ll 1, \tag{7.158}$$

which corresponds to (7.150), i.e. in the region $T \gg T_0$, a gas passes into the non-degenerate (classical) state.

Consequently, at the not very high temperature of degeneracy T_0 ($T_0 \approx$ 300–400 K), by changing the temperature of the gas from $T = 0$ to $T \gg T_0$, we can change the state of the gas from the completely degenerate state to the classical one.

7.8.3 Moderate Temperatures: $T \approx T_0$

This temperature region corresponds to the state $\eta = \mu_0 \,/\, k_0T \approx 1$, i.e. the small parameter is absent. Therefore, it is impossible to solve the problem analytically. In this case, the problem can be solved only numerically.

If we introduce the dimensionless quantities $x = \varepsilon \,/\, k_0T$, $\eta = \mu_0 \,/\, k_0T$ and take $g_0 = 2$, the general form of the equation of state (7.112) for a Fermi gas – electron gas – takes the following form:

$$\begin{cases} P = \dfrac{(2m)^{3/2}(k_0T)^{5/2}}{3\pi^2\hbar^3} \int\limits_0^\infty \dfrac{x^{3/2}\,dx}{e^{x-\eta}+1}, \\ n = \dfrac{(2mk_0T)^{3/2}}{2\pi^2\hbar^3} \int\limits_0^\infty \dfrac{x^{1/2}\mathrm{d}x}{\mathrm{e}^{x-\eta}+1}. \end{cases} \tag{7.159}$$

If the concentration of the gas n and the mass of fermions m are known, as a result of a numerical computation from the second equation of the system we can find the reduced chemical potential η at the particular temperature. Thereupon, on substituting η into the first equation of (7.159), we can find the pressure at the given temperature. By repeating the computations at different temperatures, we can find the temperature dependences of the chemical potential $\mu(T) = k_0T\eta(T)$ and pressure $P = P(T)$. Parts of the dependences $\mu = \mu(T)$ and $P = P(T)$ in the limiting cases are shown in Figs. 7.11 and 7.12.

Note that the equation of state (7.159) in the general form can be also expressed by the known Fermi integral. To do this, it is necessary to integrate (7.159) once by parts. As a result, we get

$$\begin{cases} P = \dfrac{2(2m)^{3/2}(k_0T)^{5/2}}{15\pi^2\hbar^3} F_{5/2}(\eta) \\ n = \dfrac{(2mk_0T)^{3/2}}{3\pi^2\hbar^3} F_{3/2}(\eta), \end{cases} \tag{7.160}$$

where

$$F_k(\eta) = \int^\infty \left(-\frac{\partial f}{\partial x}\right) x^k \mathrm{d}x \tag{7.161}$$

is the Fermi integral, and $f(x) = [\mathrm{e}^{x-\eta}+1]^{-1}$ is the Fermi–Dirac distribution function in dimensionless variables.

We now find the equation of the state of a Fermi gas using asymptotes of the Fermi integral in two limiting cases.

(a) *Non-degenerate and weakly degenerate limiting cases*: $A = \mathrm{e}^\eta \ll 1$. In the zeroth approximation with respect to the small parameter, the Fermi integral takes the form

$$F_k(\eta) = A_0\Gamma(\kappa+1), \tag{7.162}$$

where A_0 is the value of $\exp(\eta)$ in this approximation, which is determined by the expression (7.106), and

$$\tilde{\mathrm{A}}(k+1) = \int\limits_0^\infty x^k \mathrm{e}^{-x}\mathrm{d}x \tag{7.163}$$

is the Euler integral of the second kind or the gamma function (see Appendix A).

If we substitute asymptotes of the Fermi integral (7.162) into (7.160), we can find the value of the parameter A_0, i.e. the chemical potential [see (7.20)], and the equation of state of a non-degenerate (classical) ideal gas.

In order to calculate the chemical potential (7.108) and the equation of state (7.109) of a weakly degenerate gas, it is necessary to use the asymptote of the Fermi integral

$$F_k(\eta) = A\Gamma(k+1)\left(1 - \frac{A_0}{2^k}\right) \tag{7.164}$$

in the equation of state (7.160).

(b) *Completely and strongly degenerate limiting cases:* $e^\eta = e^{\mu/k_0T} \gg 1$. In this case, using expansion (7.134) in the zeroth approximation for the Fermi integral we get

$$F_k(\eta) = \eta_0^k = \left(\frac{\mu_0}{k_0 T}\right)^k. \tag{7.165}$$

Taking into account this asymptote of the Fermi integral in (7.160), for a completely degenerate gas we get the results deduced in Sect. 7.6.

Using the expansion (7.134) in the first approximation for the Fermi integral, we get the asymptote

$$F_k(\eta) = \eta^k \left[1 + \frac{\pi^2}{6}\frac{k(k-1)}{\eta_0^2}\right], \tag{7.166}$$

where in the second term we took $\eta = \eta_0$. On substituting this asymptote into (7.160), for a degenerate Fermi gas (electron gas) we get the results deduced in Sect. 7.7.

In conclusion, in Table 7.1 we summarise the conditions of a Fermi gas found in different statistical states.

In the next two sections, on the basis of the general statistical theory, we will consider the heat capacity and paramagnetism of an electron gas in metals and show that consideration of statistical degeneracy of an electron gas circumvents difficulties arising in classical statistics.

Table 7.1. The conditions of degeneracy of a Fermi gas

	Completely degenerate gas	Degenerate gas	Non-degenerate gas
Basic condition	$A = \exp(\eta) \gg 1$	$A = \exp(\eta) \geq 1$	$A = \exp(\eta) \ll 1$
Equivalent conditions, stemming from the basic condition	$\eta = \mu / k_0T \gg 1$ $\lambda \gg d$ $T \ll T_0$	$\eta = \mu/k_0T \approx 0$ $\lambda \geq d$ $T \leq T_0$	$-\eta = (-\mu/k_0T) \gg 1$ $\lambda \ll d$ $T \gg T_0$

7.9 Heat Capacity of Metals: First Difficulty of Classical Statistics

At the beginning of this chapter (Sect. 7.1), we saw that application of the Boltzmann classical statistics to the heat capacity of a free electron gas in metals leads to non-conformity between theory and experiment. It turned out that in the formation of the heat capacity in metals, an electron gas does not play any role. In this section we will show that the application of quantum statistics to an electron gas circumvents this non-conformity and makes it possible to understand why the presence of an electron gas does not influence the heat capacity of metals. As a result, we show that an electron gas in metals is not a usual classical gas but a quantum gas and, therefore, we cannot apply the Boltzmann distribution to it but we need to apply the Fermi–Dirac quantum statistics. We discuss all this more comprehensively below.

Assume that a crystalline lattice of a metal of volume V consists of N points at each of which one atom (ion) is found. If each atom loses one electron found at the last valence orbit, in the metal there arises a gas consisting of N electrons. The metal Na can serve as an example. If each atom of Na loses the only electron from the $3s$ level, the metal Na would consist of N positive ions and a gas consisting of the same number of free electrons.

Note that the free electron gas in metals is not associated with temperature, but due to a purely quantum effect, i.e. the ionization of atoms (the loss of electron) of a metal occurs not because of thermal motion but because of the overlap of wave functions of the valence electrons. The overlap of wave functions leads to the intermixing of valence electrons, so they do not belong to some specified point but move freely among them. For this reason, in metals a free electron gas exists at any temperature, even at the temperature of absolute zero.

Consequently, it can be regarded that in all temperature regions a metal consists of two subsystems: a crystalline lattice of ions and a free electron gas. Therefore, the heat capacity of a metal can be presented as a sum of two items:

$$C_V^{\text{met}} = C_V^{\text{lat}} + C_V^{\text{el}}. \tag{7.167}$$

We analysed the temperature dependence of the heat capacity of a crystalline lattice C_V^{lat} comprehensively in Sect. 6.4, in which we showed that the analytical expression of the heat capacity of the lattice can be found at temperatures above and below the Debye characteristic temperature θ:

$$C_V^{\text{lat}} = 3k_0N; \quad T \gg \theta \tag{7.168}$$

and

$$C_V^{\text{lat}} = \frac{12\pi^4}{5}k_0N\left(\frac{T}{\theta}\right)^3; \quad T \ll \theta, \tag{7.169}$$

where $\theta = \hbar\omega_{\max} / k_0$ is the Debye characteristic temperature, and $\omega_{\max}$ is the maximum possible frequency of the crystalline lattice. If we take the frequency $\omega_{\max} \approx 5 \times 10^{13}\,\text{s}^{-1}$, then $\theta \approx 300\,\text{K}$.

Because the electron gas is a Fermi gas, in order to find its heat capacity we apply the results obtained in Sects. 7.5 and 7.7. Then for C_V^{el} we can write the expressions in two limiting cases [see (7.111) and (7.140)]:

$$C_V^{\mathrm{el}} = \frac{3k_0 N}{2}; \quad T \gg T_0^{\mathrm{el}}, \tag{7.170}$$

corresponding to a non-degenerate electron gas and

$$C_V^{\mathrm{el}} = \frac{\pi^2}{2} k_0 N \frac{k_0 T}{\mu_0^{\mathrm{el}}} = \frac{\pi^2}{2} k_0 N \frac{T}{T_0^{\mathrm{el}}}; \quad T \ll T_0^{\mathrm{el}}, \tag{7.171}$$

corresponding to a strongly degenerate electron gas, where

$$T_0^{\mathrm{el}} = \frac{\mu_0^{\mathrm{el}}}{k_0} = \frac{\hbar^2}{2mk_0}\left(\frac{6\pi^2}{g_0} n\right)^{2/3} \tag{7.172}$$

is the temperature of degeneracy of the free electron gas in metals [see (7.126)], $n = N\,/\,V$ is its concentration, m is the effective mass of an electron in metals, μ_0^{el} is the Fermi boundary and $g_0 = 2$.

The linear dependence of the heat capacity of a strongly degenerate electron gas on temperature physically can be explained as follows: From Figs. 7.9 and 7.10, it is seen that when supplying heat externally, only electrons that are found close to the Fermi boundary at the distance $k_0 T$ receive heat and pass onto the levels lying above. Electrons that are found far below the Fermi boundary do not receive heat since the neighbouring levels are occupied (the Pauli principle). Because the number of electrons that receive heat is proportional to the width $k_0 T$, and energy of each electron is on the order of $k_0 T$, the thermal energy of the system of electrons $E \sim (k_0 T)^2 \sim T^2$. In conformity with this, it follows that $C_V \sim T$.

We evaluate the temperature of degeneracy of an electron gas T_0^{el}. To do this, assume that $m \approx 9.1 \times 10^{-28}$ g and $n \approx 5 \times 10^{22}\,\mathrm{cm}^{-3}$. Then from (7.172), we get $T_0^{\mathrm{el}} \approx 3 \times 10^4$ K. It is seen that temperature of degeneracy of a free electron gas is much higher than the Debye temperature, which characterises the vibrations of the lattice: $T_0^{\mathrm{el}} \gg \theta \approx 3 \times 10^2$ K.

In Table 7.2, we indicate the Debye temperature θ, the temperature of degeneracy of an electron gas T_0^{el} and expressions of the above-deduced asymptotes of C_V^{el} and C_V^{lat}.

From the table it is seen that the expression of the heat capacity C_V^{el} in the temperature region $\theta \ll T \ll T_0^{\mathrm{el}}$ is the same as in the region $T \ll \theta$, because

Table 7.2. The comparison the lattice and electron parts of the heat capacity metals at various temperature regions

C_v^{el}	$\frac{\pi^2}{2} k_0 N \frac{T}{T_0^{\mathrm{el}}}$		$\frac{\pi^2}{2} k_0 N \frac{T}{T_0^{\mathrm{el}}}$	
C_v^{lat}	$\frac{12\pi^4}{5} k_0 N \left(\frac{T}{\theta}\right)^3$	θ	$3k_0 N$	T_0^{e1}

$T_0^{\text{el}} \gg \theta$. The temperature region $T > T_0^{\text{el}}$ practically cannot be realized since T_0^{el} is larger than the melting temperature of metals. In other words, the classical expression of the heat capacity of an electron gas $C_V^{\text{el}} = 3k_0N\,/\,2$ is not applicable in any temperature region.

We now derive the asymptotic expressions of total heat capacities of metals measurable by experiments for different temperature regions.

7.9.1 Low Temperatures

$T \ll \theta$. It is clear that in this temperature region automatically the condition $T \ll T_0^{\text{el}}$ is fulfilled, because $\theta \ll T_0^{\text{el}}$. The total heat capacity C_V^{met} in this case ought to be equal to the sum of the expressions (7.169) and (7.171):

$$C_V^{\text{met}} = \frac{12\pi^4}{5} k_0 N \left(\frac{T}{\theta}\right)^3 + \frac{\pi^2}{2} k_0 N \left(\frac{T}{T_0^{\text{el}}}\right); \quad T \ll \theta. \tag{7.173}$$

This expression can also be presented in the form

$$C_V^{\text{met}}\,/\,T = a + bT^2, \tag{7.174}$$

where

$$a = \frac{\pi^2}{2}\frac{k_0 N}{T_0^{\text{el}}}; \quad b = \frac{12\pi^4}{5}\frac{k_0 N}{\theta^3}. \tag{7.175}$$

If, at low temperatures, we experimentally find the temperature dependence of the heat capacity $C_V^{\text{met}} = f(T)$ and construct a plot as the dependence $C_V^{\text{met}}\,/\,T$ on T^2, we get a straight line. The coordinate of the point of intersection of this straight line with the ordinate axis corresponds to the parameter a, and the slope tg φ gives the parameter b (Fig. 7.14). Thus, with the experimentally determined parameters a and b, and using (7.175), we can find the temperature of degeneracy T_0^{el} and the Debye temperature θ.

7.9.2 Region of Temperatures

$\theta \ll T \ll T_0^{\text{el}}$. From Table 7.2, it is seen that the total heat capacity of a metal in this region is determined by the expression

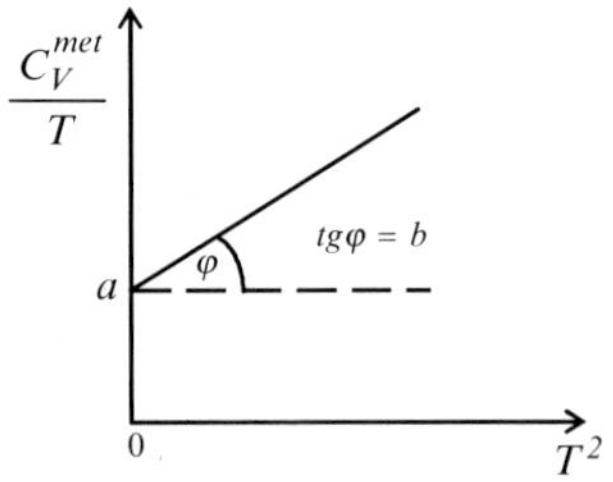

Fig. 7.14. The dependence $C_V^{\text{met}}\,/\,T$ on T^2

$$C_V^{\mathrm{met}} = 3k_0 N + \frac{\pi^2}{2} k_0 N \frac{T}{T_0^{\mathrm{el}}}. \tag{7.176}$$

We evaluate the ratio of the second term to the first term, i.e. $C_V^{\mathrm{el}} \,/\, C_V^{\mathrm{lat}}$, in the expression (7.176). Then we get

$$\frac{C_V^{\mathrm{el}}}{C_V^{\mathrm{lat}}} = \frac{\pi^2}{6} \frac{T}{T_0^{\mathrm{el}}} = \frac{\pi^2}{6} \frac{k_0 T}{\mu_0^{\mathrm{el}}} \ll 1. \tag{7.177}$$

Because the temperature of degeneracy is on the order of $T_0^{\mathrm{el}} \approx 10^4\,\mathrm{K}$, at all real temperatures and even above the Debye temperature θ, conditions $T \ll T_0^{\mathrm{el}}$ and $k_0 T \ll \mu_0^{\mathrm{el}}$ are satisfied. Therefore, at high temperatures $(T \gg \theta)$ the contribution of the electron gas to the heat capacity of metals is very small. As seen from (7.177), at a temperature $T = 500\,\mathrm{K}$ the ratio $C_V^{\mathrm{el}} \,/\, C_V^{\mathrm{lat}} \approx 0.07$, i.e. the contribution of C_V^{el}, comprises in all 7%.

Note that contributions of the electron gas and lattice to the heat capacity of a metal can be comparable only in the region of very high temperatures: $T \gg T_0^{\mathrm{el}}$. Such temperatures are non-real, because in this case the metal will not be in the solid state and the concept of a crystalline lattice loses sense.

Thus, we explained the first difficulty associated with application of the Boltzmann statistics when calculating the heat capacity of an electron gas in metals, which we noted at the beginning of this chapter (Sect. 7.1). It has now become clear that the divergence between theory and experiment is associated with the fact that in all temperature regions $(T \ll T_0^{\mathrm{el}} \approx 10^4\,\mathrm{K})$ *an electron gas is a strongly degenerate quantum gas.* Therefore, in all real temperature regions its heat capacity is determined not by (7.170) but by (7.171).

In conclusion, we note the following. We showed that at high temperatures $(T \gg \theta)$ the contribution of an electron gas to the heat capacity of a crystal is very small: $C_V^{\mathrm{el}} \ll C_V^{\mathrm{lat}}$. However, at ultra-low temperatures, C_V^{el} can be greater than C_V^{lat} (Fig. 7.15). This is associated with the fact that at $T \ll \theta$ the heat capacity $C_V^{\mathrm{lat}} \sim T^3$, whereas $C_V^{\mathrm{el}} \sim T$. Therefore, as $T \to 0$, the heat capacity C_V^{lat} tends to zero much faster than C_V^{el} (Fig. 7.15).

As is seen from the figure, as $T < T_1$ the heat capacity $C_V^{\mathrm{el}} > C_V^{\mathrm{lat}}$, where T_1 is determined from the condition $C_V^{\mathrm{el}}(T_1) = C_V^{\mathrm{lat}}(T_1)$. Hence, and also from (7.169) and (7.171), we get

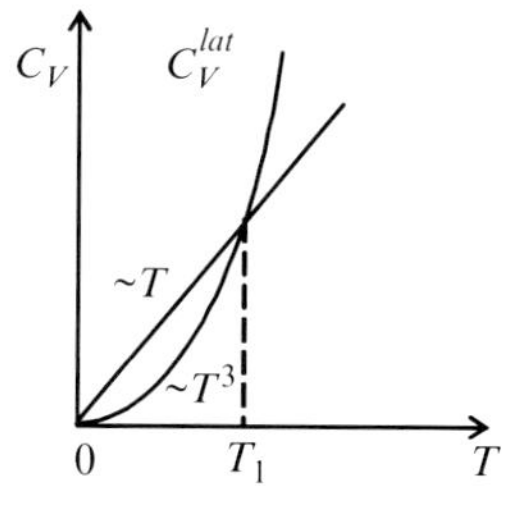

Fig. 7.15. The lattice and electron parts of the heat capacity dependence at ultra-low temperatures

$$T_1 \approx \frac{\theta}{2\pi}\left(\frac{\theta}{T_0^{\text{el}}}\right)^{1/2} . \tag{7.178}$$

If we take $\theta = 3 \times 10^2\,\text{K}$ and $T_0^{\text{el}} = 3 \times 10^4\,\text{K}$, we find $T_1 \approx 5\,\text{K}$.

7.10 Pauli Paramagnetism: Second Difficulty of Classical Statistics

The second difficulty of the Boltzmann classical statistics is associated with calculating the paramagnetic susceptibility χ of a free electron gas in metals (see Sect. 7.1). The paramagnetic susceptibility calculated on the basis of the Boltzmann statistics and its temperature dependence do not coincide with values from experiment. Thus, for instance, the experimental value of the paramagnetic susceptibility is lower by two orders of magnitude and does not depend on temperature, whereas according to the classical statistics $\chi \sim 1\,/\,T$.

This difficulty was circumvented by Pauli in 1927, after applying the new Fermi statistics to compute the paramagnetic susceptibility. He came to the conclusion that the cause of the divergence is the fact that *an electron gas is not classical but a strongly degenerate quantum gas*.

Assume that in a metal of volume V there is a gas consisting of N free electrons. Each electron possesses an intrinsic magnetic moment associated with the spin, which is equal to the Borh magneton $\mu_{\text{B}} = e\hbar\,/\,2mc$. In the absence of an external magnetic field, the intrinsic magnetic moments compensate each other (since, according to the Pauli principle in one quantum level there exist two electrons with opposite spins) and the magnetic moment of the metal as a whole equals zero.

On placing the metal in an external uniform magnetic field H, the number of electrons with spins directed along the magnetic field is more than that with the opposite spins, and therefore the electron gas possesses a paramagnetic moment. Using the Fermi statistics, we can calculate the magnetization of an electron gas in metals.

Energy of an electron in an external magnetic field H with regard to the spin has the form

$$\varepsilon = \frac{\hbar^2 k^2}{2m} + \sigma g_0 \mu_{\text{B}} H, \tag{7.179}$$

where m is the electron mass, $\mu_{\text{B}} = e\hbar\,/\,2mc = 0.93 \times 10^{-20}\,\text{erg/G}$ is the Bohr magneton, $\sigma = \mp 1\,/\,2$ (the sign "$-$" refers to electrons whose intrinsic magnetic moments are directed along the magnetic field, and "$+$" refers to electrons whose intrinsic magnetic moments are directed opposite to the field) and $g_0 = (2s+1)$ is the degree of the spin splitting degeneracy (for a free electron $s = 1\,/\,2$ and $g_0 = 2$). Energy spectra of an electron gas in the absence and in the presence of a magnetic field are schematically shown in Fig. 7.16.

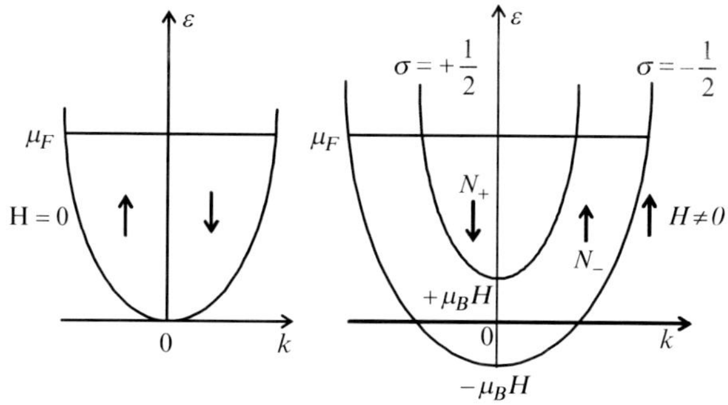

Fig. 7.16. The energy spectra of an electron gas in the absence and in the presence of a magnetic field

The total number of electrons can be found as follows:

$$N = \sum_{k,\sigma} f(\boldsymbol{k}, \sigma) = \frac{V}{(2\pi)^3} \sum_{\sigma} \int f(\boldsymbol{k}, \sigma)\, \mathrm{d}\boldsymbol{k}. \tag{7.180}$$

Here $f(\boldsymbol{k}, \sigma)$ is the Fermi–Dirac distribution function and, according to (7.11) we have changed from the summation with respect to $\boldsymbol{k}$ to the integration. If in the $\boldsymbol{k}$-space we change to the spherical coordinate system and take into account that the integral over angles equals 4π and, also, according to (7.179) change from the integration over $\mathrm{d}k$ to the integration over $\mathrm{d}\varepsilon$, (7.180) takes the form

$$N = \frac{V}{(2\pi)^2} \frac{(2m)^{3/2}}{\hbar^3} \sum_{\sigma} \int\limits_{-\sigma g_0 \mu_{\mathrm{B}} H}^{\infty} f(\varepsilon)(\varepsilon + \sigma g_0 \mu_{\mathrm{B}} H)^{1/2} \mathrm{d}\varepsilon. \tag{7.181}$$

We integrate the integral entering into (7.181) once more by parts. Then we get

$$N = \frac{2V}{3(2\pi)^2} \frac{(2m)^{3/2}}{\hbar^3} \sum_{\sigma} \int\limits_{-\sigma g_0 \mu_{\mathrm{B}} H}^{\infty} \left(-\frac{\partial f}{\partial \varepsilon}\right) (\varepsilon + \sigma g_0 \mu_{\mathrm{B}} H)^{3/2} \mathrm{d}\varepsilon. \tag{7.182}$$

If we add up with respect to $\sigma = \pm 1 / 2$ and take $g_0 = 2$, (7.182) takes the form

$$N = N_+ + N_-, \tag{7.183}$$

where

$$N_- = \frac{2V}{3(2\pi)^2} \frac{(2m)^{3/2}}{\hbar^3} \int\limits_{-\mu_{\mathrm{B}} H}^{\infty} \left(-\frac{\partial f}{\partial \varepsilon}\right) (\varepsilon + \mu_{\mathrm{B}} H)^{3/2} \mathrm{d}\varepsilon \tag{7.184}$$

is the number of electrons whose spins are parallel to the magnetic field H, and

$$N_{+} = \frac{2V}{3(2\pi)^2} \frac{(2m)^{3/2}}{\hbar^3} \int\limits_{+\mu_{\mathrm{B}}H}^{\infty} \left(-\frac{\partial f}{\partial \varepsilon}\right) (\varepsilon - \mu_{\mathrm{B}}H)^{3/2} \mathrm{d}\varepsilon \tag{7.185}$$

is the number of electrons whose spins are antiparallel to the magnetic field H.

It is evident that as a whole the paramagnetic moment of an electron gas in metal can be defined as

$$M = \mu_{\mathrm{B}}(N_{-} - N_{+}). \tag{7.186}$$

We first consider the classical (a non-degenerate electron gas) case. In this case, $f(\varepsilon) = \exp\left((\mu - \varepsilon) / k_0 T\right)$. Then, (7.186) takes the form

$$M = \frac{V\mu_{\mathrm{B}}}{4\pi^{3/2}} \frac{(2mk_0T)^{3/2}}{\hbar^3} \mathrm{e}^{\mu/k_0T}\, sh(\mu_{\mathrm{B}}H/k_0T). \tag{7.187}$$

The chemical potential μ entering into this expression is found from (7.183) with regard to (7.184) and (7.185):

$$\mathrm{e}^{\mu/k_0T} = \frac{N}{V} \frac{4\pi^{3/2}\hbar^3}{(2mk_0T)^{3/2}} ch^{-1}(\mu_{\mathrm{B}}H/k_0T). \tag{7.188}$$

If we take into account this expression in (7.187), for the paramagnetic moment per unit volume we get the well-known result

$$M = N\mu_{\mathrm{B}} th\left(\mu_{\mathrm{B}}H / k_0T\right). \tag{7.189}$$

In the region of weak magnetic fields, when $\mu_{\mathrm{B}}H \ll k_0T$, from (7.189), for the paramagnetic susceptibility we get

$$\chi = \frac{M}{VH} = \frac{n\mu_{\mathrm{B}}^2}{k_0T}, \tag{7.190}$$

where $n = N / V$ is the concentration of electrons. Note that in experiments this classical result (7.190) is just not observed.

Therefore, it is necessary to consider the case using quantum statistics. If we take into account that for a completely degenerate electron gas $(-\partial f / \partial \varepsilon) = \delta(\varepsilon - \mu_{\mathrm{F}})$, from (7.184) to (7.186) we get

$$M = \frac{\mu_{\mathrm{B}}2V}{3(2\pi)^2} \frac{(2m)^{3/2}}{\hbar^3} \mu_{\mathrm{F}}^{3/2} \left[\left(1 + \frac{\mu_{\mathrm{B}}H}{\mu_{\mathrm{F}}}\right)^{3/2} - \left(1 - \frac{\mu_{\mathrm{B}}H}{\mu_{\mathrm{F}}}\right)^{3/2}\right]. \tag{7.191}$$

In a weak magnetic field ($\mu_{\mathrm{B}}H \ll \mu_{\mathrm{F}}$), from (7.191) it follows that

$$M = \frac{V\mu_{\mathrm{B}}^2}{2\pi^2} \frac{(2m)^{3/2}}{\hbar^3} \mu_{\mathrm{F}}^{1/2} H. \tag{7.192}$$

The Fermi boundary μ_F entering into this expression is found from the expressions (7.183)–(7.185). It can be shown that the Fermi boundary μ_F, to a first approximation with respect to the parameter $\mu_B H / \mu_F \ll 1$, does not depend on the magnetic field and is determined by the known expression [see (7.116)]

$$\mu_F = \frac{\hbar^2}{2m}(3\pi^2 n)^{2/3}. \tag{7.193}$$

For the paramagnetic susceptibility $\chi_0 = M / VH$ of a completely degenerate electron gas, from (7.192) we get the simple expression

$$\chi_0 = \mu_B^2 g(\mu_F), \tag{7.194}$$

where

$$g(\mu_F) = \frac{(2m)^{3/2}}{2\pi^2\hbar^3}\mu_F^{1/2} \tag{7.195}$$

is a function of the density of quantum states on the Fermi boundary.

From (7.194), it is seen that the paramagnetic susceptibility χ_0 of a degenerate electron gas does not depend on temperature (7.194) and its ratio to χ (7.190) is given the expression

$$\frac{\chi_0}{\chi} = \frac{k_0 T}{n} g(\mu_F). \tag{7.196}$$

If we determine the concentration of electrons n from (7.193) with regard to (7.195) and substitute it into (7.196), we get

$$\frac{\chi_0}{\chi} = \frac{3}{2}\frac{k_0 T}{\mu_F} = \frac{3}{2}\frac{T}{T_0^{\text{el}}} \ll 1, \tag{7.197}$$

where $T_0^{\text{el}} = \mu_F / k_0 \approx 3 \times 10^4\,\text{K}$ is the temperature of degeneracy of the electron gas. At room temperature ($T \approx 3 \times 10^2\,\text{K}$), for ratio (7.197) we get $\chi_0 / \chi = 10^{-2}$.

Note that the ratio (7.197) is almost the same as the ratio of corresponding heat capacities of an electron gas. Indeed, if we take the ratio of (7.171) to $C_V^{\text{cl}} = 3k_0 N / 2$, we get

$$\frac{C_V^{\text{el}}}{C_V^{\text{cl}}} = \frac{\pi^2}{3}\frac{k_0 T}{\mu_F^{\text{el}}} = \frac{\pi^2}{3}\frac{T}{T_0^{\text{el}}}. \tag{7.198}$$

Thus, in all the cases we get results consistent with experiments, i.e. we are convinced that *the free electron gas in metals is not a classical gas but a strongly degenerate quantum gas.*

If at finite values of temperature with the aid of (7.134) we calculate integrals (7.184) and (7.185), for the paramagnetic susceptibility we get

$$\chi_0(T) = \mu_B^2 g(\mu_F)\left(1 - \frac{\pi^2}{12}\frac{k_0 T}{\mu_F}\right). \tag{7.199}$$

Hence, it is seen that at temperatures $k_0 T \ll \mu_F$ the paramagnetic susceptibility χ_0 depends on temperature T very weakly.

7.11 "Ultra-Relativistic" Electron Gas in Semiconductors

In this case, the energy of a particle linearly depends on the absolute value of the wave vector:

$$\varepsilon = vp = v\hbar k. \tag{7.200}$$

Photons and phonons satisfy this model of the energy spectrum. In expression (7.200), in the case of photons the velocity ought to be replaced with the velocity of light $v = c$, and for phonons with the speed of sound $v = v_0$.[4]

The conduction electrons of some complex narrow-band semiconductors (Cd_xHg_{1-x} Te; $x = 0.16$) also are described by the expression (7.200). For conduction electrons of these semiconductors, we need to replace v with the expression

$$v = (\varepsilon_g / 2m^*)^{1/2}, \quad \varepsilon = (\varepsilon_g / 2m^*)^{1/2}\, \hbar k, \tag{7.200a}$$

where ε_g is the band gap width, and m^* is the effective mass of an electron.

In this section we will consider an electron gas with the "ultra-relativistic" spectrum (7.200a). We will start from the expression of the grand thermodynamic potential (7.95). For fermions, in (7.95) we need to take the upper signs. Then, according to (7.200a), we go over from the integral over dk to the integral over dε. Then, the expression of the thermodynamic potential takes the form

$$\Omega = -\frac{Vg_0}{2\pi^2\hbar^3}\left(\frac{2m^*}{\varepsilon_g}\right)^{3/2} k_0T \int_0^\infty \ln\left(1 + \mathrm{e}^{(\mu-\varepsilon)/\,k_0T}\right)\varepsilon^2 \mathrm{d}\varepsilon. \tag{7.201}$$

If we take this expression in the system of equations (7.91) and integrate by parts, we find the general form of the thermal equation of state of a gas consisting of "ultra-relativistic" electrons:

$$\begin{cases} P = \dfrac{g_0}{6\pi^2\hbar^3}\left(\dfrac{2m^*}{\varepsilon_g}\right)^{3/2} \displaystyle\int_0^\infty \dfrac{\varepsilon^3 \mathrm{d}\varepsilon}{\mathrm{e}^{(\varepsilon-\mu)/\,k_0T} + 1}, \\ N = \dfrac{Vg_0}{2\pi^2\hbar^3}\left(\dfrac{2m^*}{\varepsilon_g}\right)^{3/2} \displaystyle\int_0^\infty \dfrac{\varepsilon^2 \mathrm{d}\varepsilon}{\mathrm{e}^{(\varepsilon-\mu)/\,k_0T} + 1}. \end{cases} \tag{7.202}$$

From (7.80), it is easy to find the mean value of energy of an "ultra-relativistic" electron gas:

$$E = \frac{Vg_0}{2\pi^2\hbar^3}\left(\frac{2m^*}{\varepsilon_g}\right)^{3/2} \int_0^\infty \frac{\varepsilon^3 \mathrm{d}\varepsilon}{\mathrm{e}^{(\varepsilon-\mu)/\,k_0T} + 1}. \tag{7.203}$$

[4] In the capacity of an example of the ultra-relativistic Bose gas in Sects. 7.14 and 7.15 photon and phonon gases will be considered, respectively.

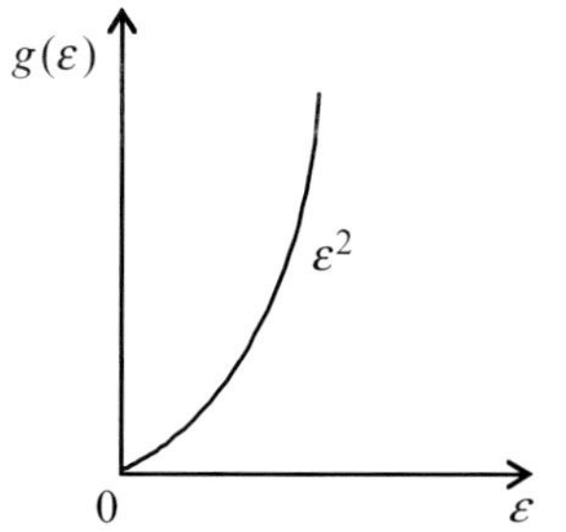

Fig. 7.17. The density of quantum states for the "ultra-relativistic" spectra

If we compare this expression with the first equation of the system (7.202), we get the relation between the energy density and pressure:

$$P = \frac{1}{3}\frac{E}{V}, \tag{7.204}$$

which does not depend on the degree of degeneracy of the electron gas [compare (7.100) with the "ultra-relativistic" case].

From a comparison of expressions of energies (7.101) and (7.203), for the density of quantum states $g(\varepsilon)$ we get the quadratic dependence (Fig. 7.17)

$$g(\varepsilon) = \frac{g_0\varepsilon^2}{2\pi^2\hbar^3}\left(\frac{2m^*}{\varepsilon_{\mathrm{g}}}\right)^{3/2} \sim \varepsilon^2. \tag{7.205}$$

It is seen that in this case the dependence of the density of states on energy $g(\varepsilon)$ is sharper compared to the usual parabolic case (Fig. 7.4).

If we integrate the expression for energy (7.203) and the second equation of (7.202) once by parts, we get the system of equations determining the caloric equation of states of an "ultra-relativistic" electron gas:

$$\begin{cases} E = \dfrac{Vg_0}{8\pi^2\hbar^3}\left(\dfrac{2m^*}{\varepsilon_{\mathrm{g}}}\right)^{3/2}\displaystyle\int_0^\infty\left(-\frac{\partial f}{\partial\varepsilon}\right)\varepsilon^4\mathrm{d}\varepsilon, \\ N = \dfrac{Vg_0}{6\pi^2\hbar^3}\left(\dfrac{2m^*}{\varepsilon_{\mathrm{g}}}\right)^{3/2}\displaystyle\int_0^\infty\left(-\frac{\partial f}{\partial\varepsilon}\right)\varepsilon^3\mathrm{d}\varepsilon. \end{cases} \tag{7.206}$$

We first consider *a completely degenerate* electron gas ($T = 0$). If we take into account that in this case $(-\partial f\,/\,\partial\varepsilon) = \delta(\varepsilon - \mu_0)$, from (7.206) we get the expressions for the Fermi boundary as

$$\mu_0 = \left(\frac{6\pi^2}{g_0}n\right)^{1/3}\hbar\left(\frac{\varepsilon_{\mathrm{g}}}{2m^*}\right)^{1/2} = \left(\frac{\varepsilon_{\mathrm{g}}}{2m^*}\right)^{1/2}p_0, \tag{7.207}$$

for the Fermi impulse as

$$p_0 = \left(\frac{6\pi^2}{g_0}n\right)^{1/3}\hbar, \tag{7.208}$$

for the zero energy as

$$E_0 = \frac{Vg_0}{8\pi^2\hbar^3}\left(\frac{2m^*}{\varepsilon_\mathrm{g}}\right)^{3/2}\mu_0^4 = \frac{3}{4}\mu_0 N \tag{7.209}$$

and for the zero pressure of the electron gas as

$$P_0 = \frac{g_0}{24\pi^2\hbar^3}\left(\frac{2m^*}{\varepsilon_\mathrm{g}}\right)^{3/2}\mu_0^4 = \frac{1}{4}n\mu_0, \tag{7.210}$$

where we used the relationship (7.204); $n = N\,/\,V$ is the concentration of electrons. Because at the Fermi boundary $\mu_0 \sim n^{1/3}$ (7.207), from (7.210) it follows that the zero pressure of a degenerate electron gas $P_0 \sim n^{4/3}$. The temperature of degeneracy in the considered case is determined by the expression $T_0 = \mu_0\,/\,k_0$.

Consider *a strongly degenerate electron gas* when $k_0T \ll \mu_0$. In this case, using (7.134) we can calculate the integrals entering into (7.206). As a result, we get

$$E = \frac{Vg_0}{8\pi^2\hbar^3}\left(\frac{2m^*}{\varepsilon_\mathrm{g}}\right)^{3/2}\mu^4\left[1 + 2\pi^2\left(\frac{k_0T}{\mu_0}\right)^2\right], \tag{7.211}$$

$$N = \frac{Vg_0}{6\pi^2\hbar^3}\left(\frac{2m^*}{\varepsilon_\mathrm{g}}\right)^{3/2}\mu^3\left[1 + \pi^2\left(\frac{k_0T}{\mu_0}\right)^2\right]. \tag{7.212}$$

Here, in the term associated with temperature we have assumed $\mu = \mu_0$. From (7.212) on determining the chemical potential

$$\mu(T) = \mu_0\left[1 - \frac{\pi^2}{3}\left(\frac{k_0T}{\mu_0}\right)^2\right], \tag{7.213}$$

and substituting it into (7.211), and restricting ourselves to the first approximation, for the caloric equation (for the total energy) we get

$$E(T) = E_0 + \frac{\pi^2}{2}\mu_0 N\left(\frac{k_0T}{\mu_0}\right)^2. \tag{7.214}$$

Hence, for the heat capacity we have

$$C_V = \pi^2 k_0 N\left(\frac{k_0T}{\mu_0}\right) = \frac{\pi^2 k_0^2 N}{\hbar}\left(\frac{2m^*}{\varepsilon_\mathrm{g}}\right)^{1/2}\left(\frac{g_0}{6\pi^2 n}\right)^{1/3} T \sim n^{-1/3}. \tag{7.215}$$

From relationship (7.204) and expression (7.214), we can obtain the thermal equation of state (pressure) ofan "ultra-relativistic" electron gas at finite temperatures:

$$P(T) = P_0 + \frac{\pi^2}{6}\mu_0 n \left(\frac{k_0 T}{\mu_0}\right)^2 , \tag{7.216}$$

where P_0 is the zero pressure determined by (7.210).

If we compare the results obtained here with those obtained in Sect. 7.6, we notice the disparity between a degenerate electron gas with the usual parabolic spectrum and an "ultra-relativistic" gas. In particular, a comparison of the expressions (7.213) and (7.136) shows that in the "ultra-relativistic" case the Fermi boundary decreases according to the temperature. This is associated with the fact that in the "ultra-relativistic" case the function of the density of quantum states depends more strongly on energy (compare Figs. 7.4 and 7.17).

7.12 Statistics of Charge Carriers in Semiconductors

According to quantum mechanics, electrons in a completely isolated atom possess a discrete energy spectrum: $1s\,2s\,2p\,3s\,3p\,3d\,4s\ldots$. At each discrete energy level, a specified number of electrons can be found, e.g. at the s-level there can be a maximum 2 electrons, at the p-level there can be 6, at the -level there can 10: $1s^2\,2s^2\,2p^6\,3s^2\,3p^6\,3d^{10}\,4s^2\ldots$.

Assume that a crystalline lattice (solid) is formed of N atoms. Then, each discrete energy level of the system (the crystalline lattice) is N-fold degenerate. Under the action of a periodic Coulomb field produced by other electrons and nuclei, this degeneracy is removed partially or completely. As a result, each discrete energy level converts into an energy band $1s^2\,2s^2\,2p^6\ldots$ of a certain width. Therefore, the energy spectrum of electrons in a crystalline lattice consists of possible energy strips, separated from each other by forbidden energy ranges - *band gaps*.

Note that the band theory of solids analytically follows from the solution of the Schrödinger equation for the motion of an electron in a periodic potential field of the crystalline lattice.

According to the band theory, all crystalline solids are divided into two classes based on the electrical conductivity in the fundamental non-excited state: *conductors* and *insulators*.

We explain this classification. In a crystal consisting of N atoms with the ordinal number Z, there exists a ZN number of electrons. In the fundamental state, these electrons occupy the lowermost energy bands. According to the Pauli principle, in each energy band there can be found a finite number of electrons. For instance, in the s-energy band, corresponding to the discrete level s, we can find a maximum of $2N$electrons, in the p-band there can be a maximum of $6N$ electrons, etc.

Depending on the distribution of electrons over the energy bands, there are two possible cases.

1. If ZN number of electrons occupy some energy bands completely and subsequent bands partially, such a crystal is called *a conductor*. For instance, in Na crystal, consisting of Na atoms with $Z = 11$, of $11\,N$ electrons, $2\,N$ electrons are positioned in the energy band $1s$, $2\,N$ are in $2s$, $6N$ are in $2p$ and N are in $3s$. Thus, the $3s$ energy band is half-filled. In Al atom with $Z = 13$, there are completely filled $1s\,2s\,2p\,3s$ bands, and the $3p$ band is filled to one-sixth. Notice that a completely filled energy band does not take part in the electrical conductivity because an external electric field cannot influence the distribution of electrons of completely filled bands. Only electrons of partially filled bands can participate in the conductivity because inside the band, energy levels are positioned infinitely close to each other and an external electric field changes the symmetry of the distribution of electrons in the $\boldsymbol{k}$-space, and thereby electrical conductivity arises in the crystal. Thus, Na and Al ought to conduct an electric current well. Remember that *a conductor* and *a metal* are used synonymously here, because most metals possess good electrical conductivity.
2. If ZN number of electrons completely fill a specified number of lowest energy bands, and the subsequent energy bands are absolutely empty, then, according to the above-mentioned reasons such a crystal is *an insulator*. The energy range between the completely filled energy band and the absolutely empty band is called *a forbidden energy band or band gap*. The completely filled band is called *the valence band* and the upper empty band is *the conduction band*.

Thus, at the absolute zero temperature ($T = 0$), all crystalline solids can be found in one of two fundamental (non-excited) states: *conductors* or *insulators*.

If the width of the band gap ε_{g} is sufficiently small ($\varepsilon_{\mathrm{g}} < 2\,\mathrm{eV}$), at finite temperatures ($T \neq 0$) some of the electrons pass from the completely filled (valence) band into the empty (conduction) band. The electrons passing into the conduction band can participate in the conductivity and therefore are called *conduction electrons*. The remaining free quantum states of the valence band behave as quasi-particles and are called *holes*. Holes are quasi-particles with a positive charge which is numerically equal to the charge of an electron. It is clear that holes are also fermions like electrons. Their effective mass is positive and distinct from the effective mass of electrons. Holes that are formed in the valence band also participate in the electrical conductivity.

Insulators that are found in such an excited state are called *semiconductors*. Thus, *semiconductors are insulators with a small width of the band gap and easily pass into the excited state*. In other words, *a semiconductor is an excited insulator*.

Consequently, *free electrons* are charge carriers in metals, and *conduction electrons* and *free holes* in the valence band are carriers in semiconductors.

Statistical properties of a free electron gas in metals were studied in Sect. 7.11. The basic distinction of metals from semiconductors consists in the fact that in metals the concentration of the conduction electrons is constant

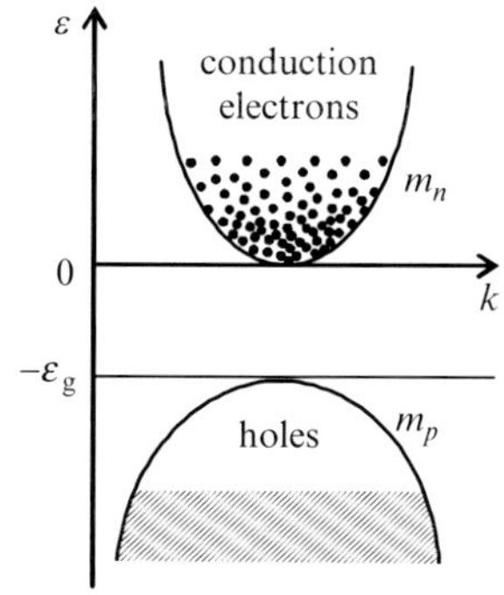

Fig. 7.18. Conduction band electrons and valence band holes, i.e. current carriers in semiconductors

whereas in semiconductors the concentration depends on temperature and on the presence or absence of impurities. In the given section we will consider the statistics of charge carriers, i.e. electron–hole systems, and analyse the temperature dependences of their chemical potential and concentration.

To do this it, is necessary to know the structure of the conduction band and the valence band, i.e. the dependence of energy on the quasi-impulse or wave vector (the dispersion law). It is a task of quantum mechanics. Note that the structure of the conduction band and the valence band can be quite complex. This depends on the construction of electron shell of the atoms forming the crystal and also on the symmetry of the crystalline lattice. Here, we will assume that energy of conduction electrons and holes depends parabolically on the wave vector (Fig. 7.18). On the basis of this model, called *a standard* or *parabolic model*, we will consider three cases: intrinsic (impurity-free) semiconductors, semi-metals and impurity semiconductors.

Intrinsic semiconductors. Consider an ideal (perfect) impurity-free semiconducting crystal. This means that the crystal consists of intrinsic atoms. The valence electrons form covalent bonds between adjacent atoms. For instance, in the tetravalent Ge or Si crystal, they are surrounded by four adjacent atoms. At the temperature of absolute zero, all four electrons participate in the formation of the bond, i.e. at $T = 0$, the valence band is completely filled and the conduction band is completely free.[5] At a finite temperature, some of the electrons participating in the formation of the covalent bond, under

[5] Note that according to the electron distribution over energy bands the crystals of germanium and silicon formally ought to be conductors. Indeed, in an atom of silicon with $Z = 14$ the level $3p^2$ is filled to $1/3$, in germanium with $Z = 32$ the level $4p^2$ is also filled to $1/3$. Consequently, at temperature of absolute zero also these crystals ought to be conductors. However, theory and numerous experiments show that Ge and Si possess the complex band structure: the band gap width for Ge equals $\varepsilon_g = 0.7\,\text{eV}$, and for Si equals $\varepsilon_g = 1.1\,\text{eV}$. So, according to the real band structure, Ge and Si are not conductors, but semiconductors.

action of the thermal motion (temperature), break away from the chemical bond and move freely inside the crystal. Holes are produced in the vacancies created by the electrons, which also move freely in the crystal. In the impulse or the k-space, this process corresponds to the fact that a specified number of electrons pass from the valence band into the conduction band, as a result of which the same number of holes arise in the valence band.

In order to consider the statistics of an electron–hole system in intrinsic semiconductors, consider a model of the band as depicted in Fig. 7.18. This model is characterised by three parameters: the effective mass of conduction electrons m_n, the effective mass of holes m_p and the width of the band gap ε_g.

If energy is taken from the bottom of the conduction band, the energy of the conduction electrons is

$$\varepsilon = \frac{\hbar^2 k^2}{2m_n}, \tag{7.217}$$

and the energy of holes is

$$\varepsilon = -\varepsilon_g - \frac{\hbar^2 k^2}{2m_p} = -\varepsilon_g - \varepsilon'. \tag{7.218}$$

The concentration of conduction electrons in the general case, according to (7.160), is

$$n = \frac{(2m_n k_0 T)^{3/2}}{3\pi^2\hbar^3} F_{3/2}(\eta), \tag{7.219}$$

where $F_{3/2}(\eta)$ is the Fermi integral determined by (7.161), and $\eta = \mu/k_0T$ is the reduced chemical potential.

If we use the asymptote of the Fermi integral (7.162), the concentration of a non-degenerate electron gas in the conduction band is

$$n = \frac{(2m_n k_0 T)^{3/2}}{4\pi^{3/2}\hbar^3} \exp(\eta). \tag{7.219a}$$

In order to find the concentration of holes, first we determine their distribution function. It is known that the probability of an electron found at a level with energy ε is given by the Fermi–Dirac distribution function $f = [1 + \exp(\varepsilon - \mu)/k_0T]^{-1}$. Then, the probability that a level with energy ε is empty (the hole), i.e. the distribution function of holes, is

$$f_p(\varepsilon) = 1 - f(\varepsilon) = [1 + \exp(\mu - \varepsilon)/k_0T]^{-1}. \tag{7.220}$$

If we substitute the expression for the energy of holes (7.218) in (7.220) and introduce the notation for the chemical potential of holes

$$\mu_p = -\varepsilon_g - \mu, \tag{7.221}$$

then distribution function of holes takes the form

$$f_p(\varepsilon') = \left[1 + \exp(\varepsilon' - \mu_p)/k_0T\right]^{-1}, \tag{7.222}$$

where ε' is energy taken from the top of the valence band. So, in order to find the distribution function of holes, it is sufficient to replace ε with ε' in the distribution function of electrons and μ with μ_p. Thus, in order to find the concentration of holes in the parabolic valence band, in (7.219) it is necessary to carry out the replacements $m_n \to m_p$ and $\eta \to \eta_p = \mu_p/k_0T$.

Consequently, the concentration of a hole gas of any degree of degeneracy is calculated by the formula

$$p = \frac{(2m_pk_0T)^{3/2}}{3\pi^2\hbar^3}F_{3/2}(\eta_p), \quad \eta_p = \mu_p/k_0T. \tag{7.223}$$

It is clear that the condition of non-degeneracy of a hole gas has the appearance

$$\exp(\eta_p) \ll 1 \quad \text{or} \quad (\varepsilon_g + \mu)/k_0T \gg 1. \tag{7.224}$$

If use the asymptote of the Fermi gas (7.162), for the concentration of a non-degenerate hole gas we get

$$p = \frac{(2m_pk_0T)^{3/2}}{4\pi^{3/2}\hbar^3}\exp(\eta_p). \tag{7.225}$$

The chemical potential of an electron–hole gas is determined from the condition of neutrality of the crystal

$$n(\mu, T) = p(\mu, T). \tag{7.226}$$

If we take into account (7.219) and (7.223) in this equation, we get

$$m_n^{3/2}F_{3/2}(\eta) = m_p^{3/2}F_{3/2}(-\eta - \varepsilon_{\mathrm{g}}^*), \tag{7.227}$$

where $\varepsilon_{\mathrm{g}}^* = \varepsilon_{\mathrm{g}}/k_0T$ is the reduced band gap width. A solution of (7.227), in principle, enables one to determine the dependence of the reduced chemical potential of an electron–hole gas as a function of the parameters of the band:

$$\eta = \eta(m_n, m_p, \varepsilon_{\mathrm{g}}^*). \tag{7.228}$$

However, in the general case (7.227) can be solved only numerically. If we take into account that a conduction electron and a hole gas in intrinsic semiconductors are usually found in the non-degenerate state, on the strength of the asymptote (7.162), expression (7.227) takes the simple form

$$m_n^{3/2}\exp(\eta) = m_p^{3/2}\exp(-\eta - \varepsilon_{\mathrm{g}}^*). \tag{7.229}$$

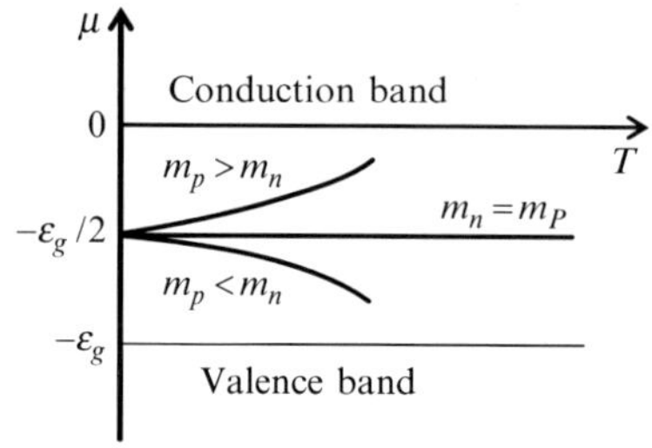

Fig. 7.19. The temperature dependence of chemical potential for intrinsic semiconductors

Hence, it is easy to find the chemical potential of an electron–hole gas:

$$\mu = -\frac{\varepsilon_g}{2} + \frac{3}{4} k_0 T \ln \frac{m_p}{m_n}. \tag{7.230}$$

From (7.230), it is seen that at $T=0$ (for any values of m_p and m_n), and also if $m_n = m_p$, at all temperatures the chemical potential passes through the middle of the band gap. If, however, the effective masses are different ($m_n \neq m_p$), with the increase of temperature starting from zero, depending on the ratio m_p/m_n the chemical potential moves to the conduction band or the valence band (Fig. 7.19).

Usually, in semiconductors $m_p > m_n$, and therefore the chemical potential moves to the conduction band. The fact that at $T = 0$ the chemical potential equals half ε_g has a deep physical meaning. It is known that the chemical potential is the free energy (the spent work) necessary to increase the number of particles by unity. In our case, because the system consists of conduction electrons and holes, in the process of passage of one electron from the valence band into the conduction band the energy ε_g is spent. As a result of this, in the system there arise two quasi-particles: a conduction electron and a hole. Therefore, at $T=0$ each arising particle accounts for a free energy equal to $\mu = -\varepsilon_g/2$.

If we substitute the value of $\exp(\eta)$ from (7.229) into (7.219a) and (7.225), we get identical dependences for the concentration of free electrons and holes in intrinsic semiconductors:

$$n_i = p_i = \frac{\left(2\sqrt{m_p m_n} k_0 T\right)^{3/2}}{4\pi^{3/2}\hbar^3} \exp\left(-\frac{\varepsilon_g}{2k_0 T}\right). \tag{7.231}$$

If from experiment we can find n_i or p_i, with the aid of (7.231) we can determine the band gap width ε_g. This is the most common classical method for the determination of ε_g.

If the conduction band and the valence band touch each other, i.e. $\varepsilon_g = 0$, (7.227) takes the form

$$m_n^{3/2} F_{3/2}(\eta) = m_p^{3/2} F_{3/2}(-\eta). \tag{7.232}$$

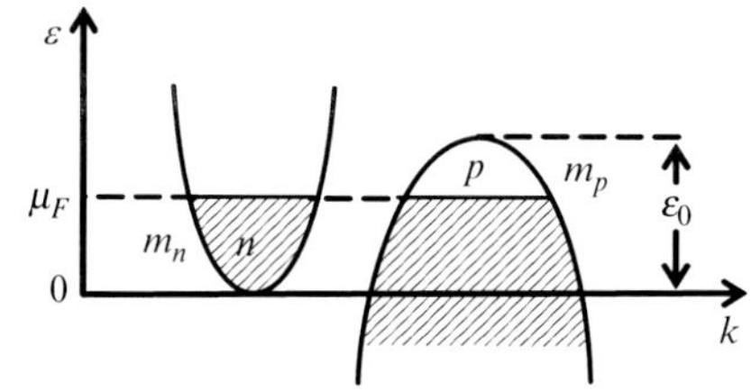

Fig. 7.20. The filling of the energy band of semimetals by electrons and holes

Because temperature does not enter into this equation, the solution of the equation does not depend on temperature, i.e. $\eta = \eta_0 = \text{const}$. Then concentration of conduction electrons (7.219) can be written down in the form

$$n_i = \frac{(2m_n k_0 T)^{3/2}}{3\pi^2\hbar^3} F_{3/2}(\eta_0) \sim T^{3/2}. \tag{7.233}$$

Thus, at $\varepsilon_g = 0$ the concentration depends on temperature not exponentially, as is shown in (7.231), but has power dependence $n \sim T^{3/2}$. It is clear the concentration of holes is

$$p_i = \frac{(2m_p k_0 T)^{3/2}}{3\pi^2\hbar^3} F_{3/2}(-\eta_0) \sim T^{3/2}. \tag{7.234}$$

As seen from (7.232), in order that the reduced chemical potential $\eta = \eta_0 = \text{const}$, it is necessary to satisfy the equality $n_i = p_i$.

Note that dependence $n_i = p_i \sim T^{3/2}$ is correct only in the case when both the conduction band and the valence band are parabolic.

Semi-metals. In the preceding paragraph, we considered semiconductors with the band gap width $\varepsilon_g \geq 0$. In these, the charge carriers arise only at finite temperatures $T \neq 0$, i.e. at $T = 0$ they do not possess any conductivity. However, there exist crystals in which the valence band and the conduction band overlap each other by a magnitude ε_0 (Fig. 7.20). Such crystals are called *semi-metals.* A crystal of bismuth is an example of a semi-metal. In such crystals, the top of the valence band is above the bottom of the conduction band and, even at $T = 0$, a part of the electrons spontaneously pass from the valence band into the conduction band, as a result of which an equal number of free electrons and holes arise. The passage of electrons from the valence band into the conduction band continues as long as the chemical potential (the Fermi level) does not become identical in both bands (as the level of a liquid in connected vessels).

Consequently, at the temperature of absolute zero, in semi-metals electrical conductivity is possible. The Fermi level in semi-metals at temperature T can be determined from the condition of neutrality $n = p$, using equality (7.227). To do this, it is necessary to carry out the replacement $-\varepsilon_g \to \varepsilon_0$ in (7.227), because, as seen from Fig. 7.20, the distance between the top of the valence band and the bottom of the conduction band equals ε_0. As a result, in the case of a semi-metal the condition of neutrality $n = p$ takes the form

$$m_n^{3/2} F_{3/2}(\eta) = m_p^{3/2} F_{3/2}(\varepsilon_0^* - \eta), \tag{7.235}$$

where $\varepsilon^* = \varepsilon_0 / k_0 T$, ε_0 is the overlap of the bands and m_n and m_p are effective masses of conduction electrons and holes, respectively.

Because at low temperatures $k_0 T \ll \varepsilon_0$, electron and hole gases are found in the degenerate state. Then, using (7.165), for the Fermi integral we get $F_k(\eta_\mathrm{F}) = \eta_\mathrm{F}^k$ and condition (7.235) takes the form

$$m_n \eta_\mathrm{F} = m_p(\varepsilon_0^* - \eta_\mathrm{F}). \tag{7.236}$$

Hence

$$\eta_\mathrm{F} = \frac{m_p}{m_n + m_p} \varepsilon_0^* \quad \text{or} \quad \mu_\mathrm{F} = \frac{m_p}{m_n + m_p} \varepsilon_0. \tag{7.237}$$

The concentration of electrons, in conformity with (7.160) and (7.165), can be written down in the form

$$n = \frac{(2 m_n k_0 T)^{3/2}}{3\pi^2 \hbar^3} \eta_\mathrm{F}^{3/2}, \tag{7.238}$$

and the concentration of holes as

$$p = \frac{(2 m_p k_0 T)^{3/2}}{3\pi^2 \hbar^3} (\varepsilon_0^* - \eta_\mathrm{F})^{3/2}. \tag{7.239}$$

On substituting (7.236) and (7.237) into these expressions, we get

$$n = p = (3\pi^2 \hbar^3)^{-1} \left(2 m_n m_p \varepsilon_0 / 2 m_n m_p \varepsilon_0 (m_n + m_p)\right)^{3/2}. \tag{7.240}$$

Consequently, in semi-metals at $T = 0$, the concentration of electrons and holes is finite and is determined by the band parameters m_n, m_p and ε_0.

Impurity semiconductors. Different defects – extraneous atoms (impurities), empty points, atoms between points, etc. – are inherent to real crystals. Here we will consider only defects associated with impurity atoms. Such defects can be artificially created, and therefore their concentration and types can be controlled. Assume that in a crystalline lattice of an element of the fourth group, for instance germanium or silicon, a minute part is replaced by atoms of elements of the third (In,Ga) or the fifth (Sb,As) group. This means that crystals of Ge or Si are doped by extraneous atoms In,Ga, Sb and As. If the concentration of impurity atoms N_i is small, i.e. the distance between them $d_i \approx N_i^{-1/3}$ is far greater than the radius of the first Bohr orbit a_0, then wave functions of impurity atoms do not overlap each other, and therefore impurity energy bands do not arise and in the band gap they create discrete energy levels. Depending on the type of impurity atoms, discrete impurity levels are positioned close to either the bottom of the conduction band or the top of the valence band.

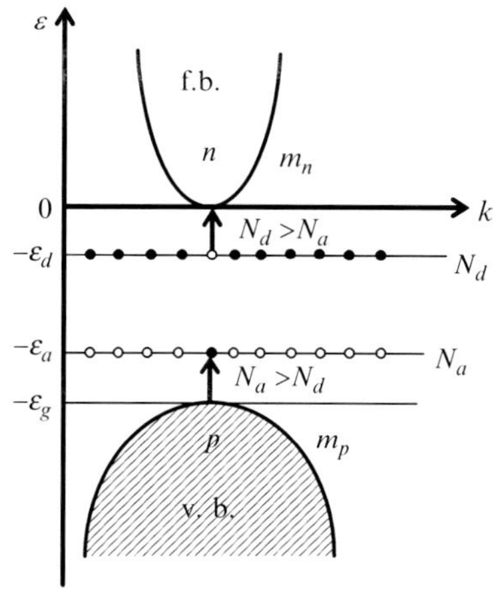

Fig. 7.21. Diagrams of bands of semiconductors with donor and acceptor impurities

If Ge and Si crystals are doped by pentavalent atoms Sb or As, four of the five electrons found in the outer electronic level of Sb or As form bonds with the adjacent atoms of Ge or Si. The remaining fifth electron is weakly bound with impurity atoms and, therefore, under action of temperature easily leaves the atom and becomes free. This means that atoms of the fifth group create discrete levels in crystals of Ge and Si close to the conduction band and the fifth valent electron easily passes into this band. Impurities of such type are called *donors*, and such crystals are *electronic semiconductors*.

If crystals of Ge and Si are doped by trivalent atoms In or Ga, then three valence electrons of In or Ga are not sufficient to form covalent bonds with adjacent atoms of Ge or Si. Therefore, they attract one electron from the adjacent covalent bond, as a result of which a non-saturated bond arises between the fundamental atoms. This means that one electron passes from the valence band to the discrete impurity level, and one free hole arises in the valence band. Impurities of such a type are called *acceptors*, and such crystals are *hole semiconductors*.

In Fig. 7.21, the band structure of semiconductors with donor and acceptor impurities is schematically presented. Because in the figure energy is taken from the bottom of the conduction band upwards, we denote the energy of donor levels by $-\varepsilon_{\mathrm{d}}$, acceptor levels by $-\varepsilon_{\mathrm{a}}$ and the band gap width by $-\varepsilon_{\mathrm{g}}$; N_{d} and N_{a} are concentrations of donor and acceptor atoms, respectively; the rest of the notations in the figure are as usual.

It is evident that even at the temperature of absolute zero, $T = 0$, weakly bound electrons pass from the donor level to the acceptor level and, depending on the ratio $N_{\mathrm{d}}/N_{\mathrm{a}}$, the donor or acceptor level is *compensated:* if $N_{\mathrm{d}} > N_{\mathrm{a}}$, acceptors are completely compensated and $N'_{\mathrm{d}} = N_{\mathrm{d}} - N_{\mathrm{a}}$ number of donors remains neutral; On the contrary, if $N_{\mathrm{a}} > N_{\mathrm{d}}$, donor atoms are compensated completely, and $N'_{\mathrm{a}} = N_{\mathrm{a}} - N_{\mathrm{d}}$ number of acceptors remains neutral.

At finite temperature ($T \neq 0$), when $N_\mathrm{d} > N_\mathrm{a}$, an electron passes from neutral donor levels into the conduction band and, when $N_\mathrm{a} > N_\mathrm{d}$, an electron passes from the valence band into the non-compensated acceptor level. In the first case, a crystal is called a *compensated electronic semiconductor*, and in the second case it is a *compensated hole semiconductor.*

Thus, at a particular finite temperature and in the thermodynamic equilibrium in a crystal, there are a specified number of free electrons in the conduction band, free holes in the valence band, positively charged donor levels ("bound holes") and negatively charged acceptor ions. It is clear that in the equilibrium state this system of charges ought to be characterised by only the chemical potential, which is found from the condition of neutrality.

The condition of neutrality: the sum of the concentration of electrons in the conduction band n and negatively ionised acceptors $n_\mathrm{a} = N_\mathrm{a}^-$ ought to equal the sum of the concentration of free holes of the valence band p and the concentration of the "bound holes" $p_\mathrm{d} = N_\mathrm{d}^+ = N_\mathrm{d} - n_\mathrm{d}$ at donor levels; here, n_d is the concentration of electrons at the donor level. Thus, the condition of neutrality of a crystal has the form

$$n + n_\mathrm{a} = p + p_\mathrm{d}. \tag{7.241}$$

Concentrations of conduction electrons n and free holes p are given by the expressions (7.219) and (7.223), respectively. In order to calculate n_a and p_d, we use the expression for the mean number of electrons at the impurity level with energy ε_i:

$$n(\varepsilon_i) = \left[1 + \beta \exp(\varepsilon_i - \mu)/k_0 T\right]^{-1}. \tag{7.242}$$

Here the factor β is associated with the degeneracy of the impurity level. If at the impurity level there is degeneracy only with respect to the spin, then $\beta = 1/2$. In this case, the mean concentration of electrons at the donor level with energy $\varepsilon_i = -\varepsilon_\mathrm{d}$ is

$$n_\mathrm{d} = N_\mathrm{d}\left[1 + \frac{1}{2}\exp\left(-(\varepsilon_\mathrm{d} + \mu)/k_0 T\right)\right]^{-1}. \tag{7.243}$$

Then the concentration p_d entering into (7.241) can be presented in the form

$$p_\mathrm{d} = N_\mathrm{d} - n_\mathrm{d} = N_\mathrm{d}\left[1 + 2\exp(\varepsilon_\mathrm{d} + \mu)/k_0 T\right]^{-1}. \tag{7.244}$$

For impurities of the acceptor type, $\beta = 2$. Then the mean concentration of electrons at the acceptor level with energy $\varepsilon_i = -\varepsilon_\mathrm{a}$ is

$$n_\mathrm{a} = N_\mathrm{a}\left[1 + 2\exp\left(-\frac{\varepsilon_\mathrm{a} + \mu}{k_0 T}\right)\right]^{-1}. \tag{7.245}$$

If we substitute (7.219), (7.223), (7.244) and (7.245) into (7.241), the condition of neutrality in the general case takes the form

$$\frac{(2m_n k_0 T)^{3/2}}{3\pi^2\hbar^3} F_{3/2}(\eta) + N_{\mathrm{a}}\left[1 + 2\exp(-\varepsilon_{\mathrm{a}}^* - \eta)\right]^{-1} =$$
$$= \frac{(2m_p k_0 T)^{3/2}}{3\pi^2\hbar^3} F_{3/2}(-\varepsilon_{\mathrm{g}}^* - \eta) + N_{\mathrm{d}}\left[1 + 2\ \exp(\varepsilon_{\mathrm{d}}^* + \eta)\right]^{-1}, \quad (7.246)$$

where

$$\varepsilon_{\mathrm{d}}^* = \varepsilon_{\mathrm{d}}/k_0 T, \varepsilon_{\mathrm{a}}^* = \varepsilon_{\mathrm{a}}/k_0 T \quad \text{и} \quad \eta = \mu/k_0 T.$$

If parameters of the energy band m_n, m_p, ε_{g}, ε_{d}, ε_{a}, N_{d} and N_{a} (see Fig. 7.21) are known, from (7.246), in principle, we can find the dependence of the reduced chemical potential on these parameters, and also on the temperature $\eta(T)$. Knowing the dependence $\eta(T)$, according to (7.219) and (7.223), we can determine the temperature dependences of the concentration of free charge carriers $n(T)$ and $p(T)$. However, the transcendental equation (7.246) in the general form is impossible to solve with reference to η. Therefore (7.246) can be solved only analytically in particular cases.

Intrinsic semiconductors. This case corresponds to the impurity-free crystal ($N_{\mathrm{d}} = N_{\mathrm{a}} = 0$), which we considered at the beginning of the section.

Electronic semiconductors. As a more comprehensive example, we consider semiconductors with donor impurities. In this case, $N_{\mathrm{a}} = 0$, $N_{\mathrm{d}} \neq 0$ and (7.246) acquires the form

$$\frac{(2m_n k_0 T)^{3/2}}{3\pi^2\hbar^3} F_{3/2}(\eta) = N_{\mathrm{d}}\left[1 + 2\ \exp(\varepsilon_{\mathrm{d}}^* + \eta)\right]^{-1}. \quad (7.247)$$

If the free electron gas in the conduction band is degenerate, even the simplified equation of neutrality (7.247) is impossible to be solved analytically; we can solve it only numerically and construct the graphic dependence $\eta(T)$.

If, however, we suppose that the electron gas is non-degenerate, and, according to (7.162) use the asymptote $F_{3/2}(\eta) = 3\sqrt{\pi}/4\ \exp(\eta)$, (7.247) can be rewritten in the form

$$\frac{(2m_n k_0 T)^{3/2}}{4\pi^{3/2}\hbar^3} \exp(\eta) = N_{\mathrm{d}}\left[1 + 2\ \exp(\varepsilon_{\mathrm{d}}^* + \eta)\right]^{-1}. \quad (7.248)$$

This is a quadratic equation with respect to $\exp(\eta)$. Solving it for the chemical potential, we get

$$\mu = -\varepsilon_{\mathrm{d}} + k_0 T\ \ln\left[\frac{1}{4}\left(\sqrt{1 + \frac{32\pi^{3/2}\hbar^3 N_{\mathrm{d}}}{(2m_n k_0 T)^{3/2}} \exp(\varepsilon_{\mathrm{d}}^*)} - 1\right)\right]. \quad (7.249)$$

Here, the solution of the quadratic equation for which the condition $\exp(\eta) > 0$ is fulfilled is conserved.

Expression (7.249) obtained for the simplest case has a very complex appearance. To obtain physically meaningful results, on the basis of (7.249) we consider concrete cases.

(1) *Region of low temperatures.* Assume that the temperature T satisfies the inequality

$$\exp\left(\frac{\varepsilon_{\mathrm{d}}}{k_0 T}\right) \gg \frac{(2m_n k_0 T)^{3/2}}{32\pi^{3/2}\hbar^3 N_{\mathrm{d}}}. \tag{7.250}$$

Then, from (7.249) we get for the chemical potential

$$\mu = -\frac{\varepsilon_{\mathrm{d}}}{2} + \frac{k_0 T}{2} \ln\left[\frac{2\pi^{3/2}\hbar^3 N_{\mathrm{d}}}{(2m_n k_0 T)^{3/2}}\right]. \tag{7.251}$$

Hence, we arrive at the following conclusions:

(a) At the temperature of absolute zero ($T = 0$), the chemical potential equals one-half ε_d. This result completely satisfies the physical sense, according to which on the passage of one electron from the donor level into the conduction band, two particles arise in the system: a free electron and a "bound hole," and energy ε_{d} is expended in this process.
(b) The sign of the second item in (7.251) depends on the value of the expression under the logarithm sign, i.e. on temperature. Thus far, the temperature satisfies the condition

$$2\pi^{3/2}\hbar^3 N_{\mathrm{d}} > (2m_n k_0 T)^{3/2}, \tag{7.252}$$

where the second term in (7.251) is positive and μ approaches the bottom of the conduction band. If the concentration of donors is sufficiently large, i.e. ($N_{\mathrm{d}} > 10^{18}\,\mathrm{cm}^{-3}$), μ can even pass into the conduction band (Fig. 7.22). At even higher temperatures, when the condition reverse to (7.252) is fulfilled, μ, which is decreasing, can pass to a level below $-\varepsilon_{\mathrm{d}}/2$. If the chemical potential passes into the conduction band, from Fig. 7.22 it is seen that an electron gas in the conduction band in the range of temperatures $\Delta T = T_2 - T_1$ can be found in the degenerate state.

It is clear that outside this range ($T < T_1$ and $T > T_2$) the chemical potential $\mu < 0$, and therefore the concentration of conduction electrons depends exponentially on the temperature.

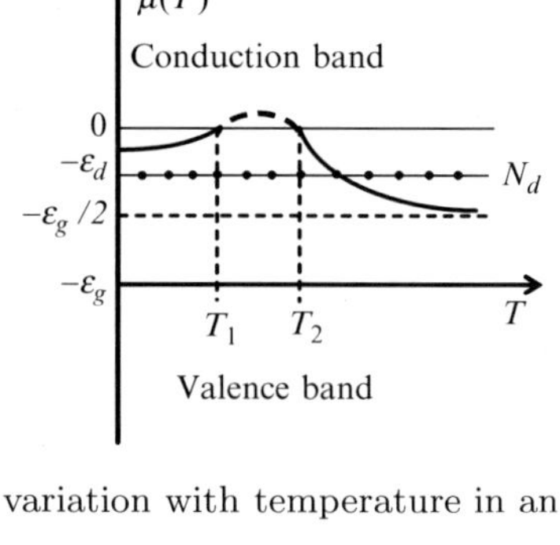

Fig. 7.22. Fermi level variation with temperature in an electron semiconductor

(2) *Region of relatively "high" temperatures.* When the condition reverse to (7.250) is fulfilled, the second term of the expression (7.249) is much smaller than unity and the square root can be expanded into a series. Then, in this region of temperatures the chemical potential takes the very simple form

$$\mu = k_0 T \ln \left[\frac{4\pi^{3/2}\hbar^3 N_{\mathrm{d}}}{(2m_n k_0 T)^{3/2}} \right] . \tag{7.253}$$

From (7.253) and the condition reverse to (7.250), it follows that in the considered temperature region the chemical potential sinks below donor levels $-\varepsilon_{\mathrm{d}}$ (Fig. 7.22).

At still higher temperatures, electrons begin to pass from the valence band into the conduction band, as a result of which the intrinsic conductivity dominates over the electronic cinductivity and the chemical potential asymptotically approaches $-\varepsilon_{\mathrm{g}}/2$ (Fig. 7.22).

If the expression of the chemical potential (7.253) is substituted into (7.219a), we get

$$n = N_{\mathrm{d}} . \tag{7.254}$$

This means that in the considered temperature region all donor levels are singly ionised, i.e. all valence electrons have passed from the donor levels into the conduction band. The concentration of conduction electrons remains constant, i.e. $n = N_{\mathrm{d}}$, as long as the intrinsic conductivity does not begin.

In a like manner, hole semiconductors ($N_{\mathrm{d}} = 0, \quad N_{\mathrm{a}} \neq 0$) also can be analysed.

In conclusion, we notice one interesting fact. From expressions of concentrations of conduction electrons (7.219a) and free holes (7.225) of non-degenerate semiconductors, it is seen that the product np does not depend on the chemical potential μ. From comparison of the product np from (7.231), we get the very simple equality

$$np = n_i^2 = p_i^2 . \tag{7.255}$$

This relationship shows that *if an electron gas in the conduction band and free hole gas in the valence band are non-degenerate, the product np does not depend on the amount of impurities and their distribution.* This result also follows from the fact that the amount of impurities and their distribution determine the value of the chemical potential μ, and the product np does not depend on μ.

7.13 Degenerate Bose Gas: Bose–Einstein Condensation

We construct a statistical theory of the thermodynamic properties of a Bose gas of volume V consisting of N bosons of the mass m. To do this, on the strength of (7.98) we write the equations of state of a Bose gas:

$$\begin{cases} P = \dfrac{2}{3}\dfrac{g_0(2m)^{3/2}}{(2\pi)^2\hbar^3}\int\limits_0^\infty \dfrac{\varepsilon^{3/2}\mathrm{d}\varepsilon}{\mathrm{e}^{(\varepsilon-\mu)/k_0T}-1}, \\ N = \dfrac{Vg_0(2m)^{3/2}}{(2\pi)^2\hbar^3}\int\limits_0^\infty \dfrac{\varepsilon^{1/2}\mathrm{d}\varepsilon}{\mathrm{e}^{(\varepsilon-\mu)/k_0T}-1}. \end{cases} \tag{7.256}$$

It is clear that, in the general form, the system of equations (7.256) at arbitrary temperature is impossible to solve analytically. Therefore, we consider only high- and low-temperature regions. If the concentration of the gas is constant, at high temperatures, the chemical potential, according to (7.20), can satisfy the condition of classicity: $A = \exp(\eta) \ll 1$. Then system of equations (7.256) can be solved by expanding it into a series with respect to the parameter $A \ll 1$. In the zeroth approximation, we obtain the well-known classical results (see Sect. 7.1). In the first approximation (the approximation of weak degeneracy), for a Bose gas as well as for a Fermi gas we obtain results that differ a little from those for a classical Boltzmann gas (see Sect. 7.5). Therefore, in this section we will consider the low-temperature region, i.e. the state of degeneracy. Qualitatively new physical results are obtained just in this case. At first, we analyse the temperature dependence of the chemical potential $\mu(T)$ of a Bose gas with a particular number of particles: $N = \text{const}$. Recall that the chemical potential of a Bose gas at any temperature can never be positive:

$$\mu(T) \leq 0. \tag{7.257}$$

It is known that at high temperatures the chemical potential takes on very large negative values. In order to elucidate how the function $\mu(T)$ behaves with the lowering of temperature, take the derivative with respect to T on each side of the second equation of the system of equations (7.256) at $N = \text{const}$. As a result, we get

$$\left(\frac{\partial\mu}{\partial T}\right)_N = -\frac{1}{T}\frac{\int\limits_0^\infty \left[\exp\left(\frac{\varepsilon-\mu}{k_0T}\right)-1\right]^{-2}(\varepsilon-\mu)\exp\left(\frac{\varepsilon-\mu}{k_0T}\right)\varepsilon^{1/2}\mathrm{d}\varepsilon}{\int\limits_0^\infty \left[\exp\left(\frac{\varepsilon-\mu}{k_0T}\right)-1\right]^{-2}\exp\left(\frac{\varepsilon-\mu}{k_0T}\right)\varepsilon^{1/2}\mathrm{d}\varepsilon}. \tag{7.258}$$

If we take into account the condition (7.257), from (7.258) it follows that

$$\left(\frac{\partial\mu}{\partial T}\right)_N < 0. \tag{7.259}$$

Hence, it is seen that with the increase in T the chemical potential $\mu(T)$, remaining negative in magnitude, decreases, and at a specified, finite temperature θ_0 becomes zero, and on further decrease in temperature, according to (7.257) and (7.259), remains equal to zero:

$$\mu(T)|_{T\leq\theta_0} = 0. \tag{7.260}$$

In Fig. 7.23, the temperature dependence of the chemical potential of a Bose gas at $N = \text{const}$ is schematically presented.

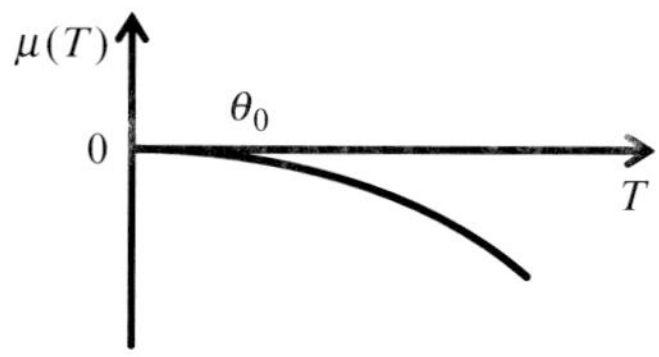

Fig. 7.23. The temperature dependence of the chemical potential of a Bose gas

Temperature θ_0 can be determined from the second equation of the system of equations (7.256) using condition (7.260):

$$N = \frac{Vg_0(2m)^{3/2}}{(2\pi)^2\hbar^3}\int\limits_0^\infty \frac{\varepsilon^{1/2}\mathrm{d}\varepsilon}{\mathrm{e}^{\varepsilon/k_0\theta_0}-1}. \tag{7.261}$$

If we change to the dimensionless variable of integration $x = \varepsilon\,/k_0\theta_0$, (7.261) takes the form

$$N = \frac{Vg_0(2mk_0\theta_0)^{3/2}}{(2\pi)^2\hbar^3}\int\limits_0^\infty \frac{x^{1/2}\mathrm{d}x}{\mathrm{e}^{x}-1}. \tag{7.262}$$

The finite integral entering into this equation (according to Appendix A) equals $\Gamma\,(3/2)\,\zeta\,(3/2) = \frac{\sqrt{\pi}}{2}\cdot 2.61 = 2.3$. Then, from (7.262) we get

$$\theta_0 = \frac{(2\pi)^{4/3}}{(2.3g_0)^{2/3}}\frac{\hbar^2}{2mk_0}\left(\frac{N}{V}\right)^{2/3}. \tag{7.263}$$

θ_0 is called *temperature of degeneracy* of a Bose gas. Note that if the masses of fermions and bosons are identical, from a comparison of (7.263) and (7.126) it follows that the temperature of degeneracy of a Bose gas θ_0 and that of a Fermi gas T_0 are of the same order.

We now analyse the thermodynamic properties of a Bose gas in the region of degeneracy $T < \theta_0$. Taking into account condition (7.260), from (7.256) in the region $T < \theta_0$ we get

$$N = \frac{Vg_0(2m)^{3/2}}{(2\pi)^2\hbar^3}\int\limits_0^\infty \frac{\varepsilon^{1/2}\mathrm{d}\varepsilon}{\mathrm{e}^{\varepsilon/k_0 T}-1};\quad T \le \theta_0. \tag{7.264}$$

Note that this expression in the region of temperatures $T < \theta_0$ contains a certain paradox. Indeed, though the left-hand side of (7.264) ($N = \mathrm{const}$) does not depend on temperature, the right-hand side depends on temperature as $\sim T^{3/2}$.

This paradox was resolved by Einstein in 1925. He paid attention to the fact that, because the function of the density of states $g(\varepsilon) \sim \varepsilon^{1/2}$, in integral (7.264) we have not taken into account the number of particles with energy

$\varepsilon = 0$. This means that the right-hand side of (7.264) determines not the total number of bosons but the number of bosons with energy different from zero ($\varepsilon > 0$). In the temperature region $T \leq \theta_0$, a part of bosons passes to the level of energy $\varepsilon = 0$, i.e. it is "condensed". This phenomenon is called *Bose–Einstein condensation*. Consequently, according to Einstein's reasoning, the right-hand side of equality (7.264) shows not the total number of bosons but the number of non-condensed bosons, i.e. the number of bosons at levels $\varepsilon > 0$. If we denote the number of these bosons by N_1, the number of "condensed" bosons at the level $\varepsilon = 0$ is

$$N_0 = N - N_1, \tag{7.265}$$

where N_1 is determined by (7.264) with regard to Einstein's idea

$$N_1 = \frac{V g_0 (2m)^{3/2}}{(2\pi)^2 \hbar^3} \int\limits_{\varepsilon > 0}^{\infty} \frac{\varepsilon^{1/2} \mathrm{d}\varepsilon}{\mathrm{e}^{\varepsilon / k_0 T} - 1}; \quad T \leq \theta_0. \tag{7.266}$$

We introduce the dimensionless integration variable $\varepsilon / k_0 \theta_0 = x$ and take into account (7.262). Then the number of non-condensed bosons equals

$$N_1 = N \left(\frac{T}{\theta_0} \right)^{3/2}; \quad T \leq \theta_0, \tag{7.267}$$

and the number of condensed bosons at the level $\varepsilon = 0$, according to (7.265) and (7.267), is

$$N_0 = N \left[1 - \left(\frac{T}{\theta_0} \right)^{3/2} \right]; \quad T \leq \theta_0. \tag{7.268}$$

It is seen that at the point $T = \theta_0$ the number of condensed bosons is $N_0 = 0$. With the lowering of temperature, beginning from θ_0, the number N_0 increases, and at the temperature of absolute zero $N_0 = N$, i.e. all bosons pass to the level $\varepsilon = 0$ and a Bose gas is completely condensed. Thus, temperature $T = \theta_0$ is the beginning of the process of condensation, i.e. temperature of degeneracy, of a Bose gas. For usual molecular Bose gases, θ_0 is very small. For instance, at $m = 10^{-24}$g, and $N/V \approx 10^{18}\,\mathrm{cm}^{-3}$, from (7.263) it follows that $\theta_0 \approx 10^{-2}\,\mathrm{K}$. This is a temperature much lower than the temperature of liquefaction of any gas. Therefore, the phenomenon of the Bose–Einstein condensation in practice is very difficult to observe.

However, it is interesting to note that in 1995 the American scientists Eric Cornell, Carl Wieman and Wolfgang Ketterle, on cooling a very rarefied gas consisting of atoms of rubidium and sodium with the aid of laser techniques to $10^{-8}\,\mathrm{K}$, achieved Bose–Einstein condensation. In 2001, they received the Nobel Prize for this work.

Recall that the Bose–Einstein condensation should not be confused with the usual condensation in the three-dimensional space. In the process of

Bose–Einstein condensation, bosons in the p-space, on losing their impulses, are accumulated at the point $p = 0$ or at the energy level $\varepsilon = 0$.

We consider the thermodynamic properties of a Bose gas in the region of condensation ($T \leq \theta_0$). To do this, we recall that in the region of condensation the chemical potential equals zero $\mu(T \leq \theta_0) = 0$. If we take this into account in (7.256) and introduce the dimensionless variable $x = \varepsilon/k_0 T$, the pressure of a Bose gas is

$$P = \frac{2}{3}\frac{g_0(2m)^{3/2}(k_0T)^{5/2}}{(2\pi)^2\hbar^3}\int\limits_0^\infty \frac{x^{3/2}\mathrm{d}x}{\mathrm{e}^x - 1}; \quad T \leq \theta_0. \tag{7.269}$$

Because the integral in this expression (according to Appendix A) equals $\Gamma(5/2)\,\zeta(5/2) \approx 1.8$,

$$P = \frac{1.2 g_0(2m)^{3/2}(k_0T)^{5/2}}{(2\pi)^2\hbar^3}; \quad T \leq \theta_0. \tag{7.270}$$

From this expression, it is seen that pressure of a degenerate Bose gas strongly depends on temperature as $P \sim T^{5/2}$, whereas it does not depend on the volume. This fact resembles the independence of pressure of a saturated vapour on volume V, and is explained by the fact that with an increase in V the temperature of condensation θ_0 rises [see (7.263)]; thereby, a larger number of bosons pass to the level with $\varepsilon = 0$ [see (7.268)], i.e. they are condensed, as a result of which the density of non-condensed bosons does not changes. Because the impulse of bosons, on passing to the level with $\varepsilon = 0$, equals zero, i.e.$p = 0$, they do not create pressure.

If we consider the expressions (7.109) for high temperatures and (7.270) for low temperatures, we get the schematic dependence of pressure on temperature $P(T)$ in all the temperature regions, which is shown in Fig. 7.24.

The energy of a Bose gas in the region of condensation ($T \leq \theta_0$) can be found from the relationship $E = 3PV/2$ and expression (7.270):

$$E = \frac{1.8 g_0 V(2m)^{3/2}(k_0T)^{5/2}}{(2\pi)^2\hbar^3}; \quad T \leq \theta_0. \tag{7.271}$$

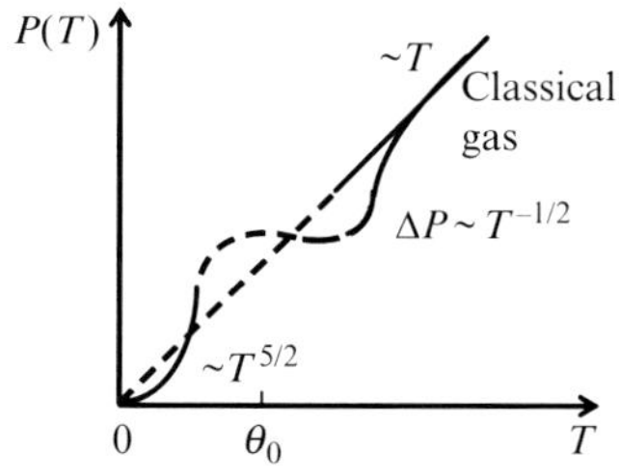

Fig. 7.24. The temperature dependence of the pressure of a Bose gas

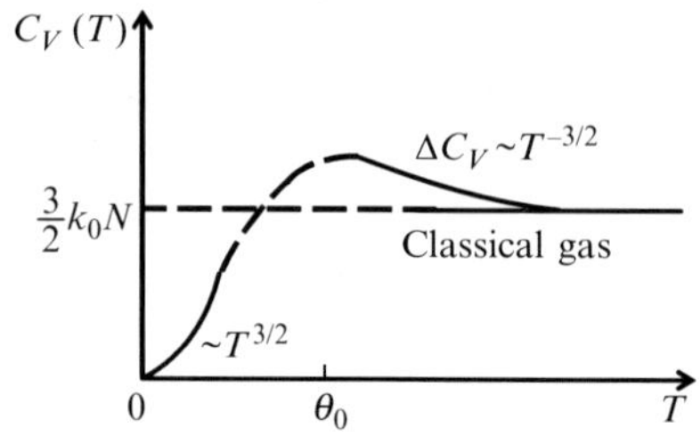

Fig. 7.25. The temperature dependence of the heat capacity of a Bose gas

Hence, for the heat capacity $C_V = (\partial E/\partial T)_V$ we have

$$C_V = \frac{4.5 g_0 V k_0 (2mk_0 T)^{3/2}}{(2\pi)^2 \hbar^3}; \quad T \leq \theta_0. \tag{7.272}$$

As is seen, for the heat capacity the Nernst principle is fulfilled: $C_V(T \to 0) \to 0$.

If we take into account expressions (7.111) and (7.172), we can determine the temperature dependence of the heat capacity of a Bose gas $C_V(T)$ in all the temperature regions, which is presented in Fig. 7.25. The portion corresponding to the high-temperature region is taken from Fig. 7.6.

7.14 Photon Gas: Third Difficulty of Classical Statistics

In Sect. 7.1, we noted the difficulties associated with the application of the Boltzmann classical statistics to a photon gas. In this section, we will show how this difficulty is circumvented.

Assume that a vessel of volume V and temperature of the inner walls T radiates inside electromagnetic waves of different frequencies $\omega(k)$; $k = 2\pi/\lambda$ is the wavenumber of these waves, and λ is the wavelength. In thermodynamic equilibrium between the electromagnetic field and walls of the vessel, there arises an energy balance, i.e. walls of the vessel in unit of time absorb the same amount of energy that they radiate. In 1900, Planck put forward the idea that this process of absorption and radiation occurs not continuously but in *quanta* and, in this way, explained all experimental observations associated with the radiation of a black body. Therefore, the year 1900 is regarded as the beginning of quantum physics. According to Planck, walls of the vessel radiating or absorbing electromagnetic waves with the frequency $\omega(k)$ radiate or absorb energy in multiples of $\hbar\omega$, i.e. $\hbar\omega$, $2\hbar\omega$, $3\hbar\omega$, $\ldots$, $n\hbar\omega$. Here, $\hbar = h/2\pi$, h being the Planck constant.

In 1905, Einstein proposed to associate a particle with energy $\varepsilon = \hbar\omega$ and impulse $p = \hbar\omega/c$ with each portion of the energy of a quantum being absorbed or irradiated. This particle was called *a photon*. According to Einstein, the process of absorption or radiation can be presented as the absorption or radiation of one or several photons by the walls of the vessel.

According to Maxwell's equations, there exists a simple relationship $\omega = ck$ (c is the velocity of light) between the frequency of an electromagnetic wave ω and the wavenumber $k = 2\pi/\lambda$. Taking this into account, the energy ε and the impulse p of a photon can be written as

$$\begin{cases} \varepsilon = \hbar\omega, \\ p = \hbar k. \end{cases} \tag{7.273}$$

Later, quantum mechanics was introduced and the *principle of dualism* was formulated. According to this principle, the relationship (7.273) can be applied to all elementary particles with the mass at rest different from zero, and they can be read in two ways: from the left to the right and from the right to the left.

From the left to the right: each elementary particle with energy ε and impulse p can be considered as a wave with frequency $\omega = \varepsilon/\hbar$ and wavenumber $k = p/\hbar = 2/\pi\lambda$.

From the right to the left: each wave with the frequency ω and wave number $k = 2\pi/\lambda$ can be considered as an elementary particle with energy $\varepsilon = \hbar\omega$ and impulse $p = \hbar k$.

The wave that represents the elementary particle is called the *de Broglie wave* and its length is determined by the impulse of the elementary particle $\lambda = h/p = h/mv$, i.e. the mass and velocity. Note that in (7.273) the first equality is called the Einstein relationship, and the second is the de Broglie formula. Remember also that quantities ε and p characterising a particle are related to the corresponding parameters of the wave ω and k by the universal constant $\hbar$.

Thus, an electromagnetic field inside a vessel can be considered as an ideal photon gas filling its volume V. A photon gas is regarded as ideal, because, as the Maxwell equations are linear and for an electromagnetic field the superposition principle is satisfied, photons do not interact with each other.

In the composition of a photon gas, there exist photons of different types, which is determined by the frequency ω and wave vector $\boldsymbol{k}$. Inside the gas, the mean number of photons of the type (ω, $\boldsymbol{k}$) corresponds to the intensity (the square of amplitude) of a wave with frequency ω.

Consequently, the study of an electromagnetic field is reduced to consideration of the statistical properties of an ideal photon gas. Difficulties that arise from the application of classical statistics to a photon gas were considered in Sect. 7.1. Therefore, here we will apply quantum statistics to a photon gas.

Note that the number of photons with identical frequency can take on any value; therefore, a photon is a boson with spin equal to unity. Hence, it follows that we need to apply the Bose–Einstein distribution to the photon gas.

Then, for the mean number of photons $\bar{n}_k$ of the type (ω, k) with energy $\varepsilon = \hbar\omega$ at temperature T, we can write

$$\bar{n}_k = \frac{1}{\mathrm{e}^{(\hbar\omega_k - \mu)/k_0 T} - 1}, \tag{7.274}$$

where T is temperature of the inner walls of the vessel which is in equilibrium with the photon gas, and μ is the chemical potential of the photon gas.

It is clear that the total number of photons $N(T)$ in the vessel is not constant but depends on temperature of the inner walls T. In order that at a pre-assigned temperature and volume a photon gas would be found in thermodynamic equilibrium with the walls of the vessel, it is necessary that the free energy F of the photon gas be a minimum, i.e. it should fulfil the condition

$$\left(\frac{\partial F}{\partial N}\right)_{V,T} = 0. \tag{7.275}$$

This means that when changing the number of photons the free energy of such a gas remains constant. This result also follows from the fact that in the state of thermodynamic equilibrium, the walls of the vessel, when absorbing or emitting a particular amount of energy, can absorb or radiate various number of photons. It is evident that during the processes of radiation or absorption of photons the law of conservation of energy is satisfied.

On the other hand, it is known that the chemical potential of the system is the change in the free energy when changing the number of particles by unity: $(\partial F/\partial N)_{V,T} = \mu(T)$. Therefore, the condition of equilibrium (7.275) can be rewritten in the form

$$\mu(T) = 0, \tag{7.276}$$

i.e. in thermodynamic equilibrium, the chemical potential of a photon gas ought to be equal to zero at all temperatures.

Thus, the following definition can be given to a photon gas: *a photon gas is an ultra-relativistic* ($\varepsilon = \hbar ck$) *quantum gas with the zero chemical potential* ($\mu(T) = 0$) *at all temperatures.*

As a result, from (7.274) and (7.276) the Bose distribution for a photon gas takes the form

$$\bar{n}_k = \frac{1}{\mathrm{e}^{\hbar\omega_k/k_0 T} - 1}. \tag{7.277}$$

Using this distribution function, we can construct the statistical theory of all thermodynamic properties of a photon gas.

The mean number of photons in a vessel of volume V at temperature T is defined by

$$N(T) = \sum_k \bar{n}_k = \frac{2V}{(2\pi)^3} \int \bar{n}_k \mathrm{d}\boldsymbol{k}, \tag{7.278}$$

where the formula of passage (7.11) is used; the factor 2 on the right-hand side of the expression takes into account the presence of two plane-polarized transverse electromagnetic waves. If we take into account that in the spherical coordinate system $\mathrm{d}\boldsymbol{k} = 4\pi k^2 \mathrm{d}k$, use $\omega = ck$ and in (7.278) pass from the integration over $\mathrm{d}k$ to the integration over $\mathrm{d}\omega$, we get

$$N(T) = \frac{V}{\pi^2 c^3} \int_0^\infty \frac{\omega^2 \mathrm{d}\omega}{\mathrm{e}^{\hbar\omega/k_0 T} - 1} = \int_0^\infty g(\omega) n(\omega) \mathrm{d}\omega, \tag{7.279}$$

where

$$g(\omega) = \frac{V}{\pi^2 c^3} \omega^2 \tag{7.280}$$

is the number of frequencies found in the unit range taken close to ω, i.e. the function of the density of frequencies (analogously see Fig. 7.17).

We introduce the dimensionless integration variable $x = \hbar\omega/k_0 T$. Then (7.279) takes the form

$$N(T) = \frac{V}{\pi^2 c^3} \left(\frac{k_0 T}{\hbar}\right)^3 \int_0^\infty \frac{x^2 \mathrm{d}x}{\mathrm{e}^x - 1}. \tag{7.281}$$

If we take into account the value of the integral (according to Appendix I) as $\Gamma(3)\xi(3) = 2.4$, for $N(T)$ finally we get

$$N(T) = \frac{2,4V}{\pi^2 c^3} \left(\frac{k_0 T}{\hbar}\right)^3 \sim T^3. \tag{7.282}$$

The total energy of a photon gas has the form

$$E(T) = \sum_k \hbar\omega(k)\bar{n}_k = \frac{2V\hbar}{(2\pi)^3} \int \omega(k)\bar{n}_k \mathrm{d}\boldsymbol{k}. \tag{7.283}$$

If we transform the integral in (7.283), for the mean total energy of a photon gas we get the expression

$$E(T) = \frac{V\hbar}{\pi^2 c^3} \int_0^\infty \frac{\omega^3 \mathrm{d}\omega}{\mathrm{e}^{\hbar\omega/k_0 T} - 1}, \tag{7.284}$$

which can be presented also in the form

$$E(T) = \int_0^\infty \rho(\omega, T) \mathrm{d}\omega, \tag{7.285}$$

where $\rho(\omega, T)$ is called *the spectral density* of the radiation and shows the energy of photons found in unit frequency range taken close to ω. From a comparison of (7.284) and (7.285) follows the known *Planck formula* for the spectral density of the radiation:

$$\rho(\omega, T) = \frac{V\hbar}{\pi^2 c^3} \frac{\omega^3}{\mathrm{e}^{\hbar\omega/k_0 T} - 1}. \tag{7.286}$$

This formula shows the distribution of energy of a photon gas over frequencies at a particular temperature.

Thus, the Planck formula follows from application of the Bose distribution to a photon gas (7.277). Thereby, the third difficulty of the classical statistics is circumvented. In the process, we come to the conclusion that *a photon gas is not a classical gas but a quantum gas obeying the Bose distribution.*

All laws of radiation of the black body follow from (7.286), for any frequency.

Consider different regions of frequencies at a particular temperature:

1. *Low (infrared) frequencies:* $\hbar\omega \ll k_0 T$. In this case, in (7.286) the exponent can be expanded into a series. As a result, we get the well-known Rayleigh–Jeans law

$$\rho(\omega, T) = \frac{V k_0 T}{\pi^2 c^3}\omega^2; \hbar\omega \ll k_0 T. \tag{7.287}$$

The Planck constant does not enter into this expression, and therefore the classical explanation can be given to it. Indeed, we can consider normal electromagnetic vibrations in volume V in terms of classical oscillators taking into account that energy $k_0 T$ corresponds to each of them. Then, multiplying the number of frequencies in the unit range of energy (7.280) by $k_0 T$, we get (7.287). Recall that (7.287) is applicable only for low infrared frequencies. If we integrate (7.287) in the range of frequencies from 0 to ∞, the integral diverges. This phenomenon bears the name of "ultraviolet" catastrophe and shows that in actuality (7.287) cannot be applied to high frequencies.

2. *High (ultraviolet) frequencies:* $\hbar\omega \gg k_0 T$. In this case, in (7.286) unity can be neglected compared to the exponent, and as a result we get the well-known *Wien's law:*

$$\rho(\omega, T) = \frac{V\hbar}{\pi^2 c^3}\omega^3 \,\mathrm{e}^{-\hbar\omega/k_0 T}; \quad \hbar\omega \gg k_0 T. \tag{7.288}$$

The fact that the Planck constant $\hbar$ is present in this expression is natural, because in the region of frequencies $\hbar\omega \gg k_0 T$ the vibrational motion of an oscillator bears quantum character with energy not $k_0 T$ but $\hbar\omega$. Note that the spectral density determined by (7.288) can be obviously explained on the principal propositions of photons. Indeed, in order to obtain formula (7.288), it is sufficient to use three functions: density of frequencies (7.280), energy of photons $\varepsilon = \hbar\omega$ and the distribution of high-frequency photons $n(\omega) = \exp(-\hbar\omega/k_0 T)$. In this explanation, the main fact is that the Boltzmann distribution is used. This is due to the fact that at all temperatures the chemical potential of a photon gas $\mu(T) = 0$, and for high-frequency photons ($\hbar\omega \gg k_0 T$) from the Bose distribution (7.274) the Boltzmann distribution $n(\omega) = \exp(-\hbar\omega/k_0 T)$ follows.

3. *Arbitrary frequencies.* We analyse the distribution of energy over frequencies at a particular temperature. If we introduce the dimensionless variable $x = \hbar\omega/k_0T$, (7.286) takes the form

$$\rho(\omega, T) = \frac{V\hbar}{\pi^2 c^3}\left(\frac{k_0 T}{\hbar}\right)\varphi(x), \tag{7.289}$$

where

$$\varphi(x) = \frac{x^3}{\mathrm{e}^x - 1}. \tag{7.290}$$

A plot of function (7.290) that characterises the frequency dependence of the spectral density at a particular temperature is shown in Fig. 7.26. As can be seen, the function $\varphi(x)$ takes on a maximum value at the specified $x = x_{\max}$, which is found from the equation

$$(3 - x_{\max})\,\mathrm{e}^{x_{\max}} = 3. \tag{7.291}$$

A solution of this transcendental equation gives $x_{\max} = 2.822$. Then, $\varphi(x_{\max}) = 1.421$.

Thus, the distribution of energy of a photon gas over the frequencies is non-uniform. The maximum density of energy corresponds to the frequency

$$\omega_{\max} = 2.822\frac{k_0}{\hbar}T, \tag{7.292}$$

which, depending on the temperature, moves to the side of high frequencies: $\omega_{\max} \sim T$.

This rule is called the *Wien displacement law* and is well satisfied in experiments. Note that by experimentally determining $\omega_{\max}$ at a particular temperature T from (7.292), we can determine the Planck constant $\hbar$.

From (7.284), we can determine the total energy of a photon gas:

$$E(T) = \frac{V\hbar}{\pi^2 c^3}\left(\frac{k_0 T}{\hbar}\right)^4 \int\limits_0^\infty \frac{x^3\,\mathrm{d}x}{\mathrm{e}^x - 1}. \tag{7.293}$$

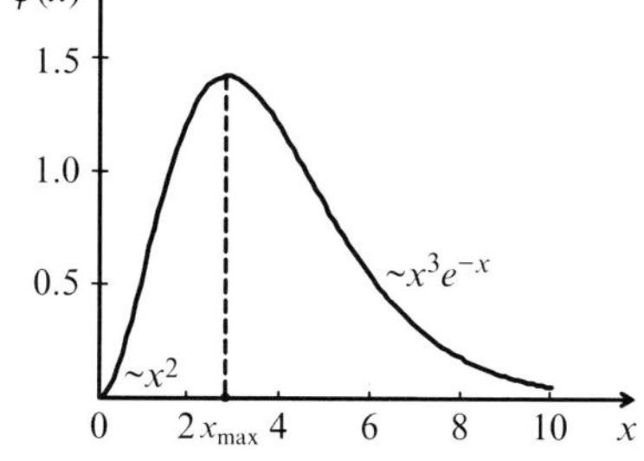

Fig. 7.26. The frequency dependence of the spectral density at a particular temperature

If we take into account that the finite integral in (7.293) equals $\Gamma(4)\zeta(4) = \pi^4/15$ (see Appendix A), for the temperature dependence of the total energy we get the well-known *Stefan–Boltzmann law*

$$E = \frac{\pi^2 V \hbar}{15c^3}\left(\frac{k_0 T}{\hbar}\right)^4 \sim T^4. \tag{7.294}$$

Hence, for the heat capacity of a photon gas $C_V = (\partial E/\partial T)_V$ we have

$$C_V = \frac{4\pi^2 V k_0^4}{15c^3\hbar^3} T^3 \sim T^3. \tag{7.295}$$

From expression $S = \int \frac{C_V}{T}\mathrm{d}T$ and (7.295), the expression of entropy of a photon gas follows:

$$S = \frac{4\pi^2 V k_0}{45c^3\hbar^3}(k_0 T)^3 \sim T^3. \tag{7.296}$$

It is seen that a photon gas satisfies the Nernst principle, $S(T \to 0) \to 0$, too.

On combining (7.294) and the relationship (7.204) for all ultra-relativistic gases, we can find the pressure of the photon gas:

$$P = \frac{\pi^2}{45c^3\hbar^3}(k_0 T)^4 \sim T^4. \tag{7.297}$$

The fact that a photon gas does not depend on volume V of the vessel resembles a saturated vapour over a liquid. This can be explained as follows: In thermodynamic equilibrium, on isothermal decrease (increase) of volume of the vessel with a photon gas, a specified number of photons are absorbed (irradiated) by the walls of the vessel, as a result of which the density of energy of a photon gas E/V does not change, and therefore the pressure remains constant and does not depend on the volume.

From the expression of pressure (7.297) follows the relationship

$$PT^{-4} = \text{const}. \tag{7.298}$$

On the other hand, for an adiabatic process ($S = \text{const}$), from (7.296) the relationship between volume and temperature follows:

$$VT^3 = \text{const}. \tag{7.299}$$

On combining the two latter relationships, we get the relation between pressure and volume:

$$PV^{4/3} = \text{const}. \tag{7.300}$$

In contrast to the isothermal process, from (7.300) it is seen that on adiabatic change in volume, the pressure of a photon gas changes. This can be explained as follows: Indeed, from (7.299) it follows that on adiabatic expansion or compression the temperature changes, and therefore the pressure of a photon gas changes.

If we define the intensity of radiation $I(T)$ as the energy radiated by unit area of the surface per second, we get

$$I(T) = \left(\frac{E}{V}\right) c, \tag{7.301}$$

where c is the velocity of light.

Substituting (7.294) into (7.301), for the intensity we have

$$I(T) = \frac{\pi^2}{15} \frac{(k_0 T)^4}{c^2 \hbar^3} \sim \sigma T^4, \tag{7.302}$$

where $\sigma = \pi^2 k_0^4 / 15 c^2 \hbar^3$ is the *Stefan–Boltzmann constant.*

It is to be noted that black-body radiation has been thoroughly studied experimentally. The results obtained are well explained theoretically, and investigations in this direction played an important role in the development of quantum physics.

In conclusion, we show that thermodynamics of a photon gas (the definition of energy, entropy and equations of state) can be immediately constructed with the aid of the grand thermodynamic potential as well. In order to obtain the expression of the grand thermodynamic potential for a photon gas, in expression (7.95) we need to consider the lower signs and take into account the relationships $\varepsilon = \hbar\omega = \hbar c k$ and $\mu = 0$ (7.276). As a result, the grand thermodynamic potential of a photon gas takes the form

$$\Omega = \frac{V k_0 T}{\pi^2 c^3} \int_0^\infty \omega^2 \ln\left(1 - \mathrm{e}^{-\hbar\omega / k_0 T}\right) \mathrm{d}\omega, \tag{7.303}$$

where we have taken $g_0 = 2$.

Integrating by parts twice the integral in (7.303) and introducing the dimensionless integration variable $x = \hbar\omega / k_0 T$, we get

$$\Omega = -V \frac{(k_0 T)^4}{3\pi^2 c^3 \hbar^3} \int_0^\infty \frac{x^3 \, \mathrm{d}x}{\mathrm{e}^x - 1} = -V \frac{\pi^2}{45} \frac{(k_0 T)^4}{(c\hbar)^3}. \tag{7.304}$$

If we use this expression in relationships $P = -\Omega / V$, $S = -(\partial\Omega / \partial T)_V$, $E = F + TS = \Omega + TS$ and $C_V = (\partial E / \partial T)_V$, we get all the above-derived results. The equality $E = \Omega + TS$ is not surprising, because at $\mu(T) = 0$ the grand thermodynamic potential Ω and the free energy F are equal: $F = \Phi - PV = \mu N + \Omega = \Omega$.

7.15 Phonon Gas

One of the systems to which quantum statistics is applied is a phonon gas. *A phonon is a quasi-particle*, corresponding to an elastic acoustic wave propagating in a crystalline lattice. In other words, *a phonon* is a quantum of a wave field created by the propagation of thermal or mechanical vibrations in a crystalline solid. A phonon can also be presented as an elementary excitation of a crystalline lattice. The phonon is a boson; therefore phonon gases obey the Bose–Einstein statistics. The concept of phonon, as in the case of a photon, also stems from *the principle of dualism* presented in the form (7.273).

In order to imagine how the concept of the phonon arises when studying the vibrational motion of a crystalline lattice, it is necessary to proceed from the principles of quantum mechanics. However, at first we recall the spectrum of frequencies of a vibrating crystal, which was comprehensively described in Chap. 6.

For simplicity, consider a simple crystalline lattice consisting of N elementary cells (in each elementary cell there is one atom). Then the number of degrees of freedom of the crystal is $3N$. In Sect. 6.1, we showed that the number of possible frequencies in such a crystal equals the number of degrees of freedom, i.e. $3N$. Because the number of possible values of the wave vector $\boldsymbol{q}$ in the range $-\pi\,/\,a \leq \boldsymbol{q} \leq \pi\,/\,a$ equals N, the entire frequency spectrum is formed by $3N$ number of frequencies (Fig. 6.4). Each frequency of this spectrum is determined by specifying two quantities: the wave vector $\boldsymbol{q}$ and the number of the branch j which takes on three values: $j = 1,\ 2,\ 3$, and is denoted as $\omega_{\boldsymbol{q}j} \equiv \omega_j(\boldsymbol{q})$.

On the other hand, in Sect. 6.2 it was shown that the total energy E of a three-dimensional lattice equals the sum of energies of $3N$ linear harmonic oscillators [see (6.57)]. In quantum mechanics (see Sect. 6.5), this result is written down in the form

$$E = \sum_{\boldsymbol{q}} \sum_{j=1}^{3} \left(n_{\boldsymbol{q}j} + \frac{1}{2} \right) \hbar\omega_j(\boldsymbol{q}), \tag{7.305}$$

where $n_{\boldsymbol{q}j} = 0, 1, 2, \ldots$ is the quantum number of an oscillator with the frequency $\omega_j(\boldsymbol{q})$. This expression can be also presented in the form

$$E = E_0 + \sum_{\boldsymbol{q}} \sum_{j=1}^{3} n_{\boldsymbol{q}j} \hbar\omega_j(\boldsymbol{q}), \tag{7.306}$$

where

$$E_0 = \frac{1}{2} \sum_{\boldsymbol{q}} \sum_{j=1}^{3} \hbar\omega_j(\boldsymbol{q}) \tag{7.307}$$

is the energy of zero vibrations.

We analyse the expression for energy (7.306). If all the oscillators are found at the fundamental level, i.e. for all $(\boldsymbol{q}j)$ the quantum number $n_{\boldsymbol{q}j} = 0$, then

$$E = E_0, \tag{7.308}$$

which corresponds to the fundamental state of the crystalline lattice at the temperature of absolute zero ($T = 0$). We can imagine this state of the crystal as an empty vessel (vacuum) of the same volume (Fig. 7.27a).

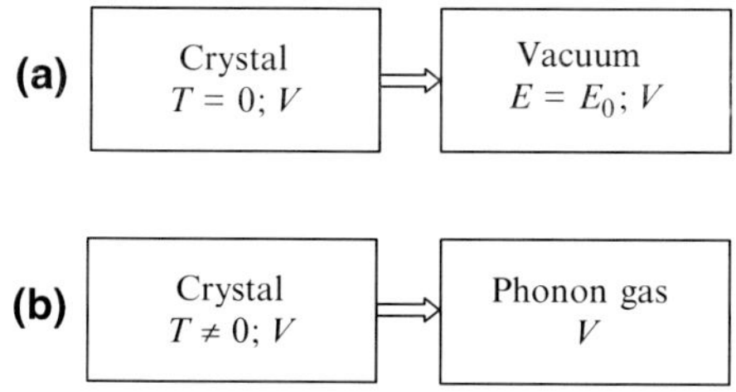

Fig. 7.27. Illustration of equivalence of the excited state of a crystal and phonon gas

At a temperature different from zero, some oscillators pass into the first excited state ($n_{\boldsymbol{q}j} = 1$) and the energy of the crystal increases by the magnitude $\hbar\omega_j(\boldsymbol{q})$. On the passage of each oscillator into the new excited state, the energy of the crystal increases by a multiple of $\hbar\omega_j(\boldsymbol{q})$. The increase in the energy of the crystal by the magnitude $\hbar\omega_j(\boldsymbol{q})$ can be presented as the appearance of one quasi-particle with energy $\varepsilon_j(\boldsymbol{q}) = \hbar\omega_j(\boldsymbol{q})$ and impulse $\boldsymbol{p} = \hbar\boldsymbol{q}$ in the vessel with volume V, which is equal to the volume of the crystalline lattice.

Such a quasi-particle is called *a phonon*:

$$\left.\begin{array}{ll} \varepsilon_j(\boldsymbol{q}) & = \hbar\omega_j(\boldsymbol{q}) \\ \boldsymbol{p} & = \hbar\boldsymbol{q} \end{array}\right\} \text{ phonon.} \tag{7.309}$$

As can be seen, *a phonon is a quantum of energy of a thermal (elastic) wave or an elementary thermal excitation of a crystal.*

Because one phonon corresponds to each frequency $\omega_j(\boldsymbol{q})$, the type of phonon is determined by the wave vector ($\boldsymbol{q}$) and the number of branch j in the frequency spectrum (Fig. 6.4). This means that in a simple crystalline lattice of N elementary cells, $3N$ types of phonons are possible. In our case (for a simple cell), these phonon are called acoustic phonons.[6]

As the temperature rises, the dynamics of filling the volume V with phonons can be described as follows: at $T = 0$ the energy of a crystal equals the energy of the fundamental state $E = E_0$, and the hypothetic vessel of volume V is empty (Fig. 7.27a). At finite temperatures ($T \neq 0$), first phonons with a small energy (low-frequency phonons) appear. As the temperature increases further, more and more new types of phonons appear and simultaneously the number of phonons that have already appeared increases. At the temperature equal to the Debye temperature, i.e. $T = \theta$, all types of phonons are excited.

[6] If the crystalline lattice is a complex one with the elementary cell, in which there are found s atoms, then there are possible $3s$ vibrational branches; therewith 3 of them are called acoustic, and $(3s - 3)$ are optical. Consequently, in a complex crystalline lattice the frequency spectrum consists of $3Ns$ frequencies, to which the same number of types of phonons corresponds. Of these phonons $3N$ are of the acoustic, and $(3s - 3)N$ are the optical type.

On further increase of temperature $T > \theta$, new types of phonons do not arise, and only the number of excited phonons increases.

In the language of classical statistics, the frequency of a harmonic vibration propagating in a crystal corresponds to each type of phonons, and their number corresponds to the intensity (the square of the amplitude).

The total energy of a crystal, according to (7.306), can be presented as

$$E = E_0 + E_{\mathrm{f}}, \tag{7.310}$$

where

$$E_{\mathrm{f}} = \sum_{\boldsymbol{q}} \sum_{j=1}^{3} n_{\boldsymbol{q}j} \hbar\omega(\boldsymbol{q}) \tag{7.311}$$

is the energy of a phonon gas. From this expression, it is seen that the oscillatory quantum number has an explicit physical sense: $n_{\boldsymbol{q}j}$ is the number of phonons with the frequency $\omega_j(\boldsymbol{q})$, i.e. the number of phonons of the type $(\boldsymbol{q}, j)$.

Because the number of phonons of one particular type can take on any value ($n_{\boldsymbol{q}j}$ takes on any value), *a phonon is a boson.* Therefore, in order to find the mean number $\bar{n}_{\boldsymbol{q}j}$ of phonons of the type $(\boldsymbol{q}, j)$, we need to use the Bose–Einstein distribution. However, first note a certain peculiarity of a phonon gas. It is known that the mean number of phonons depends on the temperature: at $T = 0$, the mean number of phonons $\bar{n}_{\boldsymbol{q}j} = 0$. With an increase of temperature in the hypothetical vessel, corresponding to the crystal, the number of phonons $N(T)$ increases (Fig. 7.27b). Because in thermodynamic equilibrium at a particular temperature T and volume V the free energy of a phonon gas ought to be minimum, the chemical potential of a phonon gas, as also in the case of a photon gas [see (7.307)], equals zero: $(\partial F \,/\, \partial N)_{V,T} = \mu(T) = 0$. If we take this into account, from the Bose–Einstein distribution it follows that the mean number of phonons with energy $\varepsilon_j(\boldsymbol{q}) = \hbar\omega_j(\boldsymbol{q})$ at temperature T is equal to

$$\bar{n}_{qj} = \frac{1}{\mathrm{e}^{\hbar\omega_j(q)/\,k_0 T} - 1}. \tag{7.312}$$

Thus, *a phonon gas is an ultra-relativistic Bose gas, the chemical potential of which is equal to zero* ($\mu(T) = 0$) *at all temperatures.* With the help of this distribution (7.312), called the Planck function, we can calculate the mean number of phonons at a particular temperature T:

$$N(T) = \sum_{\boldsymbol{q}} \sum_{j=1}^{3} n_{\boldsymbol{q}j} = \sum_{\boldsymbol{q}} \sum_{j=1}^{3} \frac{1}{\mathrm{e}^{\hbar\omega_j(\boldsymbol{q})/\,k_0 T}}, \tag{7.313}$$

and the mean energy of a phonon gas is

$$E(T) = \sum_{\boldsymbol{q}} \sum_{j=1}^{3} \hbar\omega_j(\boldsymbol{q}) n_{\boldsymbol{q}j} = \sum_{\boldsymbol{q}} \sum_{j=1}^{3} \frac{\hbar\omega_j(\boldsymbol{q})}{\mathrm{e}^{\hbar\omega_j(q)/\,k_0 T} - 1}. \tag{7.314}$$

Energy calculated with this expression is also the mean energy of the thermal motion of a crystal at temperature T.

In order to calculate the mean number of phonons $N(T)$ and the mean energy $E(T)$ of a phonon gas, we use the simple Debye model. Assume a simple crystalline lattice ($s = 1$). In this case, in the crystal only acoustic vibrations are possible (Fig. 6.4): $j = 1, 2, 3$. Also, assume that the crystal is isotropic and the branches indicated in Fig. 6.4 coincide: $\omega_1(\boldsymbol{q}) = \omega_2(\boldsymbol{q}) = \omega_3(\boldsymbol{q}) = \omega(\boldsymbol{q})$. In the Debye model, a crystal is replaced with an elastic medium and the frequency of vibrations ω depends on the wave number linearly:

$$\omega(q) = \upsilon_0 q, \tag{7.315}$$

where υ_0 is the speed of propagation of sound in the crystal (Fig. 6.5). According to this model, the wave number q and frequency ω change in the restricted range

$$0 \le q \le q_{\max}; \quad 0 \le \omega \le \omega_{\max}. \tag{7.316}$$

The maximum possible frequency $\omega_{\max}$ in the crystal is determined by formula (6.78).

On the basis of this model, the number of frequencies in the range $0 - \omega_{\max}$ equals the number of degrees of freedom of the crystalline lattice: i.e. $3N$. If we take into account that, by our supposition, the vibrational branches coincide, i.e. $(\omega_1(\boldsymbol{q}) = \omega_2(\boldsymbol{q}) = \omega_3(\boldsymbol{q}) = \omega(\boldsymbol{q}))$, in (7.313) the summation with respect to j gives the answer 3. As a result, (7.313) acquires the shape

$$N(T) = 3 \sum_{\boldsymbol{q}} \frac{1}{e^{\hbar\omega(\boldsymbol{q})/k_0 T} - 1}. \tag{7.317}$$

If we pass from the sum with respect to the quasi-discrete variable $\boldsymbol{q}$ (6.35) to the integral, we get

$$N(T) = \frac{3V}{(2\pi)^3} \int \frac{d\boldsymbol{q}}{e^{\hbar\omega(\boldsymbol{q})/k_0 T} - 1}. \tag{7.318}$$

If we pass to the spherical coordinate system and integrate over angles $d\boldsymbol{q} = 4\pi q^2 dq$ with regard to (7.315), (7.318) takes the form

$$N(T) = \frac{3V}{2\pi^2 \upsilon_0^3} \int_0^{\omega_{\max}} \frac{\omega^2 d\omega}{e^{\hbar\omega/\boldsymbol{k_0 T}} - 1}. \tag{7.319}$$

We introduce the dimensionless variable $x = \hbar\omega / k_0 T$. Then we get

$$N(T) = \frac{3V}{2\pi^2 \upsilon_0^3} \left(\frac{k_0 T}{\hbar}\right)^3 \int_0^{\theta/T} \frac{x^2 dx}{e^x - 1}, \tag{7.320}$$

where $\theta = \hbar\omega_{\max} / k_0$ is the Debye temperature [see (6.79)]

$$\theta = \frac{\hbar \upsilon_0}{k_0} \left(6\pi^2 \frac{N}{V}\right)^{1/3}. \tag{7.321}$$

If we substitute the expression of θ (7.321) into (7.320), the mean number of phonons equals

$$N(T) = 9N \left(\frac{T}{\theta}\right)^3 \int_0^{\theta/T} \frac{x^2 \mathrm{d}x}{\mathrm{e}^x - 1}. \tag{7.322}$$

At *low temperatures*, $T \ll \theta$, the upper boundary of the integral can be replaced with $\theta / T \to \infty$. Then we get

$$N(T) = 9N \left(\frac{T}{\theta}\right)^3 \int_0^{\infty} \frac{x^2 \mathrm{d}x}{\mathrm{e}^x - 1}; \quad T \ll \theta. \tag{7.323}$$

If we take into account that the integral in (7.323) equals $\Gamma(3)\zeta(3) = 2.4$ (see Appendix A), (7.323) takes the form

$$N(T) = 21.6N \left(\frac{T}{\theta}\right)^3 \sim T^3; \quad T \ll \theta. \tag{7.324}$$

At *high temperatures*, $T \gg \theta$, we can use the expansion $\mathrm{e}^x = 1 + x + \cdots$ in the integrand function. Then, (7.322) acquires the form

$$N(T) = 4.5N \left(\frac{T}{\theta}\right) \sim T; \quad T \gg \theta. \tag{7.325}$$

As can be seen, at low temperatures the mean number of phonons is proportional to the cube of temperature, and at high temperatures it varies linearly with the temperature T (Fig. 7.28).

Because the chemical potential of a phonon gas $\mu(T) = 0$, the grand thermodynamic potential and free energy coincide: $F = \Phi - PV = \mu N + \Omega = \Omega$. Therefore, if the grand thermodynamic potential is known, we can also know the free energy. In order to obtain the expression of the grand thermodynamic potential for a phonon gas (7.95), we need to take the lower signs and take into

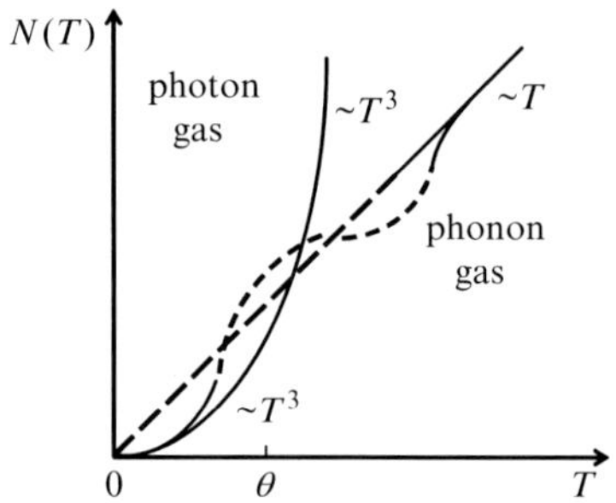

Fig. 7.28. The comparison of the temperature dependence of the mean number of photons and phonons

account relationships $\varepsilon = \hbar\omega = \hbar\upsilon_0 q$, $\mu = 0$ and $g_0 = 3$, and the upper limit of the integral to be $\omega_{\max}$. As a result, the grand thermodynamic potential Ω or the free energy F of a phonon gas takes the form[7]

$$F = \Omega = \frac{3V}{2\pi^2}\frac{k_0 T}{\upsilon_0^3}\int_0^{\omega_{\max}} \omega^2 \ln\left(1 - \mathrm{e}^{-\hbar\omega/k_0 T}\right)\mathrm{d}\omega. \tag{7.326}$$

If we integrate (7.326) once by parts, for the free energy of a phonon gas F we get

$$F = 3Nk_0 T \ln(1 - \mathrm{e}^{\theta/T}) - Nk_0 T\,(\theta\,/\,T)\,, \tag{7.327}$$

where $D\,(\theta\,/\,T)$ is the Debye function (6.113), and θ (6.70) is the Debye temperature.

Using (7.327), we can calculate entropy, mean energy and heat capacity of a phonon gas, which coincide with corresponding results derived in Sect. 6.5.

With reference to the thermal equation of state, i.e. pressure $P = -(\partial F/\partial V)_T$, we need to note the following. Because the phonon gas is obtained in the harmonic approximation, the Debye temperature for the phonon gas in this approximation does not depend on the volume V. It is known that the Debye temperature θ depends on volume only in the anharmonic approximation (see Sect. 6.3). Consequently, from (7.327) and $P = -(\partial F\,/\,\partial V)_T$, it follows that the pressure of a phonon gas ought to equal zero: i.e. $P = 0$.

Note that pressure of a solid also equals zero; it is different from zero only in the anharmonic approximation, when the Grüneisen parameter γ_G is a finite quantity, since $P \sim \gamma_G$ [see formula (6.161)].

These coincidences show that the concepts of phonon and phonon gas are very convenient to use when formulating the theory of thermodynamic properties of crystals. In that case, a vibrating crystalline lattice is replaced with an ideal phonon gas. Theoretical results obtained for the phonon gas are concerned with the thermal properties of a crystal. The concept of a phonon is also convenient when formulating the theory of thermal conductivity and the interaction of conduction electrons with a crystalline lattice. In these cases, instead of a vibrating crystal it is sufficient to take an ideal phonon gas and consider the interaction of an electron with the phonon gas (radiation or emission of a phonon by an electron).

In conclusion, note the basic distinctions between photons and phonons:

- A photon is a real particle; a phonon is a quasi-particle.
- The frequency of a photon changes in the limit $0 \le \omega \le \infty$, and the frequency of a phonon is restricted to the range $0 \le \omega \le \omega_{\max}$.

---

[7] Notice that the expression of free energy (7.326) coincides with formula (6.151) without consideration of the zero energy E_0. Hence it follows that thermodynamic parameters calculated for a phonon gas ought to coincide with thermodynamic parameters of a vibrating (exciting) crystal.

- The impulse of a photon $\boldsymbol{p} = \hbar\boldsymbol{k}$ is single-valued and changes in the infinite range $0 \leq \boldsymbol{p} \leq \infty$; the wave vector of a phonon $\boldsymbol{q}$ and, therefore, its impulse $\boldsymbol{p} = \hbar\boldsymbol{q}$ are not single-valued, since $\boldsymbol{q}$ is determined with accuracy to an arbitrary vector of a reciprocal lattice $\boldsymbol{b}_g$, i.e. $\omega(\boldsymbol{q}) = \omega(\boldsymbol{q} + \boldsymbol{b}_g)$;
- A photon can exist in vacuum whereas a phonon exists only inside a lattice; outside a crystal a phonon does not exist.
- The mean value of the total number of photons over the entire temperature region changes as $N(T) \sim T^3$, whereas the mean number of phonons in the Debye approximation in different temperature regions depends on T in different ways (Fig. 7.28):

$$\begin{aligned} &N(T) \sim T^3; \quad T \ll \theta, \\ &N(T) \sim T; \quad T \gg \theta; \end{aligned} \tag{7.328}$$

- The total energy of a photon gas over the entire temperature region equally depends on temperature, whereas in the Debye approximation over different temperature regions the total energy of a phonon gas depends on T in different ways:

$$\begin{aligned} &E(T) \sim T^4; \quad T \ll \theta, \\ &E(T) \sim T; \quad T \gg \theta; \end{aligned} \tag{7.329}$$

- In contrast to a photon gas, the pressure of a phonon gas equals zero.

Electron Gas in Quantizing Magnetic Field

Summary. In this chapter, a statistical theory of thermodynamic properties of an electron gas taking into account the energy-spectrum quantization in an external magnetic field is expounded. On the basis of the grand thermodynamic potential, the chemical potential, thermal equation of state, entropy and heat capacity of an electron gas are found. The Landau diamagnetism is considered. It is shown that all results of the quantum theory in the quasi-classical approximation pass into the known classical ones.

8.1 Motion of Electron in External Uniform Magnetic Field: Quantization of Energy Spectrum

According to classical mechanics, a charged particle in a constant uniform magnetic field H under the action of the Lorentz force moves in a circular helix with the axis along the magnetic field and with radius

$$r = \frac{v_\perp}{\omega_\mathrm{c}}, \tag{8.1}$$

where $v_\perp$ is the component of velocity in the plane perpendicular to the magnetic field, and

$$\omega_\mathrm{c} = \frac{eH}{mc} \tag{8.2}$$

is the cyclic frequency of rotation of a particle in this plane, called also *the cyclotron frequency*; c is the velocity of light; e and m are the charge and mass of a particle, respectively. The particle at the same time participates in two motions: a uniform rotation with the angular velocity ω_c in the plane perpendicular to the field and translational motion in the direction of a magnetic field. If the velocity of the particle along the field equals zero, the particle performs only a circular motion.

Note that the longitudinal translational motion along the direction of a magnetic field is infinite and is always classical. However, the rotational motion in the plane perpendicular to a magnetic field is not always classical. It is classical only in the case when its action $mv_{\perp}r$ is much more of the Planck constant $\hbar$:

$$mv_{\perp}r >> \hbar \text{ or r} >> \lambda, \tag{8.3}$$

where $\lambda = h/mv_{\perp}$ is the de Broglie wavelength.

If we take into account (8.1), the condition of classicity (8.3) can be rewritten in the form

$$mv_{\perp}^2 >> \hbar\omega_c \; or \; k_0T >> \hbar\omega_c, \tag{8.4}$$

where k_0 is the Boltzmann constant, and T is the absolute temperature.

In the case when adduced conditions of classicity of the rotational motion are not fulfilled, i.e. when $r \leq \lambda$ or $k_0T \leq \hbar\omega_c$, the motion is quantum and the problem should be solved on the basis of the Schrödinger equation.

The problem of the motion of a charged particle (an electron) in an external uniform magnetic field on the basis of quantum mechanics was solved for the first time by Landau when determining the diamagnetic susceptibility of a free electron gas in metals. He showed that the spectrum of an electron in a uniform magnetic field becomes partially discrete, i.e. the circular motion of an electron in the plane perpendicular to the field is quantized. On the basis of the obtained spectrum, the diamagnetic susceptibility of a free electron gas was calculated (according to classical notions, in a free electron gas diamagnetism is absent) and it was shown that the diamagnetic susceptibility of an electron gas equals 1/3 of the paramagnetic susceptibility, due to the spin of electrons. Subsequently, the presence of diamagnetic properties in the free electron gas was named *the Landau diamagnetism.*

Here in brief we adduce the Landau solution. We consider the motion of a conduction electron with the effective mass m in an external constant uniform magnetic field H. We disregard the spin of an electron. Then the Hamiltonian of an electron in a magnetic field in the effective mass approximation has the appearance

$$\hat{\mathscr{H}} = \frac{1}{2m}\left(\hat{\boldsymbol{p}} + \frac{e}{c}\boldsymbol{A}\right)^2, \tag{8.5}$$

where $\hat{\boldsymbol{p}} = -i\hbar\nabla$ is the operator of the impulse, e is the charge magnitude of an electron, and $\boldsymbol{A}$ is the vector-potential of a magnetic field.

The z-coordinate axis is directed along the external magnetic field. Then, $H_x = H_y = 0, H_z = H$. To such a magnetic field, different calibrations of the vector-potential $\boldsymbol{A}$ correspond. Choose the following calibration:

$$\boldsymbol{A}_x = 0, \boldsymbol{A}_y = Hx, \boldsymbol{A}_z = 0, \tag{8.6}$$

then Hamiltonian (8.5) takes the form

$$\hat{\mathscr{H}} = \frac{1}{2m}\left[\hat{p}_x^2 + \left(\hat{p}_y^2 + m\omega_c x\right)^2 + \hat{p}_z^2\right], \tag{8.7}$$

where ω_c is cyclotron frequency (8.2).

Inasmuch as Hamiltonian (8.7) commutates with operators $\hat{p}_y$ and $\hat{p}_z$, there are conserved y- and z- components of the impulse which have the following eigenvalues[1]:

$$p_y = \hbar k_y; \quad p_z = \hbar k_z, \tag{8.8}$$

where k_y and k_z are corresponding components of the wave vector $\boldsymbol{k}$. Therefore the motion of an electron in y- and z-directions is described by the plane wave, and a solution of the Schrödinger equation ε

$$\hat{\mathscr{H}}\,\Psi = \varepsilon\,\Psi \tag{8.9}$$

can be found in the form

$$\Psi(r) = \varphi(x)\exp[\mathrm{i}(k_y y + k_z z)]. \tag{8.10}$$

Substituting this solution into (8.9), with regard to (8.7), we get an equation for the unknown function $\varphi(x)$:

$$-\frac{\hbar^2}{2m}\frac{\mathrm{d}^2\varphi}{\mathrm{d}x^2} + \frac{1}{2}m\omega_\mathrm{c}^2(x-x_0)^2\varphi = \varepsilon_N\varphi, \tag{8.11}$$

where the following notations are introduced

$$x_0 = -\frac{\hbar k_y}{\omega_\mathrm{c} m}, \varepsilon_N = \varepsilon - \frac{\hbar^2 k_z^2}{2m}. \tag{8.12}$$

Notice that (8.11) is the equation of a linear harmonic oscillator with the frequency $\omega_\mathrm{c} = eH/mc$ with the centre $x = x_0$, eigenfunctions and eigenvalues of which are

$$\varphi(x-x_0) = \frac{1}{\sqrt{R}}H_N\left(\frac{x-x_0}{R}\right)\exp\left[-\frac{1}{2}\left(\frac{x-x_0}{R}\right)^2\right], \tag{8.13}$$

$$\varepsilon_N = \left(N+\frac{1}{2}\right)\hbar\omega_\mathrm{c}, \tag{8.14}$$

where $N = 0, 1, 2, 3\ldots$ is the oscillatory quantum number, H_N is the Hermite polynomial of order N,

$$R = \left(\frac{\hbar}{\omega_\mathrm{c} m}\right)^{1/2} = \left(\frac{\hbar c}{eH}\right)^{1/2} \tag{8.15}$$

is so-called *the magnetic length.*

As a result, substituting (8.14) into (8.12), for energy and the wave function of an electron in a magnetic field directed along the z-axis, we get the following expressions:

[1] If we would choose another calibration, e.g. $A_x = -Hy$, $A_y = A_z = 0$, p_x and p_z components would be conserved.

$$\varepsilon \equiv \varepsilon(N, k_z) = \left(N + \frac{1}{2}\right) \hbar\omega_{\mathrm{c}} + \frac{\hbar^2 k_z^2}{2m}, \tag{8.16}$$

$$\Psi_\alpha(r) = \varphi_N(x - x_0) \exp\left[\mathrm{i}(k_y y + k_z z)\right], \tag{8.17}$$

where $\alpha \equiv (N, k_y, k_z)$ is a set of quantum numbers, determining the state of an electron in a magnetic field, $\varphi_N(x - x_0)$ is the normalized wave-function of a linear harmonic oscillator with the quantum number N, vibrating near the equilibrium position

$$x_0 = -\frac{\hbar k_y}{\omega_{\mathrm{c}} m} = -R^2 k_y. \tag{8.18}$$

Energy spectrum (8.16) can be also presented in the form

$$\varepsilon \equiv \varepsilon(N, k_z) = (2N + 1)\,\mu H + \frac{\hbar^2 k_z^2}{2m}, \tag{8.19}$$

where $\mu = e\hbar/2mc$ is the effective Bohr magneton.

The spectrum of an electron in a magnetic field, as is seen from (8.19), becomes partially discrete. Energy (8.19) consists of two parts: one is continuously depending on k_z, corresponding to the motion of an electron along the magnetic field, and the second is discrete, corresponding to the quantization of the circular motion in the plane perpendicular to the magnetic field.

The influence of quantization of the motion in a magnetic field on the energy spectrum of an electron on the basis of (8.19) is schematically shown in Fig. 8.1a. For comparison the same is found in (Fig. 8.1b), where the dependence of energy on the wave vector $\varepsilon(\boldsymbol{k})$ in the absence of a magnetic field is adduced. It is seen that in the presence of a magnetic field, in the spectrum discrete levels separated from each other by the energy distance $2\mu H$ appear. These are called *the Landau levels.*

For the given level, energy continuously depends only on k_z, i.e. one-dimensional parabolic energy bands appear. It is also seen that the bottom

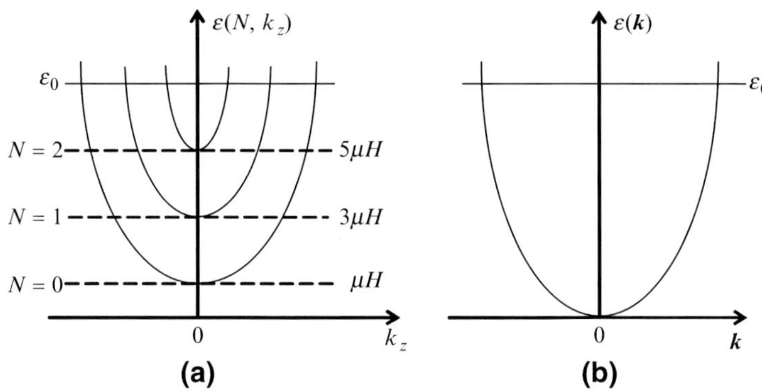

Fig. 8.1. Energy as a function of wave vector: **a)** in the absence of magnetic field, **b)** in the presence of magnetic field, discrete Landau levels and one-dimensional bands appear

of the conduction band ascends by the magnitude of μH and, thereby, the forbidden bandwidth increases.

Using the correspondence principle, the quantization of motion in the quasi-classical approximation can be obviously imagined. According to classical physics, any values of energy $\varepsilon_\perp$, associated with the motion in the plane perpendicular to the magnetic field, are permissible. Inasmuch as $\varepsilon_\perp$ is related to the radius of the orbit r by the simple relationship $\varepsilon_\perp = mv_\perp^2/2 = m\omega_c^2 r^2/2$, any circular orbits with the radius r are possible. From the point of view of quantum mechanics, according to (8.16), $\varepsilon_\perp = (N + 1/2)\,\hbar\omega_c$, i.e. energy of the transverse motion cannot take on any values. If we bring into correspondence these two expressions for energy, we see that only orbits with radii

$$r_N = (2N+1)^{1/2} R, \quad N = 0, 1, 2, 3, \ldots \tag{8.20}$$

are possible. Hence it is seen that magnetic length (8.15) is the radius of the first possible orbit $R = r_0$. Thus, by quasi-classical notions an electron can rotate around the magnetic field in discrete orbits with radii (8.20). For an electron to pass from one orbit to an adjacent one with a larger radius, it is necessary to expend the energy $\hbar\omega_c$.

We have seen that in the $\boldsymbol{r}$-space the quantization corresponds to the presence of discrete orbits, and that for the given magnetic field there is an orbit with a minimum radius R. It is of interest to consider how the distribution of states of an electron in the $\boldsymbol{k}$-space, when quantizing the motion, changes. It is known that in the simple parabolic case, states of an electron with energies less than ε_0 in the $\boldsymbol{k}$-space in the absence of a magnetic field continuously fill a sphere with the radius $k = (1/\hbar)(2m\varepsilon_0)^{1/2}$. In a magnetic field directed along the z-axis, the part of energy $\hbar^2 k_z^2/2m$ does not change, and to the energy $\hbar^2 k_\perp^2/2m$, the expression $\varepsilon_\perp = (N + 1/2)\,\hbar\omega_c$ corresponds. From this correspondence, it follows that only discrete values of $k_\perp$ are possible, namely

$$k_\perp = k_{\perp N} = (2N+1)^{1/2} R^{-1}, \quad N = 0, 1, 2, 3, \ldots \tag{8.21}$$

where R is the magnetic length (8.15).

This means that all states continuously filling the volume inside the sphere $\varepsilon_0 = const$, in the presence of a magnetic field, are found only on surfaces of co-axial discrete cylinders with the axis parallel to k_z with radii (8.21) (see Fig. 8.2a). The height of the possible cylinder with number N for the pre-assigned value of energy of an electron ε_0, according to (8.19), equals[2]

$$k_{zN} = \left[2(2m)^{1/2}\Big/\hbar\right]\,[\varepsilon_0 - (2N+1)\mu H]^{1/2}. \tag{8.22}$$

Inasmuch as the height ought to be a real quantity, a maximum number of possible cylinders, corresponding to energies less than ε_0, equals the integer part of the fraction

[2] The factor 2 takes into account the positive and negative directions of k_z.

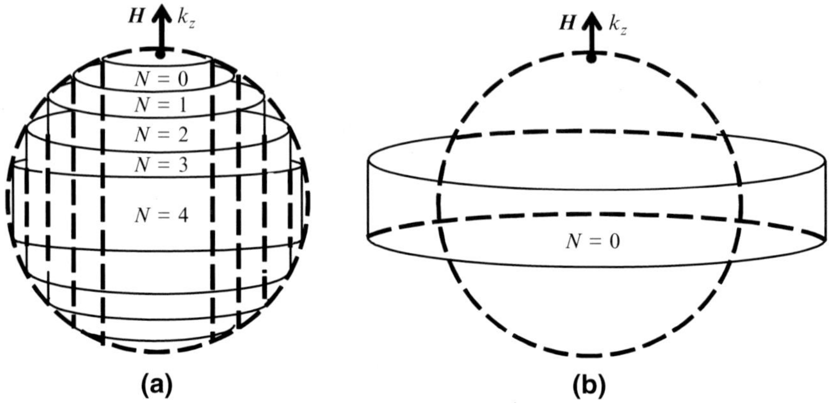

Fig. 8.2. Quantum state distribution variation in magnetic field

$$\left[\frac{\varepsilon_0 - \mu H}{2\mu H}\right] = N_{\max}. \tag{8.23}$$

Hence it is seen that the number of cylinders inside the sphere $\varepsilon_0 = \text{const}$ decreases with the growth of the magnetic field.

Thus, according to quasi-classical notions, in the presence of a quantizing magnetic field all states of an electron with energies less than ε_0 in the $\boldsymbol{k}$-space are found on surfaces of co-axial cylinders with the axis along k_z, with radii (8.21) and heights (8.22) (see Fig. 8.2a), the number of which is determined by (8.23).

It should be noted that such a quasi-classical notion of the distribution of states in a magnetic field holds true only in the case when inside the given isoenergetic surface $\varepsilon_0 = \text{const}$ if only several_cylinders are placed, i.e. $N_{\max} > 1$. At strong magnetic fields in the indicated sphere, in all only one cylinder is placed and in the limit $H \to \infty$ the radius of this cylinder $k_{\perp 0} = R^{-1} \to \infty$ becomes larger than that of the radius of the sphere $k = (1/\hbar)(2m\varepsilon_0)^{1/2}$, and its height tends to zero, i.e. all quantum states in the limiting strong magnetic field – *the quantum limit* – are found on a surface of the narrow ring with the large radius $k_{\perp 0} = R^{-1} = (eH/\hbar c)^{1/2}$ (see Fig. 8.2b).

8.2 Density of Quantum States in Strong Magnetic Field

Now we will determine the density of quantum states of an electron in a magnetic field. In the preceding section/chapter we showed that one quantum state in a magnetic field is determined by two quasi-continuous (k_y, k_z) numbers and one discrete quantum number N.

Therefore, the total number of quantum states of a unit of volume in a magnetic field equals

$$Z = \frac{2}{V} \sum_{Nk_y k_z} \to \frac{2L_2 L_3}{V(2\pi)^2} \sum_N \int \mathrm{d}k_y \mathrm{d}k_z, \tag{8.24}$$

where L_2 and L_3 are corresponding linear dimensions of a system of volume $V = L_1 L_2 L_3$, in which an electron gas is found, and the factor 2 takes into account degeneracy with respect to the spin.

From (8.19) it is seen that energy of a conduction electron depends only on two quantum numbers (N, k_z), i.e. with respect to k_y there is degeneracy. By virtue of this, the integral over $\mathrm{d}k_y$ in (8.24) can be taken, if we take into account (8.18):

$$\int \mathrm{d}k_y = \frac{1}{R^2} \int\limits_{-L/2}^{L/2} \mathrm{d}x_0 = \frac{L_1}{R^2}. \tag{8.25}$$

From the latter two formulae, after passing to the integral with respect to energy, we have

$$Z = \frac{4}{(2\pi R)^2} \sum_N \int \frac{\mathrm{d}k_z(\varepsilon, N)}{\mathrm{d}\varepsilon} \mathrm{d}\varepsilon = \int g_H(\varepsilon) \mathrm{d}\varepsilon, \tag{8.26}$$

where

$$g_H(\varepsilon) = \frac{4}{(2\pi R)^2} \sum_N \frac{\mathrm{d}k_z(\varepsilon, N)}{\mathrm{d}\varepsilon} \tag{8.27}$$

is the density of quantum states in a magnetic field; the factor 2 in (8.26) and in (8.27) takes into account the fact that ε is an even function of k_z.

In order to find the explicit form of $g_H(\varepsilon)$, it is necessary to begin with a concrete form of the dispersion law $\varepsilon(k_z, N,)$, where the spin splitting is disregarded. Then the summation with respect to spin is reduced to the factor 2 and by virtue of (8.16) density of states (8.27) takes the form

$$g_H(\varepsilon) = \frac{4}{(2\pi R)^2} \frac{(2m)^{1/2}}{\hbar} \sum_N \left[\varepsilon - (N + 1/2)\hbar\omega_{\mathrm{c}}\right]^{-1/2}. \tag{8.28}$$

The summation in (8.28) is carried out with respect to all integer values of N, for which the radicand expression is positive.

Note that if in weak magnetic fields ($\hbar\omega_{\mathrm{c}} \ll \varepsilon$) in (8.28) from the summation with respect to N pass to the integral in the limits from 0 to $(\varepsilon - 1/2\hbar\omega_{\mathrm{c}}) / \hbar\omega_{\mathrm{c}} = N_{\mathrm{max}}$, we get the known result (7.102) for the density of states without a magnetic field.

From (8.28) it is seen that the density of states has a certain peculiarity: every time, when energy coincides with one of the Landau levels, it is converted to infinity. The behaviour of $g_H(\varepsilon)$ is schematically shown in Fig. 8.3. Continuously distributed quantum states in the k-space in the presence of a

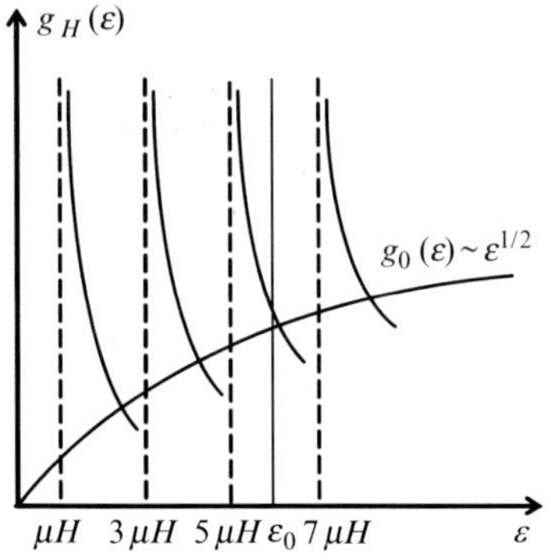

Fig. 8.3. Quantum state densities in magnetic field, $g_H(\varepsilon)$. State densities in the absence of magnetic field, $g_0(\varepsilon)$ is also shown for comparison

magnetic field basically group at Landau levels, but because of that the total number of states is conserved. The fact that many quantum states account for one Landau level is associated with the fact that each level in the magnetic field is degenerate with respect to the quantum number k_y. The degeneracy multiplicity, according to (8.25), equals

$$\sum_{k_y} \to \frac{L_2}{2\pi} \int \mathrm{d}k_y = \frac{L_1 L_2}{2\pi R^2} = \frac{L_1 L_2}{2\pi} \frac{eH}{\hbar c}. \tag{8.29}$$

It should be noted that oscillations of the magnetic susceptibility and other physical properties of conducting crystals in a strong magnetic field are due to just such a singular behaviour of density of states (8.25).

8.3 Grand Thermodynamic Potential and Statistics of Electron Gas in Quantizing Magnetic Field

To determine the criterion of degeneracy of an electron gas in a quantizing magnetic field, it is necessary to find the relation of its chemical potential to concentration and temperature. To do this, it is needed to know the explicit form of the grand thermodynamic potential Ω_e as a function of volume, temperature, chemical potential and magnetic field: $\Omega_\mathrm{e} = \Omega_\mathrm{e}(V, T, \zeta, H)$. The thermodynamic relationship for Ω_e in a magnetic field, according to (2.208), has the appearance

$$\mathrm{d}\Omega_\mathrm{e} = -S\,\mathrm{d}T - P\,\mathrm{d}V - N_\mathrm{e}\mathrm{d}\zeta - VM\,\mathrm{d}H, \tag{8.30}$$

where N_e is the number of free electrons, V and ζ are volume and chemical potential of an electron gas, respectively[3], M is the magnetization, and the rest of the notations are generally accepted.

[3] In this chapter, we denote the chemical potential by ζ, in order not to confuse it with the intrinsic magnetic moment μ.

From (8.30) it is seen that if the explicit form of the function $\Omega_e = \Omega_e(T, V, \zeta, H)$ is known, the total number of electrons N_e can be found as follows

$$N_e = -(\partial\Omega_e/\partial\zeta)_{T,V,H} \,. \tag{8.31}$$

Grand thermodynamic potential for fermions (electrons) (7.81) in our case has the appearance

$$\Omega_e = \Omega_e(T, V, \zeta, H) = -2k_0T \sum_{N,k_y,k_z} \ln\left[1 + \exp\left(\frac{\zeta - \varepsilon(N, k_z)}{k_0T}\right)\right], \tag{8.32}$$

since one quantum state of an electron is determined by three quantum numbers $\boldsymbol{k} \to (N, k_y, k_z)$; $\varepsilon(N, k_z)$ is given by formula (8.19), the factor 2 takes into account degeneracy with respect to the spin.

Assume that an electron gas occupies volume $V = L_1L_2L_3$, where L_i are linear dimensions of volume along corresponding coordinate axes.

According to the known rule (8.24), in (8.32) we pass from the summation with respect to k_y and k_z to the integration over $\mathrm{d}k_y$ and $\mathrm{d}k_z$ and take into account (8.25). Then from the integral over $\mathrm{d}k_z$ we pass to the integral over energy $\mathrm{d}\varepsilon$. As a result, (8.32) takes the form

$$\Omega_e = -\frac{4k_0TV}{(2\pi R)^2} \sum_N \int_{\varepsilon_N}^{\infty} \frac{\mathrm{d}k_z(\varepsilon, N)}{\mathrm{d}\varepsilon} \ln\left[1 + \exp\left(\frac{\zeta - \varepsilon}{k_0T}\right)\right] \mathrm{d}\varepsilon, \tag{8.33}$$

where the lower boundary of the integral $\varepsilon_N = (N + 1/2)\hbar\omega_c$, according to spectrum (8.19), is a root of the equation $k_z(\varepsilon, \varepsilon_N) = 0$. The factor 2 shows that two values of k_z correspond to one value of energy (Fig. 8.1).

Integrate up (8.33) once by parts. Then we get

$$\Omega_e = -\frac{4V}{(2\pi R)^2} \sum_N \int_{\varepsilon_N}^{\infty} k_z(\varepsilon, N) f(\varepsilon)\mathrm{d}\varepsilon, \tag{8.34}$$

where $f(\varepsilon) = [1 + \exp(\varepsilon - \zeta)/k_0T]^{-1}$ is the Fermi distribution function.

If we take into account that $(\partial f/\partial\zeta) = -(\partial f/\partial\varepsilon)$, then, according to (8.31) and (8.34), the concentration of an electron gas $n = N_e/V$ is

$$n = \frac{4V}{(2\pi R)^2} \sum_N \int_{\varepsilon_N}^{\infty} \left(-\frac{\partial f}{\partial\varepsilon}\right) k_z(\varepsilon, N)\mathrm{d}\varepsilon. \tag{8.35}$$

Using (8.19), for $k_z(\varepsilon, N)$ we can write

$$k_z(\varepsilon, N) = \frac{\sqrt{2m}}{\hbar}(\varepsilon - \varepsilon_N)^{1/2}. \tag{8.36}$$

Substituting (8.36) into (8.35), we get

$$n = \frac{4(2m)^{1/2}}{\hbar(2\pi R)^2} \sum_N \int_{\varepsilon_N}^{\infty} \left(-\frac{\partial f}{\partial \varepsilon}\right) (\varepsilon - \varepsilon_N)^{1/2} \mathrm{d}\varepsilon. \quad (8.37)$$

Consider particular cases.

1. *Non-degenerate electron gas.* In this case, the distribution function has the appearance $f(\varepsilon) = \exp\left[(\zeta - \varepsilon)/k_0 T\right]$. Then (8.37) can be rewritten in the form

$$n = \frac{4(2m)^{1/2}}{\hbar(2\pi R)^2 k_0 T} \mathrm{e}^{\zeta/k_0 T} \sum_N \int_{\varepsilon_N}^{\infty} (\varepsilon - \varepsilon_N)^{1/2} \mathrm{e}^{-\varepsilon/k_0 T} \mathrm{d}\varepsilon. \quad (8.38)$$

Introduce the notation $\varepsilon - \varepsilon_N = \varepsilon'$ and pass to the dimensionless integration variable $x = \varepsilon'/k_0 T$. Then the integral over dx is easily fulfilled and as a result we get

$$n = \frac{2(2\pi m k_0 T)^{1/2}}{\hbar(2\pi R)^2} e^{\zeta/k_0 T} \sum_{N=0}^{\infty} e^{-(2N+1)\mu H/k_0 T}. \quad (8.39)$$

The summation with respect to N gives

$$\sum_{N=0}^{\infty} \mathrm{e}^{-(2N+1)\mu H/k_0 T} = \left[2 sh\left(\mu H/k_0 T\right)\right]^{-1}. \quad (8.40)$$

Taking this into account, from (8.39) for the chemical potential we get

$$\mathrm{e}^{\zeta/k_0 T} = \frac{n\hbar(2\pi R)^2 \left[sh\left(\mu H/k_0 T\right)\right]}{(2\pi m k_0 T)^{1/2}}. \quad (8.41)$$

This expression for a non-degenerate gas, presented in the form

$$\mathrm{e}^{\zeta/k_0 T} = 4n \frac{\pi^{3/2}\hbar^3}{(2m k_0 T)^{3/2}} \frac{sh\left(\mu H/k_0 T\right)}{\mu H/k_0 T}, \quad (8.42)$$

holds true for any value of a strong field, including a quantizing magnetic field.

In the quasi-classical approximation, when the energy of the thermal motion $k_0 T$ is larger than the difference between two adjacent Landau levels ($k_0 T \gg \hbar\omega_\mathrm{c} = 2\mu H$), from (8.42) follows the known expression for the chemical potential (7.20).

In the region of strong magnetic fields, when $2\mu H \gg k_0 T$, the hyperbolic sine in (8.42) can be replaced by an exponent. Then we get

$$\mathrm{e}^{(\zeta - \mu H)/k_0 T} = \frac{2\pi^{3/2}\hbar^3 n}{(2m k_0 T)^{3/2}} \frac{k_0 T}{\mu H}. \quad (8.43)$$

Notice that (8.43) can be obtained immediately from (8.39), restricting to the first term of the summation $N = 0$. Such a state, when electrons are found at the zero Landau level, bears the name *the quantum limit*. Thus, for a non-degenerate electron gas the quantum limit is realized as $2\mu H \gg k_0 T$, and the criterion of the absence of degeneracy $\exp\left[(\zeta - \mu H)/k_0 T\right] \ll 1$, according to (8.42), has the explicit appearance

$$\left[\frac{4\pi^{3/2}\hbar^3 n}{(2mk_0T)^{3/2}}\right](k_0T/2\mu H) \ll 1. \tag{8.44}$$

It is seen that in the quantum limit an increase in a magnetic field promotes the fulfilment of the criterion of the degeneracy absence of an electron gas. This is also seen from (8.43), if it is rewritten in the form

$$\zeta - \mu H = k_0 T \ln\left[\frac{4\pi^{3/2}\hbar^3 n}{(2mk_0T)^{3/2}}\right] - k_0 T \ln\frac{2\mu H}{k_0 T}. \tag{8.45}$$

Hence it follows that the distance between the chemical potential level and the bottom of the conduction band in a magnetic field $(\zeta - \mu H)$ at the expense of the latter term in (8.45) grows with an increase in a magnetic field.

2. *Degenerate electron gas.* For any degree of degeneracy from (8.37) it is impossible to find analytically the chemical potential $\zeta = \zeta(n, H, T)$. Therefore, consider the other limiting case of *a degenerate electron gas* with the pre-assigned concentration n. In this case, replace $(-\partial f/\partial\varepsilon)$ with the $\delta(\varepsilon - \zeta)$-function and from (8.37) we get

$$n = \frac{(2m)^{3/2}}{\pi^2\hbar^3}\mu H \sum_N \left[\zeta(H) - (2N+1)\mu H\right]^{1/2}. \tag{8.46}$$

As $H \to 0$ in (8.46) from the sum with respect to N we can pass to the integration and easily obtain the Fermi boundary $\zeta(0)$ in the absence of a magnetic field (7.116).

The latter formula gives the possibility to determine $\zeta(n, H)$ for a degenerate electron gas at any value of the magnetic field. A simple analytical dependence $\zeta(n, H)$ can be obtained only in the region of the quantum limit, when electrons are found on the lower-most Landau parabola with the number $N = 0$. In the case of the quantum limit $\zeta(H) < 3\mu H$, and in (8.16) it is needed to restrict only the term with $N = 0$. Then from (8.46) it is easy to obtain

$$\zeta(n, H) = \mu H\left[1 + 3\left(\zeta(0)/3\mu H\right)^3\right], \tag{8.47}$$

where $\zeta(0) = (\hbar^2/2m)(3\pi^2 n)^{2/3}$ is the Fermi boundary at $H = 0$ (7.116).

It is seen that in the region of the quantum limit $(\zeta(H) < 3\mu H)$ the relation of the Fermi boundary in a magnetic field $\zeta(H)$ to the Fermi boundary $\zeta(0)$ at $H = 0$ can be found. From (8.47) it is easy to show that in this region $\zeta(n, H)$ is always more of $\zeta(0)$ (Fig. 8.4). As $H \to \infty$, the Fermi boundary $\zeta(H) \to \mu H$

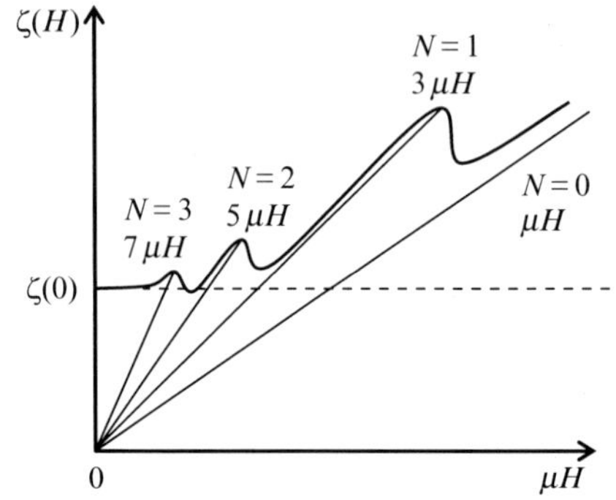

Fig. 8.4. Fermi level of electron gas with specified concentration as a function of magnetic field

and all electrons are "condensed" at the Landau level with $N = 0$. Inasmuch as all levels in a magnetic field, including the level with $N = 0$, are degenerate with multiplicity (8.29), the number of quantum states at the level $N = 0$ is sufficient to find in it all electrons. Such an extreme distribution of electrons corresponds to the picture in Fig. 8.2b, where the radius of a cylinder, on the surface of which electrons are positioned, is larger than the radius of the Fermi sphere at $H = 0$. In the limit $H \to \infty$ the height of this cylinder tends to zero, and the radius $k_{\perp 0} = R^{-1} \to \infty$. In such a state, the motion of an electron along the magnetic field is completely absent, i.e. $k_z = 0$. Notice that the described picture of the distribution takes place at the pre-assigned concentration of an electron gas $n = \text{const}$.

Formula (8.47) is just in the range of a magnetic field, restricted from below by the quantum limit condition $\zeta(H) < 3\mu H$

$$3\mu H > \left(\frac{3}{2}\right)^{1/3} \zeta(0), \tag{8.48}$$

and from the above by the strong-degeneracy criterion $[\zeta(H) - \mu(H)] \gg k_0 T$

$$\frac{\zeta(0)}{k_0 T} \left(\frac{\zeta(0)}{3\mu H}\right)^2 \gg 1. \tag{8.49}$$

From the latter inequality it is seen that in the quantum limit an increase in a magnetic field leads to the degeneracy-criterion violation, i.e. promotes the degeneracy removal. This circumstance is associated with the fact that the density of state is proportional to H (8.28). Therefore, with the growth of a magnetic field at the bottom of the conduction band more electrons are placed and thereby the Fermi boundary energy decreases.

If the latter two conditions are combined, we get inequalities determining the region of a magnetic field, where an electron gas is strongly degenerate and the quantum limit takes place

$$\left(\frac{3}{2}\right)^{1/3}\frac{\zeta(0)}{3\mu} < H \ll \frac{\zeta(0)}{3\mu}\left(\frac{\zeta(0)}{k_0T}\right)^{1/2}, \tag{8.50}$$

i.e. in this region formula (8.47) is right.

Using formula (7.134) to calculate the temperature smearing of the Fermi boundary, with accuracy to $(k_0T)^2$, we get

$$\zeta(H,T) = \zeta(H,0) + \frac{\pi^2}{12}\frac{(k_0T)^2}{\zeta(H)-\mu H}. \tag{8.51}$$

Notice that the temperature correction in (8.51) and (7.136) are distinct by the sign. This is associated with the fact that in the quantum limit the density of states with the growth of energy decreases (the right branch in Fig. 8.3), since the electron gas behaves as though it is one-dimensional. Therefore, electrons that are found in the energy range k_0T below the Fermi boundary which get excited by the heat motion occupy a wider range of energy than k_0T, and, consequently, the mean position of the Fermi level $\zeta(H,T)$ is found above the absolute zero temperature. Inasmuch as in the absence of a magnetic field the density of states $g_0(\varepsilon)$ grows with energy (Fig. 8.3), $\zeta(T)$ ought to be found below $\zeta(0)$ [see (7.136)].

We have found the relation (8.47) between the Fermi boundary in a magnetic field $\zeta(H)$ and one without a magnetic field $\zeta(0)$ in the quantum limit, when below the boundary $\zeta(H)$ there is in all one Landau level with $N=0$. In that case, when two and more Landau levels are found below the Fermi boundary, from (8.46) it is difficult to find this relation analytically for the whole region of the magnetic field. However, this can be done for pre-assigned values of H_N, at which the Fermi boundary $\zeta(H)$ coincides with some Landau level. Assume that the number of this level is N. Then $\zeta(H_N) = (2N+1)\mu H_N$. Substituting this value into (8.46) we get

$$n = \frac{(2m)^{3/2}}{\pi^2\hbar^3}\mu H_N \sum_{N'=0}^{N}\left[2(N-N')\mu H_N\right]^{1/2}, \tag{8.52}$$

which gives the relation between the Fermi boundary $\zeta(0)$ at $H=0$ and the Fermi boundary for magnetic fields H_N, at which it coincides with the Landau level with number N:

$$\zeta(H_N) = \frac{N+1/2}{\left(\frac{3}{2}\sum_{k=0}^{N}\sqrt{k}\right)^{2/3}}\zeta(0), \quad N \neq 0. \tag{8.53}$$

Hence it can be evaluated that $\zeta(H_1) = (3/2)^{1/3}\zeta(0) \approx 1{,}146\zeta(0)$, $\zeta(H_2) \approx 1{,}058\zeta(0)$, etc. It is easy to show that for large N relationship (8.53) gives $\zeta(H_N) \approx \zeta(0)$. The behaviour of the function $\zeta(H)$ with regard to (8.47) is schematically shown in Fig. 8.4. It is seen that the Fermi boundary is a pulsating function of the magnetic field: the Fermi boundary does not much expand every time, when some Landau level intersects it. After that all Landau levels, except the zero one, go out of the Fermi boundary, i.e. $\zeta(H) < 3\mu H$, with a further increase in the magnetic field, where the Fermi boundary monotonically grows and asymptotically approaches the zero Landau level with energy $\varepsilon_0 = \mu H$.

8.4 Thermodynamic Properties of Electron Gas in Quantizing Magnetic Field

In this section on the basis of relationship (8.30) and the expression of the grand thermodynamic potential (8.34), we consider the basic thermodynamic properties of an electron gas in a quantizing magnetic field. Finding the explicit form of the equation of state, we calculate entropy and heat capacity.

1. *Thermal equation of state.* From (8.30) it is seen that knowing the explicit form of $\Omega_e = \Omega_e(T, V, \zeta, H)$, pressure can be determined as follows:

$$P = -\left(\partial\Omega_e/\partial V\right)_{\zeta,H,T}. \tag{8.54}$$

Hence and from (8.34) it follows that

$$P = \frac{4}{(2\pi R)^2}\sum_N \int_{\varepsilon_N}^{\infty} k_z(\varepsilon, N) f(\varepsilon)\mathrm{d}\varepsilon. \tag{8.55}$$

If we use (8.36), (8.55) takes the form

$$P = \frac{4(2m)^{1/2}}{\hbar(2\pi R)^2}\sum_N \int_{\varepsilon_N}^{\infty} (\varepsilon - \varepsilon_N)^{1/2} f(\varepsilon)\mathrm{d}\varepsilon. \tag{8.56}$$

Integrate this expression once by parts. Then we get

$$P = \frac{8(2m)^{1/2}}{3\hbar(2\pi R)^2}\sum_N \int_{\varepsilon_N}^{\infty} (\varepsilon - \varepsilon_N)^{3/2} \left(-\frac{\partial f}{\partial \varepsilon}\right)\mathrm{d}\varepsilon. \tag{8.57}$$

For an arbitrary degree of degeneracy of an electron gas the integral in (8.57) cannot be analytically calculated. Therefore consider particular cases.

Non-degenerate electron gas. If we take into account that $f(\varepsilon) = \exp((\zeta - \varepsilon)/k_0T)$ and $(-\partial f/\partial\varepsilon) = (1/k_0T)\exp((\zeta - \varepsilon)/k_0T)$, carry out the change of the integration variable $(\varepsilon - \varepsilon_N) = \varepsilon'$, and also integrate over $\mathrm{d}\varepsilon'$, then the integral in (8.57) acquires the shape:

$$\int_{\varepsilon_N}^{\infty} (\varepsilon - \varepsilon_N)^{3/2} \left(-\frac{\partial f}{\partial \varepsilon}\right)\mathrm{d}\varepsilon = \frac{3\sqrt{\pi}}{4}(k_0T)^{3/2}\exp\left(\frac{\zeta - \varepsilon_N}{k_0T}\right). \tag{8.58}$$

Taking into account (8.58) in (8.57), we have

$$P = \frac{8(2m)^{1/2}}{3\hbar(2\pi R)^2}\frac{3\sqrt{\pi}}{4}(k_0T)^{3/2}\mathrm{e}^{\zeta/k_0T}\sum_{N=0}^{\infty}\mathrm{e}^{-(2N+1)\mu H/k_0T}. \tag{8.59}$$

The summation with respect to N in (8.59), according to (8.40), is easily fulfilled and as a result, for the thermal equation of the state of a non-degenerate electron gas $P = P(T, N, \zeta)$ in indicated variables we get

$$P = \frac{(2\pi m)^{1/2}(k_0T)^{3/2}}{\hbar(2\pi R)^2}\frac{\exp(\zeta/k_0T)}{sh\,(\mu H/k_0T)}. \tag{8.60}$$

If substituted the expression for $\exp(\zeta/k_0T)$ from (8.42) into (8.60), we get pressure as a function of the concentration and temperature

$$P = nk_0T. \tag{8.61}$$

It is seen that pressure of a non-degenerate electron gas does not depend on the magnetic field and has a known classical appearance.

Degenerate electron gas. In this case, if in the zeroth approximation with respect to temperature we take into account that $(-\partial f/\partial\varepsilon) = \delta(\varepsilon - \zeta_{\mathrm{F}})$, from (8.57) for the zero pressure (at $T = 0$) we get

$$P_0 = \frac{8(2m)^{1/2}}{3\hbar(2\pi R)^2}\sum_{N=0}^{N_0}[\zeta_{\mathrm{F}} - (2N+1)\mu H]^{3/2}, \tag{8.62}$$

where $N_0 = (\zeta - \mu H)/2\mu H$.

In the quasi-classical approximation, when many Landau levels ($\zeta_{\mathrm{F}}/2\mu H \gg 1$) are found below the Fermi boundary energy, we can replace the sum with respect to N with the integral and integrate up in indicated limits. Then we get the known expression for the zero pressure of a Fermi gas in the absence of the magnetic field [see (7.115)].

Applying the Euler summation formula (4.85) to (8.62), the quantum correction to the zero pressure, associated with discreteness of the spectrum, can be obtained.

In *the quantum limit*, when all electrons are found at the zero Landau level, i.e. when $\mu H < \zeta_{\mathrm{F}} < 3\mu H$, in sum (8.62) it can be restricted to the term with $N = 0$ and from (8.62) it follows that

$$P_0 = \frac{8(2m)^{1/2}}{3\hbar(2\pi R)^2}(\zeta_{\mathrm{F}} - \mu H)^{3/2}. \tag{8.63}$$

Substitute the expression for $(\zeta_{\mathrm{F}} - \mu H)$ from (8.47) into (8.63). Then for the thermal equation of the state of a degenerate electron gas in the quantum limit we get

$$P_0 = \frac{\pi^2}{3}\frac{\hbar^4c^3}{me^2}\frac{n^3}{H^2} \approx \frac{n^3}{H^2}. \tag{8.64}$$

Hence it is seen that in contradistinction to the non-degenerate case, the zero pressure of a degenerate electron gas in the quantum limit strongly depends on the concentration n and magnetic field $P_0 \sim n^3/H^2$.

Applying the approximate formula (7.134) to the integral in (8.57), the temperature correction to the zero pressure of a degenerate electron gas can be obtained.

2. *The caloric equation of state or mean energy* of an electron gas is defined as follows

$$E = 3 \times 2 \sum_{N,k_y k_z} (\varepsilon - \varepsilon_N) f(\varepsilon_{Nk_z}). \tag{8.65}$$

Here the factor 2 takes into account degeneracy with respect to the spin, and the factor 3 appears owing to equivalency of all three directions of a magnetic field in the isotropic space.

We pass from the summation with respect to quasi-continuous quantum numbers k_y and k_z to integrals, according to the known rule (8.24), and take into account (8.25). Then (8.65) takes the form:

$$E = \frac{6V}{(2\pi R)^2} \sum_N \int (\varepsilon - \varepsilon_N) f(\varepsilon) \mathrm{d}k_z. \tag{8.66}$$

From the integration over $\mathrm{d}k_z$ we pass to the integration over $\mathrm{d}\varepsilon$ and multiply by 2, taking into account that two values $\pm k_z$ correspond to one value of ε. As a result, we have

$$E = \frac{12V}{(2\pi R)^2} \sum_N \int_{\varepsilon_N}^{\infty} (\varepsilon - \varepsilon_N) f(\varepsilon) \left(\frac{\mathrm{d}k_z}{\mathrm{d}\varepsilon}\right) \mathrm{d}\varepsilon. \tag{8.67}$$

According to (8.36), we can write

$$\frac{\mathrm{d}k_z}{\mathrm{d}\varepsilon} = \frac{(2m)^{1/2}}{2\hbar} (\varepsilon - \varepsilon_N)^{1/2}. \tag{8.68}$$

Taking this into account in (8.67), for the caloric equation of state (the mean energy) we get

$$E = \frac{6V}{\hbar(2\pi R)^2} \sum_N \int_{\varepsilon_N}^{\infty} (\varepsilon - \varepsilon_N)^{1/2} f(\varepsilon) \mathrm{d}\varepsilon. \tag{8.69}$$

Comparing this expression with (8.56), we get the known relationship, relating the thermal equation of the state of a free electron gas to the caloric one

$$P = \frac{2}{3}\frac{E}{V}. \tag{8.70}$$

3. *Entropy.* According to (8.30), entropy of an electron gas is defined as follows:

$$S = -\left(\partial\Omega_{\mathrm{e}}/\partial T\right)_{V,\zeta,H} . \tag{8.71}$$

We started with the expression (8.34) for the grand thermodynamic potential Ω_{e}. From this expression it is seen that only the Fermi distribution function $f(\varepsilon)$ depends on temperature. If we take into account that

$$\left(\frac{\partial f}{\partial T}\right)_{\zeta,H} = \frac{(\varepsilon-\zeta)}{T}\left(\frac{\partial f}{\partial \varepsilon}\right), \tag{8.72}$$

from (8.34) and (8.71) for entropy of an electron gas in a quantizing magnetic field we get

$$S = \frac{4V}{(2\pi R)^2 T}\sum_N \int_{\varepsilon_N}^{\infty} k_z(\varepsilon,\varepsilon_N)(\varepsilon-\zeta)\left(-\frac{\partial f}{\partial \varepsilon}\right)\mathrm{d}\varepsilon \tag{8.73}$$

or, if we take into account (8.36), we have

$$S = \frac{4V}{(2\pi R)^2 T}\frac{(2m)^{1/2}}{\hbar}\sum_N \int_{\varepsilon_N}^{\infty} (\varepsilon-\varepsilon_N)^{1/2}(\varepsilon-\zeta)\left(-\frac{\partial f}{\partial \varepsilon}\right)\mathrm{d}\varepsilon . \tag{8.74}$$

On the basis of this general formula, different particular cases can be considered.

4. *The heat capacity* of an electron gas in a quantizing magnetic field can be calculated, using the expression either for entropy (8.74), as $C_V = T\left(\partial S/\partial T\right)_V$, or for energy (8.69), as $C_V = \left(\partial E/\partial T\right)_V$. We use the second variant.

In the case of a non-degenerate electron gas, by virtue of (8.61) and (8.70), the heat capacity has a surprisingly simple appearance:

$$C_V = \frac{3}{2}k_0 N_{\mathrm{e}} . \tag{8.75}$$

Consequently, in the case of a non-degenerate electron gas a quantizing magnetic field (discreteness of the energy spectrum) influences neither the equation of state nor the heat capacity. To all appearance, it is associated with the fact that the high temperatures, at which discreteness of the spectrum becomes unessential, are required in order that an electron gas be non-degenerate.

To calculate heat capacity of a degenerate electron gas (8.69) integrate up by parts:

$$E = \frac{4V(2m)^{1/2}}{\hbar(2\pi R)^2}\sum_N \int_{\varepsilon_N}^{\infty} \left[\varepsilon-(2N+1)\mu H\right]^{3/2}\left(-\frac{\partial f}{\partial \varepsilon}\right)\mathrm{d}\varepsilon . \tag{8.76}$$

In order to find the temperature correction to energy, apply formula (7.134). As a result, we have

$$E = E_0 + \frac{\pi^2}{2}(k_0T)^2\frac{V(2m)^{1/2}}{\hbar(2\pi R)^2}\sum_N [\zeta_F - (2N+1)\mu H]^{-1/2}, \tag{8.77}$$

where $E_0 = 3P_0V/2$ is the zero energy at $T = 0$, and P_0 is determined by (8.62).

From (8.77) for the heat capacity of a degenerate electron gas in a quantizing magnetic field we get the expression:

$$C_V = \pi^2 k_0^2 T\frac{V(2m)^{1/2}}{\hbar(2\pi R)^2}\sum_N [\zeta_F - (2N+1)\mu H]^{-1/2}, \tag{8.78}$$

which can be presented as

$$C_V = \frac{\pi^2}{4}k_0^2 TVg_H(\zeta_F), \tag{8.79}$$

where $g_H(\zeta_F)$ is the density of quantum states at the Fermi level [see (8.28)].

From these formulae it is seen that with the change in the magnetic field, every time the Landau level $\varepsilon_N = (2N+1)\mu H$ coincides with the Fermi boundary the heat capacity experiences a sharp jump, i.e. has a peculiarity.

In weak magnetic fields ($\zeta_F >> \mu H$) in (8.78) from the summation with respect to N we can pass to the integration and show that in the quasi-classical approximation (8.78) coincides with the known expression (7.140).

In the quantum limit ($\zeta_F < 3\mu H$) in (8.78) it can be restricted to the term with $N = 0$ and the heat capacity takes the form:

$$C_V = \pi^2 k_0^2 T\frac{V(2m)^{1/2}}{\hbar(2\pi R)^2}(\zeta_F - \mu H)^{-1/2}. \tag{8.80}$$

If we used (8.47), for the heat capacity of a degenerate electron gas in the quantum limit we get

$$C_V = \frac{k_0^2 TV(2m)^3}{(2\pi)^2\hbar^6 n}(\mu H)^2 \sim H^2. \tag{8.81}$$

It is seen that, in contradistinction to the non-degenerate case, the heat capacity of a degenerate electron gas in the quantum limit strongly depends on the magnetic field.

8.5 Landau Diamagnetism

According to classical mechanics, free electrons in metals placed in a uniform magnetic field move in spiral trajectories, and in the plane perpendicular to the magnetic field follow a cyclotron orbit. The motion of each electron in the circular orbit creates a closed electric current, and thereby a magnetic moment, directed against the external field, arises. As a result of this, a free electron gas in metal ought to possess *diamagnetism*, which is observed on test.

However, if it is attempted to calculate the diamagnetic moment of an electron gas in the quasi-classical approximation, zero is obtained. Indeed, the Hamilton function of a free electron in a magnetic field is distinct from that without a magnetic field by the fact that in it the impulse $\boldsymbol{P}$ is replaced with $\boldsymbol{P}' = (\boldsymbol{P} + e/c\boldsymbol{A})$, where e is the electron charge magnitude, $\boldsymbol{A}$ is the vector potential of the magnetic field $\boldsymbol{H} = \mathrm{rot}\boldsymbol{A}$. If we calculate the statistical integral, it is easy to show that it and, consequently, the free energy F do not depend on the magnetic field H, inasmuch as integrals with respect to components of the impulse P'_x, P'_y, P'_z are also taken in the limits from $-\infty$ to $+\infty$. Hence it follows that the diamagnetic magnetization $\boldsymbol{M} = -1/V\,(\partial F/\partial \boldsymbol{H})_{T,V} = 0$.

Note that this quasi-classical result ($\boldsymbol{M} = 0$) was obtained by Nils Bohr as far back as in 1911. He showed that magnetic moments, formed by circular orbits of free electrons in the magnetic field, are compensated by magnetic moments of non-closed orbits close to the surface of metal.

The further exposition of the given section will show that the absence of diamagnetism does not depend on whether an electron gas obeys the Boltzmann classical statistics or the Fermi quantum statistics. The main question is whether the motion of an electron is classical or quantum.

In 1930 L.D. Landau showed that if the motion of an electron is considered on the basis of quantum mechanics and quantization of energy of free electrons in a magnetic field is taken into account (8.16), the diamagnetic susceptibility of an electron gas does not equal zero. Here we expound the theory of Landau diamagnetism on the basis of energy spectrum (8.19).

From (8.30) it follows that the magnetization of an electron gas M can be found originating from the explicit form of the grand thermodynamic potential $\Omega_\mathrm{e} = \Omega_\mathrm{e}(T, V, \zeta, H)$ as follows:

$$M = -\frac{1}{V}\left(\frac{\partial \Omega_\mathrm{e}}{\partial H}\right)_{T,V,\zeta}. \tag{8.82}$$

If we substitute (8.36) into (8.34) and once integrate up by parts, for the grand thermodynamic potential we get:

$$\Omega_\mathrm{e} = -\frac{8V(2m)^{1/2}}{3\hbar(2\pi R)^2}\sum_N \int\limits_{\varepsilon_N}^{\infty} (\varepsilon - \varepsilon_N)^{3/2}\left(-\frac{\partial f}{\partial \varepsilon}\right)\mathrm{d}\varepsilon. \tag{8.83}$$

For any degree of degeneracy of an electron gas it is impossible to find the analytical form of Ω_e; therefore the classical and the quantum statistics are considered separately.

Non-degenerate electron gas. In this case, the distribution function has the appearance $f(\varepsilon) = \exp\left[(\zeta - \varepsilon)/k_0 T\right]$. If in (8.83) we take into account that $(-\partial f/\partial\varepsilon) = -(1/k_0T)\exp[(\zeta-\varepsilon)/k_0T]$ and introduce the notation $\varepsilon' = \varepsilon - \varepsilon_N$, the integral with respect to ε' is easily calculated (see Appendix A). As a result, we get

$$\Omega_e = -\frac{2V(2\pi m)^{1/2}(k_0T)^{3/2}}{\hbar(2\pi R)^2}e^{\zeta/k_0T}\sum_{N=0}^{\infty}e^{-(2N+1)\mu H/k_0T}. \tag{8.84}$$

The summation with respect to N in (8.84), according to (8.40), is easily fulfilled and Ω_e has the appearance

$$\Omega_e = -\frac{V(2\pi m)^{1/2}(k_0T)^{3/2}}{\hbar(2\pi R)^2}\frac{\exp(\zeta/k_0T)}{sh(\mu H/k_0T)}. \tag{8.85}$$

In the quasi-classical approximation, when inequality (8.4) analogous to $\mu H \ll k_0T$ is satisfied, in (8.85) $sh(\mu H/k_0T) \approx \mu H/k_0T$ can be replaced, and if we take into account (8.15), we see that Ω_e does not depend on the magnetic field and coincides with the classical value (7.16). Thereby the magnetization in this case, according to (8.82), equals zero. This result can be obtained if in (8.84) from the sum with respect to N it is passed to the integral.

Note that to realize this case the simultaneous fulfilment of two inequalities is required:

$$k_0T \gg \mu H \text{ and } k_0T \gg \zeta_0, \tag{8.86}$$

which can also be presented in the form

$$\lambda \ll R \text{ and } \lambda \ll d, \tag{8.87}$$

where ζ_0 is the Fermi boundary energy (7.116), λ is the de Broglie wavelength, $d = n^{-1/3}$ is the mean distance between electrons, and R is the magnetic length.

From (8.82) and (8.85) for the magnetization of a non-degenerate electron gas in an arbitrary quantizing magnetic field we get the following expression

$$M = \frac{(2\pi m)^{1/2}(k_0T)^{3/2}}{(2\pi)^2\hbar^2}\frac{e}{c}\frac{\exp(\zeta/k_0T)}{sh(\mu H/k_0T)} \times [1 - \mu H/k_0T\, cth(\mu H/k_0T)] \tag{8.88}$$

If we substitute the expression of the chemical potential (8.41) into (8.88), for the diamagnetic magnetization of a non-degenerate electron gas we finally get

$$M = -n\mu\left[cth(\mu H/k_0T) - \frac{1}{(\mu H/k_0T)}\right] \tag{8.89}$$

or

$$M = -n\mu L(\mu H/k_0T), \tag{8.90}$$

where $L(\mu H/k_0T)$ is the Langevin function (4.175).

In a weak magnetic field ($\mu H \ll k_0T$), if it is restricted to the zeroth, i.e. the quasi-classical approximation, and one item kept in an expansion of a hyperbolic cotangent $cth(\mu H/k_0T) \approx k_0T/\mu H$, then $L(\mu H/k_0T) = 0$ and, consequently, the magnetization equals zero. The magnetization is distinct from zero only in the approximation when $cth(\mu H/k_0T) = (k_0T/\mu H) + \mu H/3k_0T$. Then from (8.89) we have:

$$M = -\frac{n\mu^2 H}{3k_0T}; \quad \mu H \ll k_0T. \tag{8.91}$$

The diamagnetic susceptibility $\chi_{\text{dia}} = M/H$, conformably, equals

$$\chi_{\text{dia}} = -\frac{m\mu^2}{3k_0T}. \tag{8.92}$$

If we compare (8.92) with the paramagnetic susceptibility χ_{para}, associated with the spin of an electron (7.190), we get the Landau result

$$\frac{\chi_{\text{dia}}}{\chi_{\text{para}}} = -\frac{1}{3}\left(\frac{\mu}{\mu_{\text{B}}}\right)^2 = -\frac{1}{3}\left(\frac{m_0}{m}\right)^2, \tag{8.93}$$

where m_0 is the mass of a free electron m is the effective mass of an electron in metal or semiconductor. Usually in semiconductors $m < m_0$; therefore diamagnetism of an electron gas in them dominates.

In the quantum limit, when the magnetic field satisfies the condition $\mu H \gg k_0T$, all electrons are found at the first Landau level and from (8.89) we get a simple result

$$M = -n\mu; \quad \mu H \gg k_0T. \tag{8.94}$$

The diamagnetic susceptibility in this case depends on the magnetic field

$$\chi_{\text{dia}} = -\frac{n\mu}{H}. \tag{8.95}$$

From the expression (7.189) in the strong magnetic field $\mu_{\text{B}}H \gg k_0T$ we find the paramagnetic susceptibility and determine the relationship

$$\frac{\chi_{\text{dia}}}{\chi_{\text{para}}} = -\frac{\mu}{\mu_{\text{B}}} = -\frac{m_0}{m}. \tag{8.96}$$

From the comparison (8.93) and (8.96) it is seen that in the strong magnetic field the ratio $\chi_{\text{dia}}/\chi_{\text{para}}$ is distinct from the case of the weak magnetic field. The obtained result (8.94) is true when fulfilling the condition: $\zeta_0 \ll k_0T \ll \mu H$, which can also be presented as $R \ll \lambda \ll d$.

Degenerate electron gas. In this case in the zeroth approximation with respect to temperature $(-\partial f/\partial\varepsilon) = \delta(\varepsilon - \zeta_{\text{F}})$ and from (8.83) for the grand thermodynamic potential of a degenerate electron gas Ω_{e} we get

$$\Omega_{\text{e}} = -\frac{8}{3}\frac{V(2m)^{1/2}}{(2\pi R)^2}\sum_{N=0}^{N_0}[\zeta_{\text{F}} - (2N+1)\mu H]^{3/2}, \tag{8.97}$$

where $N_0 = (\zeta_F - \mu H)/2\mu H$.

In the general form for an arbitrary value of the magnetic field it is impossible to conduct the summation in (8.97). Therefore we consider different limiting cases.

In weak magnetic fields, when $\mu H \ll \zeta_F$, i.e. at large N_0, in the zeroth quasi-classical approximation in (8.97) the summation with respect to N we can replace the integral in indicated limits and regard that $\zeta_F = \zeta_0$. Then in the zeroth approximation for the grand thermodynamic potential of a completely degenerate electron gas we get the following result:

$$\Omega_e = -\frac{8}{15}\frac{V(2m)^{3/2}}{(2\pi)^2\hbar^3}\zeta_0^{5/2}, \tag{8.98}$$

not depending on the magnetic field and coinciding with the known expression (7.120).

Inasmuch as $\Omega_e(0)$ does not depend on the magnetic field, the magnetization in the quasi-classical approximation equals zero.

Note that independent of the kind of statistics (an electron gas is degenerate or non-denerate) in the quasi-classical approximation, diamagnetism is absent. So, the question lies not in the statistics, but in the nature of the motion of an electron in the magnetic field: if the motion is classical, diamagnetism is absent, and if the motion is quantum, diamagnetism of a free electron gas exists.

In order to find the quantum correction to the grand thermodynamic potential, we calculate the sum in (8.97) with the aid of the Euler summation formula (4.85) and everywhere neglect μH compared with ζ_0. As a result, we get

$$\Omega_e = \Omega_e'^{(0)} + \frac{V(2m)^{3/2}}{3\hbar^3(2\pi)^2}\zeta_0^{5/2}(\mu H)^2, \tag{8.99}$$

where $\Omega_e'^{(0)}$ does not depend on the magnetic field part of the grand thermodynamic potential.

From (8.82) and (8.99) for the diamagnetic magnetization we have

$$M_{dia} = -\frac{(2m)^{3/2}}{6\pi^2\hbar^3}\mu^2 H\zeta_0^{5/2}. \tag{8.100}$$

If we compare (8.100) with paramagnetic susceptibility (7.192), for the ratio $M_{dia}/M_{para} = \chi_{dia}/\chi_{para}$ we get the same result as for a non-degenerate electron gas in a weak magnetic field (8.93).

In the quantum limit, when $\zeta_F < 3\mu H$, in sum (8.97) it can be restricted only to one term ($N = 0$). Then we have

$$\Omega_e = -\frac{8}{3}\frac{V(2m)^{3/2}}{3\hbar^3(2\pi^2)}\frac{eH}{\hbar c}(\zeta_F - \mu H)^{3/2}. \tag{8.101}$$

Hence, according to (8.82), find the magnetization M, where for $(\zeta_F - \mu H)$ we use (8.47). As a result, we get

$$M = \frac{8}{3} \frac{(2m)^{3/2}}{\hbar(2\pi^2)} \frac{e}{\hbar c} \zeta_0^{3/2} \left[\left(\frac{\zeta_0}{3\mu H} \right)^3 - \frac{1}{2} \right] . \tag{8.102}$$

If we leave out the first term in the square bracket and use the expression for the boundary Fermi energy ζ_0 (7.193) at $H = 0$, for the magnetization we get the same result as in the case of a non-degenerate electron gas in the quantum limit: $M = -n\mu$ (8.94).

9

Non-Equilibrium Electron Gas in Solids

Summary. An ideal Fermi gas, in particular an electron gas in an equilibrium state, is dealt with in Chap. 7. There the state equation is found, thermodynamic coefficients are calculated and Pauli's paramagnetism due to the free electron spin in metals is investigated.

In this chapter electron gas in metals and semiconductors is dealt with in a nonequilibrium state. Nonequilibrium processes associated with charge carriers' motion in a crystal under external disturbances such as electric field, temperature gradient, magnetic field, etc. are referred to as *electron transport phenomena* or *kinetic effects*. They include electric conductivity, thermal conductivity, thermoelectric, galvanomagnetic and thermomagnetic effects.

If the values governing transport phenomena, i.e. electric current density, heat flux, electric field strength, etc. do not depend on time, the charge or energy transport process is referred to as stationary. Here, we shall discuss stationary transport phenomena only. The definition and classification of these phenomena are presented below.

In the classical case, the Boltzmann equation for a nonequilibrium charge carrier distribution function which accounts for the interaction with a crystal lattice is used to construct a microscopic theory of transport phenomena.

9.1 Boltzmann Equation and Its Applicability Conditions

9.1.1 Nonequilibrium Distribution Function

The conduction electrons at thermodynamic equilibrium are described by the equilibrium Fermi–Dirac distribution function. In the presence of an external electric field, temperature or concentration gradients, or other effects, the electron gas is in a nonequilibrium state; therefore it cannot be described by an equilibrium distribution function $f_0(\boldsymbol{k})$. In a quasiclassical approximation in the nonequilibrium case, the distribution function $f(\boldsymbol{k}, \boldsymbol{r}, t)$ can be introduced while giving it a physical meaning, namely the local concentration near a point

$\boldsymbol{r}$ at time t of electrons in the $\boldsymbol{k}$-state. Knowing the nonequilibrium distribution function $f(\boldsymbol{k}, \boldsymbol{r}, t)$ it is possible to calculate the current density

$$\boldsymbol{j}(\boldsymbol{r}, t) = -\frac{2e}{V}\sum_{k} v(\boldsymbol{k}) f(\boldsymbol{k}, \boldsymbol{r}, t) = -\frac{2e}{(2\pi)^3}\int v(\boldsymbol{k}) f(\boldsymbol{k}, \boldsymbol{r}, t)\mathrm{d}\boldsymbol{k} \tag{9.1}$$

and the heat energy flux density

$$\boldsymbol{w}(\boldsymbol{r}, t) = \frac{2}{V}\sum_{\boldsymbol{k}} (\varepsilon(\boldsymbol{k}) - \mu) v(\boldsymbol{k}) f(\boldsymbol{k}, \boldsymbol{r}, t) = \frac{2}{(2\pi)^3}\int (\varepsilon(\boldsymbol{k}) - \mu) v(\boldsymbol{k}) f(\boldsymbol{k}, \boldsymbol{r}, t)\mathrm{d}\boldsymbol{k}, \tag{9.2}$$

which makes it possible to find an explicit transport equation form

$$\begin{aligned} j_i &= \sigma_{ik} E_k - \beta_{ik} \nabla_k T, \\ w_i &= \gamma_{ik} E_k - \kappa_{ik} \nabla_k T \quad i, k = x, y, z \end{aligned} \tag{9.3}$$

Here $E_k = -\nabla_k \varphi$, $w_i = w_i^* - \varphi j_i$, $\varphi = \varphi_0 - (\mu/e)$ is the electrochemical potential, φ_0 the electrostatic potential, μ the chemical potential, $(-e)$ the electron charge, and $w_i{}^*$ the component of energy flux density transported by the conduction electrons. The value φj_i derived from the total energy density is associated with the fact that every electron transports energy $e\varphi$ and, consequently, w_i is none other than the flux density of kinetic or heat energies transported by the conduction electrons.

The tensor components entering into (9.3) are the magnetic field functions H and they satisfy the following relations from the symmetry principle of kinetic coefficients:

$$\begin{aligned} \sigma_{ik}(H) &= \sigma_{ki}(-H) \\ \kappa_{ik}(H) &= \kappa_{ki}(-H) \\ \gamma_{ik}(H) &= T\beta_{ki}(-H). \end{aligned} \tag{9.4}$$

The advantage of writing the relations in the form of (9.3) is that the coefficients involved are derived immediately from the Boltzmann equation solution.

In the expressions (9.1) and (9.2), the factor 2 appears due to electron spin, $v(\boldsymbol{k})$ is the electron velocity in the $\boldsymbol{k}$-state with energy $\varepsilon(\boldsymbol{k})$. If in (9.1) and (9.2) $f(\boldsymbol{k}, \boldsymbol{r}, t)$ is replaced by $f_0(\boldsymbol{k}) = f_0(\varepsilon(\boldsymbol{k}))$, then by virtue of energy evenness $\varepsilon(\boldsymbol{k}) = \varepsilon(-\boldsymbol{k})$ and velocity oddness $v(-\boldsymbol{k}) = -v(\boldsymbol{k})$, the electric current and energy flux are equal to zero, i.e. in the equilibrium state there are no currents and fluxes.

In order to calculate $\boldsymbol{j}$ and $\boldsymbol{w}$ it is necessary to know the nonequilibrium distribution function $f(\boldsymbol{k}, \boldsymbol{r}, t)$ that can be derived from the Boltzmann equation.

9.1.2 Boltzmann Equation

By definition, the nonequilibrium distribution function $f(\boldsymbol{k}, \boldsymbol{r}, t)$ is the number of electrons which at time t are in a unit volume near the point $\boldsymbol{r}$ and have a wave vector $\boldsymbol{k}$. Except for $(\partial f/\partial t)$, their number can vary due to the following three physical processes: diffusion associated with temperature or concentration gradients, leading to a variation with respect to $\boldsymbol{r}$; acceleration by external fields causing a variation of the wave electron vector $\boldsymbol{k}$; electron *scattering* by phonons or by other lattice defects which also cause a variation in $\boldsymbol{k}$.

The total effect of all these processes must be equal to the distribution function variation per unit time, i.e. we can write

$$\partial f/\partial t = (\partial f/\partial t)_{\text{diff}} + (\partial f/\partial t)_{\text{field}} + (\partial f/\partial t)_{\text{scat}} \tag{9.5}$$

This is a symbolic writing of the Boltzmann equation where the terms of the right-hand side correspond to the variation rate due to the above processes. In order to find an explicit form of the Boltzmann equation (9.5) we shall consider these processes separately.

In the presence of a temperature or concentration gradient diffusion takes place in the conductor; therefore the charge carriers arrive at the special region near the point $\boldsymbol{r}$ and then leave it. Of interest is the variation in the electron concentration with wave vector $\boldsymbol{k}$. The electrons have the velocity $\upsilon(\boldsymbol{k}) = \hbar^{-1}\nabla_k\varepsilon(\boldsymbol{k})$. The concentration of these $\boldsymbol{k}$-electrons near the point $\boldsymbol{r}$ at time t is $f(\boldsymbol{k}, \boldsymbol{r}, t)$. After a time interval Δt, these electrons will be near the point $\boldsymbol{r} + \Delta\boldsymbol{r}$, acquiring the concentration $f(\boldsymbol{k}, \boldsymbol{r} + \Delta r, t + \Delta t)$. In accordance with the Liouville theorem on the invariance of the phase space volume the charge carrier concentration in the vicinity of the point $\boldsymbol{r}$ at time t is equal to their concentration in the vicinity of the point $\boldsymbol{r} + \Delta\boldsymbol{r}$ at time $t + \Delta t$, i.e.

$$f(\boldsymbol{k}, \boldsymbol{r}, t) = f(\boldsymbol{k}, \boldsymbol{r} + \Delta\boldsymbol{r}, t + \Delta t). \tag{9.6}$$

Decomposing the right-hand side near the point $\boldsymbol{r}$, taking account of $\Delta\boldsymbol{r} = \upsilon(\boldsymbol{k})\Delta t$ and making $\Delta t \to 0$, we obtain

$$(\partial f/\partial t)_{\text{diff}} = -\upsilon(\boldsymbol{k})(\partial f/\partial \boldsymbol{r}) = -\upsilon(\boldsymbol{k})\nabla_r f \tag{9.7}$$

for the rate of the nonequilibrium distribution function variation because of diffusion.

The external force $\boldsymbol{F}$, accelerating the electrons, varies their wave vector such that $\dot{\boldsymbol{k}} = \hbar^{-1}F$, and consequently the number of $\boldsymbol{k}$-electrons near the point $\boldsymbol{r}$. If these considerations are also repeated for the $\boldsymbol{k}$-space it is possible to write

$$f(\boldsymbol{k}, \boldsymbol{r}, t) = f(\boldsymbol{k} + \Delta\boldsymbol{k}, \boldsymbol{r}, t + \Delta t) \tag{9.8}$$

which yields

$$(\partial f/\partial t)_{\text{field}} = -\dot{\boldsymbol{k}}(\partial f/\partial \boldsymbol{k}) = -\hbar^{-1}\boldsymbol{F}\nabla_k f \tag{9.9}$$

for the rate of variation of $f(\boldsymbol{k},\boldsymbol{r},t)$ owing to the external force $\boldsymbol{F}$.

Let us assume that a conductor is in an external electric field $\boldsymbol{E}_0$ and magnetic field H. Then a Lorentz force

$$\boldsymbol{F} = -e\left(\boldsymbol{E}_0 + \frac{1}{c}\left[\boldsymbol{v}(\boldsymbol{k})\mathscr{H}\right]\right), \tag{9.10}$$

where $\boldsymbol{v}(\boldsymbol{k}) = \hbar^{-1}\nabla_k\varepsilon(\boldsymbol{k})$ is the electron velocity and c the velocity of light, acts upon an electron with charge $(-e)$.

The Boltzmann equation (9.5) for conduction electrons while accounting for (9.7), (9.9) and (9.10) has the form

$$\frac{\partial f}{\partial t} + \boldsymbol{v}(\boldsymbol{k})\nabla_r f - \frac{e}{\hbar}\left(\boldsymbol{E}_0 + \frac{1}{c}\left[\boldsymbol{v}(\boldsymbol{k})\mathscr{H}\right]\right)\nabla_k f = \left(\frac{\partial f}{\partial t}\right)_{\text{scat}} \tag{9.11}$$

In order to find $\left(\frac{\partial f}{\partial t}\right)_{\text{scat}}$ we shall introduce a function $W(\boldsymbol{k},\boldsymbol{k}')$ which is the per-unit-time probability of electron transition from the state $\boldsymbol{k}$ to the state $\boldsymbol{k}'$ as a result of scattering by a lattice defect. The transitions from state $\boldsymbol{k}$ to every possible $\boldsymbol{k}'$-states will take place if there is an electron in the $\boldsymbol{k}$-state and the $\boldsymbol{k}'$-states are vacant (Pauli's exclusion principle). Therefore, because of the transitions $\boldsymbol{k} \to \boldsymbol{k}'$ the number of $\boldsymbol{k}$-electrons decreases, i.e. leaving the $\boldsymbol{k}$-state, in unit time by $\sum\limits_{k} W(\boldsymbol{k},\boldsymbol{k}')f(\boldsymbol{k})(1-f(\boldsymbol{k}'))$. The number of $\boldsymbol{k}$-electrons, i.e. the function $f(\boldsymbol{k},\boldsymbol{r})$, increases due to electron transition from every possible $\boldsymbol{k}'$-state to the $\boldsymbol{k}$-state by $\sum\limits_{k}^{\prime} W(\boldsymbol{k}',\boldsymbol{k})f(\boldsymbol{k}')(1-f(\boldsymbol{k}))$ per unit time. The difference between the numbers of electrons arriving and leaving gives the rate of variation of the nonequilibrium distribution function

$$(\partial f/\partial t)_{\text{scat}} = \sum_{k'}\left\{W(\boldsymbol{k}',\boldsymbol{k})f(\boldsymbol{k}')(1-f(\boldsymbol{k})) - W(\boldsymbol{k},\boldsymbol{k}')f(\boldsymbol{k})(1-f(\boldsymbol{k}'))\right\}. \tag{9.12}$$

Further, we shall consider the stationary case where the external fields do not explicitly depend on time and, consequently, $\mathrm{d}f/\mathrm{d}t = 0$. In the stationary case, from (9.11) and (9.12) we can obtain an explicit form of the Boltzmann equation for the conduction electrons in electric and magnetic fields:

$$\begin{aligned}\boldsymbol{v}(\boldsymbol{k})\nabla_r f &- \frac{e}{\hbar}\left(\boldsymbol{E}_0 + \frac{1}{c}\left[\boldsymbol{v}(\boldsymbol{k})\mathscr{H}\right]\right)\nabla_k f \\ &= \sum_{k'}\left\{W(\boldsymbol{k}',\boldsymbol{k})f(\boldsymbol{k}')(1-f(\boldsymbol{k})) - W(\boldsymbol{k},\boldsymbol{k}')f(\boldsymbol{k})(1-f(\boldsymbol{k}'))\right\}\end{aligned} \tag{9.13}$$

Converting the summation with respect to $\boldsymbol{k}'$ to integration we shall see that the Boltzmann equation (9.13) is an integro-differential equation. The function $W(\boldsymbol{k}, \boldsymbol{k}')$ entering into the collision part of the Boltzmann equation depends on the nature of the conduction electron–scatterer interaction, and for a particular mechanism it is governed by the quantum theory of scattering.

Thus the calculation of the currents (9.1) and (9.2) is reduced to two independent problems, namely the calculation of $W(\boldsymbol{k}, \boldsymbol{k}')$ for different scattering mechanisms and the solution of the Boltzmann equation (9.13) in order to find the nonequilibrium distribution function $f(\boldsymbol{k}, \boldsymbol{r})$. Of course, these problems are not solved in a general form. Some approximate solutions of the Boltzmann equation will be discussed in subsequent paragraphs, but now we shall dwell upon the applicability criteria of the Boltzmann equation itself.

9.1.3 Applicability Conditions of the Boltzmann Equation

From both the definition of the nonequilibrium distribution function $f(\boldsymbol{k}, \boldsymbol{r})$ and the derivation of the Boltzmann equation (9.11) it is seen that it is valid only in the quasiclassical case where the notion of trajectory is used and the states of the electron gas are specified in a phase space. The quasiclassical condition imposes a limitation on the generalized force, namely the electrochemical potential gradient $-\nabla(\varphi_0 - (\mu/e))$, as well as on the magnetic field strength H, and it defines the applicability limit of the Boltzmann equation. We shall now dwell upon it.

The quasiclassical condition runs as follows: the properties of a free system can be considered using classical mechanics if the de Broglie wavelengths λ of the particles are small as compared with the sizes L that are typical of the given problem, i.e. $\lambda \ll L$. If the system is not free and a certain force F acts upon it, the quasiclassical condition has the form

$$(\lambda/2\pi)\,|\mathrm{d}\lambda/\mathrm{d}x| \ll \lambda \tag{9.14}$$

i.e. the electron wavelength λ must vary slightly over a distance equal to the wavelength itself. Accounting for $\lambda = h/p(x)$ where h is the Planck constant and $p(x)$ the electron impulse in the external field $U(x)$, the quasiclassical condition (9.14) can be written as follows:

$$(mh/p^3)\,|F| \ll 1, \tag{9.15}$$

where m is the effective mass of conduction electron, and F is the force acting upon the electron. In the general case where there are dynamic (electric field) and static (temperature or concentration gradients) disturbances, the force F depends on the electrochemical potential gradient

$$\boldsymbol{F} = e\nabla(\phi_0 - (\mu/e)). \tag{9.16}$$

When $\lambda = h/p$ and $p^2 \approx m\overline{\varepsilon}$ are taken into account, (9.15) reduces to the applicability condition of the Boltzmann equation in general form

$$\lambda\,|F| \ll \overline{\varepsilon}, \tag{9.17}$$

which has an explicit physical meaning: the energy gained by a electron due to the action of the disturbing force F at a distance λ must be much less than the mean electron energy $\overline{\varepsilon}$.

From (9.17), for the electric field E_0 it is easy to obtain the condition

$$\lambda e E_0 \ll \overline{\varepsilon} \tag{9.18}$$

for the Boltzmann equation to be applicable. In nondegenerate semiconductors, $\overline{\varepsilon} \approx k_0 T$, while in degenerate ones, $\overline{\varepsilon}$ *as* μ_F. At $T = 100\,\mathrm{K}$ and $\lambda = 10^{-7}\,\mathrm{cm}$, we obtain from (9.18) $E_0 \ll 10^5\,\mathrm{V/cm}$ for nondegenerate semiconductors. Thus, the Boltzmann equation is valid up to very strong electric fields.

In the case of nonuniform semiconductors, the condition (9.17) while accounting for (9.16) also determines the chemical potential gradient

$$(\partial \mu/\partial x)\lambda \ll \overline{\varepsilon} \tag{9.19}$$

when the Boltzmann equation is valid. If the nonuniformity is caused by a temperature gradient, then (9.19) assumes the form

$$(\partial \mu/\partial T)\lambda \nabla T \ll \overline{\varepsilon}.$$

Finding $\partial \mu/\partial T$ from (7.20) and (7.136) we shall obtain respectively

$$|\mu/k_0 T - 3/2|\,\lambda\,|\nabla T| \ll T \tag{9.20}$$

for a nondegenerate electron gas, and

$$(k_0 T/\mu_F)\lambda\,|\nabla T| \ll T_0 \tag{9.21}$$

for a degenerate electron gas, where $T_0 = \mu_F/k_0 T$ is the degeneracy temperature of the electron gas. Conditions similar to (9.19)–(9.21) can also be obtained from (9.19) for the concentration gradient ∇n.

Now we shall consider the limitations imposed on the magnetic field strength H for applying the Boltzmann equation. An electron moving in a uniform magnetic field is under the action of the Lorentz force $F = (e/c)v_\perp H$ along a helix with axis parallel to the magnetic field of radius

$$r = v_\perp/\Omega \tag{9.22}$$

where $v_\perp$ is the velocity component in a plane perpendicular to the magnetic field,

$$\Omega = eH/mc \tag{9.23}$$

is the cyclic frequency of the electron revolution in this plane and is referred to as the cyclotron frequency, and m is the effective electron mass.

Substituting the Lorentz force $F = (e/c)v_\perp H$ into the general condition (9.17) and accounting for $\lambda \approx \hbar/mv_\perp$ we obtain the following applicability criterion of the Boltzmann equation in the magnetic field

$$\hbar\Omega \ll k_0 T, \quad \text{or} \quad e\hbar H/mc \ll k_0 T \tag{9.24}$$

for nondegenerate semiconductors, and

$$\hbar\Omega \ll \mu_F, \text{or}\, e\hbar H/mc \ll \mu_F \tag{9.25}$$

for degenerate semiconductors. It is easy to see that these conditions can be rewritten in the form

$$r \gg \lambda \tag{9.26}$$

In the case of degenerate semiconductors, λ in (9.26) is to be replaced by the wavelength for electrons on the Fermi surface, $\lambda_F \approx \hbar/mv_F$.

The last inequality is nothing more than the quasiclassical condition: *The Boltzmann equation is applicable in magnetic fields where the wavelength λ is much less than the radius r of cyclotron electron orbit.*

A quantomechanical analysis of the motion of charge carriers in a magnetic field shows that their motions are quantized in a plane perpendicular to the direction of the magnetic field and the energy separation between neighbouring discrete levels (Landau levels) is equal to $\hbar\Omega$ (see Chap. 8).

In magnetic fields satisfying (9.24) or (9.25) spectrum quantization becomes insignificant. These magnetic fields are referred to as classical or non-quantizing magnetic fields. In such a region of magnetic fields the Boltzmann equation can be applied to calculate the kinetic effects.

In sufficiently strong magnetic fields and at low temperatures, the conditions (9.24) and (9.25) cannot be fulfilled while $\hbar\Omega \geq k_0 T$ or $\hbar\Omega \geq \mu_F$ can take place (same for $\lambda \geq r$), and spectrum discreteness becomes important. Such magnetic fields are referred to as quantizing fields. In quantizing magnetic fields, the Boltzmann equation is not valid and the problem should be solved by methods of the quantum transport theory.

There is another condition limiting the scope of applicability of the Boltzmann equation (9.13) which is associated with charge carriers' scattering by various lattice defects. If Δt denotes scattering duration (collision time with defects), then according to the uncertainty principle ($\Delta\varepsilon\Delta t \geq \hbar$) an uncertainty $\Delta\varepsilon \geq \hbar/\Delta t$ appears in the energy. It is possible to use a distribution function that varies significantly in an energy interval on the order of $k_0 T$ only in the case where $\Delta\varepsilon < k_0 T$. This means that the collision duration must be relatively large: $\Delta t \gg \hbar/k_0 T$. On the other hand, the collision time Δt cannot be too large since the right-hand side of the Boltzmann equation (9.13) suggests that the interaction of the charge carriers with the crystal lattice occurs as separate scattering events with the probability $W(\boldsymbol{k}, \boldsymbol{k}')$ and this interaction does not change the electron energy (basically the charge carriers move freely in the crystal though sometimes scattered by the lattice defects). Therefore, Δt must be less than the mean free time τ (i.e. the relaxation time). As a result, we come to the inequality

$$\tau \gg \hbar/k_0 T, \tag{9.27}$$

which is the condition for the collision duration to be small compared with the time between two successive collisions.

Since the mobility of the charge carriers is $u = e\bar{\tau}/m$ the criterion (9.27) can be expressed as

$$u \gg e\hbar/mk_0T. \tag{9.28}$$

The above condition is well fulfilled for a wide class of semiconductors, except for substances with an abnormally small mobility of charge carriers.

The criterion (9.27) can also be written in another form. If both sides are multiplied by the mean velocity of the charge carriers $\bar{\upsilon}$, and accounting for $k_0T/\bar{\upsilon} = \bar{p}$ and $h/\bar{p} \approx \lambda$ we shall obtain

$$\bar{l} \gg \lambda, \tag{9.29}$$

where $\bar{l} = \bar{\upsilon}\bar{\tau}$ is the mean free path. Then it is possible to say that *the Boltzmann equation is applicable if the mean free path of the charge carriers is much larger than the de Broglie wavelength.*

9.2 Solution of Boltzmann Equation in Relaxation Time Approximation

Let us consider a conductor with an arbitrary spherically symmetric band. That is, we assume that the energy of charge carriers ε is an arbitrary function of the wave vector of magnitude $\boldsymbol{k}$:

$$\varepsilon(k_x, k_y, k_y) = \varepsilon(k). \tag{9.30}$$

This property is inherent in the dispersion relation of the conduction electrons for metals and many semiconductor compounds of types $A^{\mathrm{III}}B^{\mathrm{V}}$ and $A^{\mathrm{II}}B^{\mathrm{VI}}$.

9.2.1 Relaxation Time

First we transfer the collision term of the Boltzmann equation (9.13). To do this, we shall use the principle of detailed equilibrium, according to which, in the state of equilibrium where $f(\boldsymbol{k}, \boldsymbol{r}) = f_0(\boldsymbol{k})$, in unit time the number of electrons coming into the $\boldsymbol{k}$-state from the $\boldsymbol{k}'$-state is equal to that of electrons coming out from the $\boldsymbol{k}$-state into the $\boldsymbol{k}'$-state:

$$W(\boldsymbol{k}', \boldsymbol{k})f_0(\boldsymbol{k}')(1 - f_0(\boldsymbol{k})) = W(\boldsymbol{k}, \boldsymbol{k}')f_0(\boldsymbol{k})(1 - f_0(\boldsymbol{k}')). \tag{9.31}$$

Using the explicit form of the Fermi–Dirac distribution function (7.77) we shall obtain from (9.31) the following expression:

$$\frac{W(\boldsymbol{k}', \boldsymbol{k})}{W(\boldsymbol{k}, \boldsymbol{k}')} = \exp\left(\frac{\varepsilon(\boldsymbol{k}') - \varepsilon(\boldsymbol{k})}{k_0T}\right). \tag{9.32}$$

For $\varepsilon(\boldsymbol{k}) < \varepsilon(k')$, as seen from (9.32), $W(\boldsymbol{k}, \boldsymbol{k}') < W(\boldsymbol{k}', \boldsymbol{k})$, i.e. the scattering $\boldsymbol{k} \to \boldsymbol{k}'$ is less probable for an increase in the energy than for a decrease. The smaller probability of transition from states of smaller energies is compensated by their greater occupation $f_0(\boldsymbol{k})$, which provides detailed equilibrium.

Solution of the Boltzmann equation (9.13) can be presented in the following form:

$$f(\boldsymbol{k}) = f_0(\boldsymbol{k}) + f_1(\boldsymbol{k}), \tag{9.33}$$

where $f_0(\boldsymbol{k}) \equiv f_0(\varepsilon(\boldsymbol{k}))$ is the equilibrium function, while $f_1(\boldsymbol{k})$ is a nonequilibrium supplement to be defined. It is obvious that $f_1(\boldsymbol{k})$ is an odd function of $\boldsymbol{k}$ while $f_0(\boldsymbol{k})$ is an even function of $\boldsymbol{k}$. In addition, $f_1(\boldsymbol{k})$ must be proportional to the electric field and the temperature gradient, since these forces result in a deviation from the equilibrium conduction electron distribution in the $\boldsymbol{k}$-state.

Almost in all real cases the deviation from equilibrium is small, i.e.

$$|f_1| = |f - f_0| \ll f_0. \tag{9.34}$$

In the absence of quantization, when the inequality (9.24) takes place the probability of the transition $W(\boldsymbol{k}, \boldsymbol{k}')$ does not depend on the magnetic field. It is reasonable to assume that this function does not depend on the electric field and the temperature gradient either, i.e. $W(\boldsymbol{k}, \boldsymbol{k}')$ is one and the same for both the equilibrium and nonequilibrium states. Then it is possible to resort to the equation of detailed equilibrium (9.31) to transform the scattering term (9.12) in the Boltzmann equation.

Let us substitute (9.33) into (9.12) and use (9.31). Accounting for (9.34) we confine ourselves to terms that are linear with respect to $f_1(\boldsymbol{k})$. As a result, we obtain

$$(\partial f/\partial t)_{\text{scatt}} = -f_1(\boldsymbol{k})/\tau(\boldsymbol{k}), \tag{9.35}$$

where the notation

$$\frac{1}{\tau(\boldsymbol{k})} = \sum_{k'} W(\boldsymbol{k}, \boldsymbol{k}') \left\{ \frac{1 - f_0(\varepsilon')}{1 - f_0(\varepsilon)} - \frac{f_0(\varepsilon)}{f_0(\varepsilon')} \frac{f_1(\boldsymbol{k}')}{f_1(\boldsymbol{k})} \right\}, \tag{9.36}$$

has been used.

Magnitude $\tau(\boldsymbol{k})$ is referred to as relaxation time, the meaning of which can be explained as follows. Assume that the external forces violate the equilibrium of the conduction electron system. If at time $t = 0$ the external field is switched off, $F = 0$, and the temperature gradient is canceled, $\nabla_r f = 0$, the nonequilibrium function $f(\boldsymbol{k}, t)$ will tend to an equilibrium function $f_0(\boldsymbol{k})$. In this case, $f(\boldsymbol{k}, t)$ satisfies the equation

$$\frac{\partial f(\boldsymbol{k}, t)}{\partial t} + \frac{f(\boldsymbol{k}, t) - f_0(\boldsymbol{k})}{\tau(\boldsymbol{k})} = 0, \tag{9.37}$$

which is derived from (9.11), taking into account (9.35).

From (9.37) we obtain the following law of the nonequilibrium function variation,

$$(f - f_0)_t = (f - f_0)_{t=0} \exp(-t/\tau). \tag{9.38}$$

It is seen that the relaxation time τ characterizes the speed of recovery of the equilibrium state.

Substituting (9.35) into the right-hand side of (9.13) we obtain the following simple form of the Boltzmann equation,

$$\boldsymbol{v}(\boldsymbol{k})\nabla_r f - \frac{e}{\hbar}\left(\boldsymbol{E}_0 + \frac{1}{c}\left[\boldsymbol{v}(\boldsymbol{k})\mathscr{H}\right]\right)\nabla_k f = -\frac{f - f_0}{\tau(\boldsymbol{k})}. \tag{9.39}$$

Thus, introducing the notion of relaxation time we formally reduce the integro-differential equation (9.13) to the differential equation (9.39) for the nonequilibrium distribution function f. However, in the general case this is a seeming simplification of the Boltzmann equation (9.13). Indeed, the ratio of two values of the unknown function f_1 enters into the right-hand side of (9.39) in terms of $\tau(\boldsymbol{k})$ of (2.36) under the summation (integration) sign. In the case where the ratio $f_1(\boldsymbol{k}')/f_1(\boldsymbol{k}')$ does not depend on both the type and the magnitude of the disturbance (electric field or temperature gradient) which causes a deviation of the distribution function from the equilibrium function, the value $\tau(\boldsymbol{k})$ is a characteristic of the conductor and has a reasonable physical sense.

9.2.2 Solution of the Boltzmann Equation in the Absence of Magnetic Field

In this case, (9.39) has the form

$$\boldsymbol{v}(\boldsymbol{k})\nabla_r f - e\boldsymbol{E}_0/\hbar\nabla_k f = -\frac{f - f_0}{\tau(\boldsymbol{k})}, \tag{9.40}$$

from which it is possible to find a nonequilibrium function, assuming that $\tau(\boldsymbol{k})$ does not depend on the electric field and temperature gradient. To do this, we substitute (9.33) into (9.40) and confine ourselves to the function $f_0(\boldsymbol{k})$ on the left-hand side, assuming that in the presence of a temperature gradient the equilibrium function $f_0(\boldsymbol{k})$ depends on the local temperature $T(\boldsymbol{r})$ and chemical potential $\mu(\boldsymbol{r})$, i.e.

$$f_0(\boldsymbol{k}, \boldsymbol{r}) = \left[1 + \exp\left(\frac{\varepsilon(\boldsymbol{k}) - \mu(\boldsymbol{r})}{k_0 T(\boldsymbol{r})}\right)\right]^{-1}. \tag{9.41}$$

Taking into account

$$\nabla_k f_0(\boldsymbol{k}) = \left(\frac{\partial f_0}{\partial \varepsilon}\right)\left(\frac{\partial \varepsilon}{\partial \boldsymbol{k}}\right) = \left(\frac{\partial f_0}{\partial \varepsilon}\right)\hbar\vec{\boldsymbol{v}}(\boldsymbol{k}) \tag{9.42}$$

and

$$\nabla_r f_0(\boldsymbol{k}, \boldsymbol{r}) = k_0 T \left(\frac{\partial f_0}{\partial \varepsilon}\right) \nabla_r \left(\frac{\varepsilon - \mu(\boldsymbol{r})}{k_0 T(\boldsymbol{r})}\right) \tag{9.43}$$

we obtain, as a result, the expression

$$f(\boldsymbol{k}) = f_0(\boldsymbol{k}) - \tau(\boldsymbol{k})(\boldsymbol{v}(\boldsymbol{k})\boldsymbol{\phi}_0(\varepsilon)) \left(\frac{\partial f_0}{\partial \varepsilon}\right) \tag{9.44}$$

for the nonequilibrium but stationary distribution function $f(k)$ in linear approximation with respect to electric field and temperature gradient, where

$$\boldsymbol{\phi}_0(\varepsilon) = -e\boldsymbol{E}_0 + k_0 T \nabla_r \left(\frac{\varepsilon - \mu}{k_0 T}\right) = -e\boldsymbol{E} - \frac{\varepsilon - \mu}{k_0 T} k_0 \nabla T \tag{9.45}$$

is the generalized disturbing force (dynamic and static) causing a deviation from the equilibrium distribution, $\boldsymbol{E} = \boldsymbol{E}_0 + (1/e)\nabla\mu = -\nabla(\varphi_0 - (\mu/e))$ being the electrochemical potential gradient.

Now consider again the expression for the relaxation time. Substituting (9.44) into (9.36) and accounting for

$$f_0(\varepsilon)(1 - f_0(\varepsilon)) = -\frac{1}{k_0 T} \left(\frac{\partial f_0}{\partial \varepsilon}\right) \tag{9.46}$$

gives the following integral equation for $\tau(\boldsymbol{k})$:

$$\frac{1}{\tau(\boldsymbol{k})} = \sum_{k'} W(\boldsymbol{k}, \boldsymbol{k}') \frac{1 - f_0(\varepsilon')}{1 - f_0(\varepsilon)} \left\{1 - \frac{\tau(\boldsymbol{k}')}{\tau(\boldsymbol{k})} \frac{(v(\boldsymbol{k}')\boldsymbol{\phi}_0(\varepsilon'))}{(v(\boldsymbol{k})\boldsymbol{\phi}_0(\varepsilon))}\right\}. \tag{9.47}$$

As we are considering spherically symmetric bands, the scattering must be isotropic, i.e. the transition probability $W(\boldsymbol{k}, \boldsymbol{k}')$ does not depend on $\boldsymbol{k}$ and $\boldsymbol{k}'$ separately but on both k and k' and the angle between them, i.e.

$$W(\boldsymbol{k}, \boldsymbol{k}') = W(k, k', (\boldsymbol{k}\boldsymbol{k}')). \tag{9.48}$$

In the case of isotropic scattering it is obvious that the relaxation time τ will only depend on the magnitude $k = |\boldsymbol{k}|$. Moreover, for the isotropic dispersion relation (9.30) the directions of the electron velocity v and wave vector $\boldsymbol{k}$ coincide and

$$\boldsymbol{v}(\boldsymbol{k}) = \frac{\hbar \boldsymbol{k}}{m(k)}, \tag{9.49}$$

where $m(k)$ is the effective electron mass.

Let us take into account these consequences of the band isotropy and apply spherical coordinates in (9.47). If the polar axis is directed along $\boldsymbol{k}$, and the polar angles of the vectors $\boldsymbol{k}'$ and $\phi_0(\varepsilon')$ are denoted by (θ, ϕ) and (α, β) respectively, then, as is well known, we have

$$\cos(\boldsymbol{k}', \boldsymbol{\phi}_0(\varepsilon')) = \cos\theta\cos\alpha + \sin\theta\sin\alpha\cos(\phi - \beta). \tag{9.50}$$

Then, from (9.47) it is easy to obtain

$$\frac{1}{\tau(k)} = \sum_{k'} W(\boldsymbol{k},\boldsymbol{k}')\frac{1-f_0(\varepsilon')}{1-f_0(\varepsilon)}\left\{1 - \frac{\tau(\boldsymbol{k}')}{\tau(\boldsymbol{k})}\frac{(\boldsymbol{v}(\boldsymbol{k}')\phi_0(\varepsilon'))}{(\boldsymbol{v}(\boldsymbol{k})\phi_0(\varepsilon))}\frac{\boldsymbol{v}(\boldsymbol{k})\boldsymbol{v}(\boldsymbol{k}')}{v^2(\boldsymbol{k})}\right\}. \tag{9.51}$$

Note that this integral relation, from which it is possible to define $\tau(k)$, is obtained on two assumptions: the smallness of the deviation from the equilibrium distribution (9.34) and the isotropy of the dispersion relation (9.30). However, (9.51) is an inconvenient relation since it contains an equilibrium-disturbing force $\phi_0(\varepsilon)$ (electric field $\boldsymbol{E}$ and temperature gradient ∇T) given by (9.45). Consequently, $\tau(k)$ as derived from (9.50) will not be universal; it will depend on the disturbance type and magnitude ($\boldsymbol{E}$ and ∇T). Therefore, it is necessary to make another assumption referring to the scattering nature. Assuming that the charge carriers' scattering caused by lattice defects is purely elastic, $\varepsilon(\boldsymbol{k}') = \varepsilon(k)$, and consequently $k = k'$, i.e. the charge carriers exchange during scattering by the lattice only by impulses, (9.51) simplifies considerably and $\tau(k)$ becomes a universal function characterizing the system relaxation. In this case, the relaxation speed $\bar{\tau}^{-1}(k)$ does not depend on the force ($\boldsymbol{E}$ or ∇T) causing the deviation of the distribution from the equilibrium function but characterizes the conductor itself. Taking into consideration the scattering elasticity $\varepsilon' = \varepsilon$ as well as (9.50), we obtain from (9.51) a simple expression for the relaxation time[1]:

$$\frac{1}{\tau(k)} = \sum_{k'} W(\boldsymbol{k},\boldsymbol{k}')\left(1 - \frac{\boldsymbol{k}\boldsymbol{k}'}{k^2}\right) \tag{9.52}$$

This simple expression for the relaxation time is valid when four conditions are satisfied, namely, smallness of deviation from equilibrium (9.34), isotropy of spectrum (9.30), scattering probability (9.48) and scattering elasticity $\varepsilon' = \varepsilon$. In the case where these conditions are fulfilled simultaneously the determination of the nonequilibrium distribution function is reduced to two independent problems: the calculation of $\tau(k)$ from (9.52) and the solution of the differential Boltzmann equation (9.39).

Section 9.4 is devoted to the calculation of $\tau(k)$ for different scattering mechanisms. But here we shall present the solution of the Boltzmann equation (9.39) in the relaxation time approximation.

We have found the solution of the Boltzmann equation (9.39) in the absence of magnetic field ($H = 0$). It has the form of (9.44). Using this solution we shall first find the constraints to be imposed on the electric field and temperature gradient to satisfy the condition for linear approximation or

[1] Note that the relaxation time can also be introduced in the case of inelastic scattering if the transition probability satisfies the condition $W(\boldsymbol{k},\boldsymbol{k}') = W(\boldsymbol{k},-\boldsymbol{k}')$ which takes place in the case of scattering by nonpolar optical phonons without accounting for the dispersion. This conclusion follows immediately from (2.22).

the condition for smallness of deviation of the distribution function $f(\boldsymbol{k})$ from the equilibrium function $f_0(\boldsymbol{k})$. If only the electric field is present from (9.44) and (9.45) we have

$$f(\boldsymbol{k}) = f_0(\boldsymbol{k}) - e\tau(\boldsymbol{k})(\boldsymbol{v}\boldsymbol{E}_0)\,(-\partial f_0/\partial\varepsilon). \tag{9.53}$$

For a nondegenerate electron gas, $f_0(\varepsilon) = \exp((\mu - \varepsilon)/k_0T)$ and we obtain from (9.53) an explicit form of the condition for small deviations (9.34):

$$\bar{l}eE_0 \ll k_0T, \tag{9.54}$$

where $\bar{l} = \bar{v}\bar{\tau}$ is the mean free path.

It is seen from (9.54) that the energy acquired by an electron in the electric field for one mean free path must be much less than the mean thermal energy k_0T to satisfy an approximation that is linear in E_0 or the nonequilibrium addition $f_1 < f_0$. A comparison of (9.54) with the applicability condition for the Boltzmann equation (9.18) shows that the field for which linear approximation is valid must be weaker by a factor $\lambda/\bar{l}$. Since in a nondegenerate semiconductor $\lambda/\bar{l} \approx 10^{-1}$, linear approximation is valid up to rather large fields $E_0 \approx 10^4\,\mathrm{V/cm}$.

Accounting for $k_0T \sim m\overline{v}^2$ the condition (9.54) can be rewritten as follows:

$$v_\mathrm{d} \ll \overline{v}, \tag{9.55}$$

where $v_\mathrm{d} = (e\tau/m)E_0$ is the electron drift velocity and $\overline{v}$ is the mean value of the thermal velocity.

In strong degenerate electron gas, instead of (9.54), (9.55) and (9.53), we obtain the following inequalities

$$l(\mu)eE_0 \ll \mu \quad \text{or} \quad v_\mathrm{d} \ll v(\mu), \tag{9.56}$$

where $l(\mu) = \tau(\mu)v(\mu)$ is the mean free path of an electron on the Fermi level, $v_\mathrm{d} = (e\tau(\mu)/m(\mu))E_0$ is the drift velocity, and $v(\mu)$ is the electron velocity on the Fermi surface.

We now consider the conditions to be imposed on the temperature gradient. In the presence of a temperature gradient, from (9.44) and (9.45) we obtain

$$f(\boldsymbol{k}) = f_0(\boldsymbol{k}) - \tau(\boldsymbol{k})\frac{\varepsilon - \mu}{T}(\boldsymbol{v}\nabla T)\left(-\frac{\partial f_0}{\partial\varepsilon}\right). \tag{9.57}$$

For nondegenerate semiconductors, (9.57) and (9.34) yield the following inequality governing the applicability of the linear approximation of the solution with respect to temperature gradient

$$|\mu/k_0T - 1|\,\bar{l}\,|\nabla T| \ll T. \tag{9.58}$$

As in a nondegenerate semiconductor, we usually have $|(\mu/k_0T) - 1| \sim 10$, so it can be said that the temperature variation over a length $\sim 10\bar{l}$ must be much smaller than the temperature T itself to satisfy the condition of the deviation from the equilibrium distribution (9.34).

In the conclusion of this subsection, we shall demonstrate the difference of stationary distribution from equilibrium distribution in the $\boldsymbol{k}$-space. To do this, the solution (9.44) will be presented as follows:

$$f(\boldsymbol{k}) = f_0(\boldsymbol{k}) - \frac{\tau(k)}{\hbar}\phi_0(\varepsilon)\left(\frac{\partial f_0}{\partial \boldsymbol{k}}\right). \tag{9.59}$$

Consider the action of the electric field and temperature gradient separately. In the presence of an electric field $\phi_0 = -e\boldsymbol{E}_0$, and, according to (9.59), the stationary distribution $f(\boldsymbol{k})$ is related to the equilibrium distribution as follows:

$$f(\boldsymbol{k}) = f_0(\boldsymbol{k}) + \frac{\tau(\boldsymbol{k})}{\hbar}e\boldsymbol{E}_0\left(\frac{\partial f_0}{\partial \boldsymbol{k}}\right) \tag{9.60}$$

or

$$f(\boldsymbol{k}) \approx f_0\left(\boldsymbol{k} + \tau(\boldsymbol{k})e\boldsymbol{E}_0/\hbar\right) = f_0(\boldsymbol{k} - \Delta\boldsymbol{k}). \tag{9.61}$$

It is seen that the stationary distribution $f(\boldsymbol{k})$ is identical with the equilibrium distribution $f_0(\boldsymbol{k})$ except that the coordinate origin in the $\boldsymbol{k}$-space is shifted from the point $\boldsymbol{k} = 0$ to the point $\boldsymbol{k} - \Delta\boldsymbol{k} = -\tau e\boldsymbol{E}_0/\hbar$. Figure 9.1 shows schematically $f(\boldsymbol{k})$ and $f_0(\boldsymbol{k})$ in the case of a degenerate electron gas. This result is clear from the standpoint of the influence of the electric field on each quantum state. Indeed, it follows from $\dot{\boldsymbol{p}} = \boldsymbol{F}$ that the variation speed $\boldsymbol{k}$ under the action of the force $\boldsymbol{F} = -e\boldsymbol{E}_0$ is the same for all electrons. Therefore, the distribution in the $\boldsymbol{k}$-space will drift in the field, holding its shape invariant, at a constant speed $\mathrm{d}\boldsymbol{k}/\mathrm{d}t = -e\boldsymbol{E}_0/\hbar$. However, scattering processes that tend to return the system to the equilibrium state restrict this drift by $\Delta\boldsymbol{k} = -e\boldsymbol{E}_0\tau/\hbar$ and establish the stationary (time-independent) electron distribution in the $\boldsymbol{k}$-space (dash line in Fig. 9.1). As seen from Fig. 9.1, the stationary distribution $f(\boldsymbol{k})$, in contrast to the equilibrium distribution $f_0(\boldsymbol{k})$, is nonsymmetric with respect to the point $\boldsymbol{k} = 0$. Therefore this state corresponds to a finite constant current in a conductor.

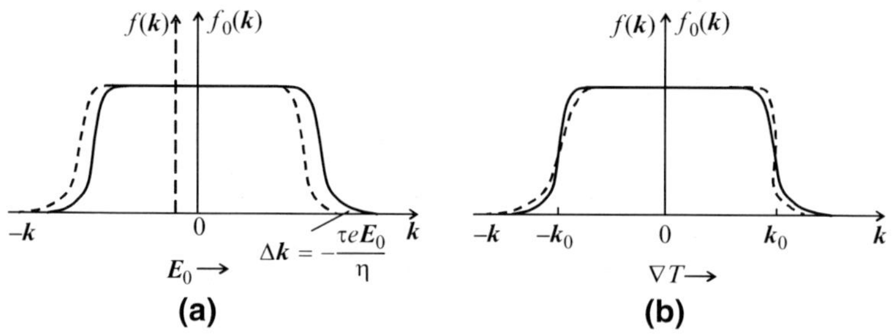

Fig. 9.1. Equilibrium (solid line) and nonequilibrium but stationary (dash line) distribution functions for a degenerate electron gas in $\boldsymbol{k}$ - space: a) in the presence of electric field, b) in the presence of temperature gradient

Note, however, that the mean energy of the electron system in a stationary state is equal to the energy in an equilibrium state. Actually, using (9.60) we have

$$\int \varepsilon(\boldsymbol{k}) f(\boldsymbol{k}) \mathrm{d}\boldsymbol{k} = \int \varepsilon(\boldsymbol{k}) f_0(\boldsymbol{k}) \mathrm{d}\boldsymbol{k}, \tag{9.62}$$

since as $\varepsilon(\boldsymbol{k})$ is an even function of $\boldsymbol{k}$ the addend of (9.60) does not make a contribution to the integral (9.62). Physically, this is associated with the fact that the electrons, moving in the direction of the electric field and accelerating, reduce (due to their negative charge) their wave vector by $\Delta\boldsymbol{k}$. The electrons moving against the field increase their $\boldsymbol{k}$ by the same value (see Fig.9.1). As $\varepsilon(\boldsymbol{k})$ is an even function of $\boldsymbol{k}$, on average the electron system energy does not vary.

In the presence of the temperature gradient $\phi_0(\varepsilon) = -\left(\frac{\varepsilon-\mu}{T}\right)\nabla T$, according to (9.59) the stationary distribution is

$$f(\boldsymbol{k}) = f_0(\boldsymbol{k}) + \frac{(\varepsilon-\mu)\tau(\boldsymbol{k})}{\hbar}\frac{\nabla T}{T}\frac{\partial f_0}{\partial \boldsymbol{k}}, \tag{9.63}$$

which can be presented in the following form:

$$f(\boldsymbol{k}) \approx f_0\left(k + \frac{(\varepsilon-\mu)\tau(k)}{\hbar}\frac{\nabla T}{T}\right) = f_0(\boldsymbol{k} - \Delta\boldsymbol{k}), \tag{9.64}$$

where

$$\Delta\boldsymbol{k} = -\frac{(\varepsilon-\mu)\tau(\boldsymbol{k})}{T\hbar}\nabla T. \tag{9.65}$$

In the case of a degenerate electron gas with a parabolic dispersion relation $\mu = \hbar^2 k_0^2/2m$, from (9.65) we have

$$\Delta\boldsymbol{k} = -\frac{\hbar\tau(\boldsymbol{k})}{2mT}(k^2 - k_0^2)\nabla T, \tag{9.66}$$

where m is the effective conduction electron mass, and $\boldsymbol{k}_0$ is the wave vector of an electron on the Fermi surface.

The function (9.64) is formally similar to (9.61). However, in contrast to the electric field in the presence of a temperature gradient the shift $\Delta\boldsymbol{k}$ of the stationary distribution relative to the equilibrium distribution $f_0(\boldsymbol{k})$ depends on the value $\boldsymbol{k}$. In particular, for $\boldsymbol{k} = \boldsymbol{k}_0$, $\Delta\boldsymbol{k} = 0$ and $f(\boldsymbol{k}_0) = f_0(\boldsymbol{k}_0)$. For electrons with $\boldsymbol{k} > \boldsymbol{k}_0$ the distribution shift is $\nabla\boldsymbol{k} < 0$ if the electron moves along the temperature gradient ∇T, and $\Delta\boldsymbol{k} > 0$ if the electron moves against ∇T, and vice verse for electrons with wave vector $\boldsymbol{k} < \boldsymbol{k}_0$: the shift is $\Delta\boldsymbol{k} > 0$ for motion along the gradient and $\Delta\boldsymbol{k} < 0$ if the electrons move against ∇T. An equilibrium distribution $f_0(\boldsymbol{k})$ and a stationary one $f(\boldsymbol{k})$ for a degenerate electron gas in the presence of the temperature gradient are shown schematically in Fig. 9.1b. As seen from this figure, in the presence of the temperature gradient the smearing width of the Fermi level in the distribution of electrons moving against ∇T becomes greater (left part of Fig. 9.1b), for the electrons moving along ∇T the smearing width of the Fermi level is smaller (right part of Fig. 9.1b) than the width for zero gradient of the temperature, which is the same for all electrons.

This result can be understood from the following obvious considerations. A crystal electron which is at the point $\boldsymbol{r}_0$ moving in the direction of increasing temperature (in the direction of ∇T) has to experience its last collision at the point where the temperature is less than $T(\boldsymbol{r})$ by $\tau v \nabla T$. This suggests that the distribution width for such electrons will be relatively smaller. Similarly, an electron at the point $\boldsymbol{r}$ moving against ∇T must experience the last collision at a higher temperature. Therefore, the distribution width for such electrons will be greater than the mean width.

Thus, a temperature gradient causes violation of the symmetric electron distribution with respect to the point $\boldsymbol{k} = 0$ (see Fig. 9.1b) and thereby causes a finite electric current proportional to ∇T.

Note that only for degenerate electron gases, as seen from Fig. 9.1, the behaviour of the distribution deviation would be different from the equilibrium one depending on the causes of this deviation, namely: an electric field or a temperature gradient. When the electron gas is nondegenerate, $(\varepsilon - \mu) > 0$. Therefore, as seen from (9.61) and (9.64), in contrast to the degenerate case a shift of the stationary electron gas distribution relative to an equilibrium one is of the same nature for disturbance by an electric field or by a temperature gradient. In both cases, the shift occurs in a direction opposite to the action of $\boldsymbol{E}_0$ or ∇T (Fig. 9.2). In Fig. 9.2 the equilibrium (solid line) and stationary (dash line) distribution functions for nondegenerate electron gas in $\boldsymbol{k}$-space are represented.

9.2.3 Solution of Boltzmann Equation with an Arbitrary Nonquantizing Magnetic Field

We have shown that in the absence of a magnetic field in the τ -approximation the Boltzmann equation solution (9.39) has the form of (9.44). In this solution the vector $\Phi_0(\varepsilon)$ possesses an explicit physical meaning, namely a generalized force disturbing the equilibrium electron distribution in the $\boldsymbol{k}$-space. Obviously, in a magnetic field H this force will be different. The disturbing force, changed by the magnetic field, will be denoted as $\Phi(\varepsilon) = \Phi(\varepsilon, \boldsymbol{E}_0, \nabla T, \mathscr{H})$. Then, the general solution of equation (9.39) will be

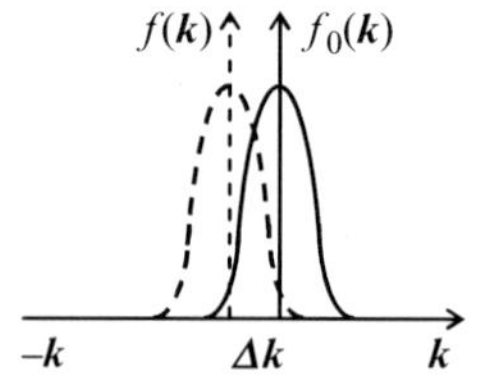

Fig. 9.2. Equilibrium (solid line) and stationary (dash line) distribution function for a nondegenerate electron gas in $\boldsymbol{k}$-space

$$f(\boldsymbol{k}) = f_0(\boldsymbol{k}) - \tau(\boldsymbol{k})(\boldsymbol{v}(\boldsymbol{k})\Phi(\varepsilon))(\partial f_0/\partial\varepsilon). \tag{9.67}$$

Thus, the problem is reduced to finding the disturbing force $\Phi_0(\varepsilon)$ in a magnetic field. To do this, substitute (9.67) into equation (9.39) and take into account that it is possible to confine oneself to the equilibrium function in the terms $\nabla_r f$ and $\boldsymbol{E}_0\nabla_k f$, while in the term involving the magnetic field, $[vH]$, it is necessary to retain the nonequilibrium addition since, according to (9.42), in a zero approximation this term is identical to zero. As a result, in a first nonvanishing approximation for $\Phi_0(\varepsilon)$, we obtain the following equation:

$$(\boldsymbol{v}\Phi_0) + \frac{e}{\hbar c}\,[\boldsymbol{v}\mathscr{H}]\,\nabla_\kappa(\tau\boldsymbol{v}\Phi) = (\boldsymbol{v}\Phi), \tag{9.68}$$

where Φ_0 is defined by (9.45).

To solve the equation (9.68), it is convenient to introduce the vector

$$\boldsymbol{P}(\varepsilon) = \tau(k)\Phi(\varepsilon), \tag{9.69}$$

which has the significance of the impulse of a disturbing force in the magnetic field.

Then, $\boldsymbol{P}(\varepsilon)$ will satisfy the equation

$$(\boldsymbol{v}\Phi_0) + \frac{e}{\hbar c}\,[\boldsymbol{v}\boldsymbol{H}]\,\nabla_\kappa(\boldsymbol{v}\boldsymbol{P}) = \frac{1}{\tau(k)}(\boldsymbol{v}\boldsymbol{P}). \tag{9.70}$$

Let us transform the expression $\nabla_\kappa(\boldsymbol{v}\boldsymbol{P}(\varepsilon))$ using the known formula of the vector analysis

$$\nabla_\kappa(\boldsymbol{v}\boldsymbol{P}) = (\boldsymbol{P}\nabla_\kappa)\boldsymbol{v} + (\boldsymbol{v}\nabla_\kappa)\boldsymbol{P} + [\boldsymbol{v}\,\mathrm{rot}\,\boldsymbol{P}] + [\boldsymbol{P}\,\mathrm{rot}\,\boldsymbol{v}]\,. \tag{9.71}$$

The last term in (9.71) vanishes since $\mathrm{rot}\,\boldsymbol{v} = (1/\hbar)\mathrm{rot}\,\mathrm{grad}\,\varepsilon(\boldsymbol{k}) \equiv 0$. It is known that in the case of a spherically symmetric zone the velocity $\boldsymbol{v}$ is directed along the $\boldsymbol{k}$-vector and is determined by the expression (9.49). Taking this into consideration it is easy to show that the first term of the right-hand side of (9.71) is

$$(\boldsymbol{P}\nabla_\kappa)\boldsymbol{v} = \frac{2m(k)}{k}\frac{\partial}{\partial k}\left(\frac{1}{m(k)}\right)(\boldsymbol{k}\boldsymbol{P})\boldsymbol{v} + \frac{\hbar}{m(k)}\boldsymbol{P}. \tag{9.72}$$

By means of direct calculation it is possible to obtain

$$(\boldsymbol{v}\nabla_k)\boldsymbol{P} + [\boldsymbol{v}\,\mathrm{rot}\,\boldsymbol{P}] = \left(k_x\frac{\partial P_x}{\partial\varepsilon} + k_y\frac{\partial P_y}{\partial\varepsilon} + k_z\frac{\partial P_z}{\partial\varepsilon}\right)\frac{\hbar^2}{m(\varepsilon)}\boldsymbol{v}. \tag{9.73}$$

Substituting (9.72) and (9.73) into (9.70) and considering that all terms containing the velocity $\boldsymbol{v}$ vanish on scalar multiplication by $[vH]$ we obtain the equation

$$(\boldsymbol{v}\Phi_0) + \frac{e\tau(k)}{cm(k)}\,([\boldsymbol{v}\mathscr{H}]\,\Phi) = (\boldsymbol{v}\Phi), \tag{9.74}$$

where, according to (9.69), we have gone on from $\boldsymbol{P}(\varepsilon)$ to the sought vector $\Phi(\varepsilon)$.

The value $m(k)$ entering into equation (9.74) is the effective mass of charge carriers in the case of an isotropic band:

$$m(k) = \hbar^2 k \frac{\partial k}{\partial \varepsilon(k)}. \tag{9.75}$$

Assuming $v \neq 0$ we obtain from (9.74) an ordinary vector equation for $\Phi(\varepsilon)$:

$$\Phi_0 + \frac{e}{c}\frac{\tau(k)}{m(k)}[H\Phi] = \Phi. \tag{9.76}$$

To define $\Phi(\varepsilon)$ from this equation we shall proceed as follows. Equation (9.76) is multiplied by H scalarly to give

$$(H\Phi_0) = (H\Phi). \tag{9.77}$$

Next, (9.76) is multiplied by H vectorially from the left:

$$[H\Phi_0] = \frac{e}{c}\frac{\tau(k)}{m(k)} H^2\Phi - \frac{e}{c}\frac{\tau(k)}{m(k)}(H\Phi)H + [H\Phi]. \tag{9.78}$$

From the last three equations it is easy to derive the final expression for the disturbing force in the magnetic filed $\Phi(\varepsilon)$:

$$\Phi(\varepsilon) = \frac{1}{1+\nu^2}\left\{\Phi_0 + \frac{e}{c}\frac{\tau}{m}[H\Phi_0] + \left(\frac{e\tau}{cm}\right)^2 H(H\Phi_0)\right\}. \tag{9.79}$$

Here

$$\nu = \Omega\tau, \tag{9.80}$$

where $\Omega = eH/cm(k)$ is the cyclotron frequency.

The function (9.79), together with (9.67), gives the general linear solution of the Boltzmann equation with an arbitrary nonquantizing magnetic field for an arbitrary isotropic band in the relaxation time approximation. This solution in the above approximation forms the basis of the theory of electron transport phenomena in the classical domain. The applicability limits of this solution are defined by the inequalities (9.24) and (9.26).

Now we shall take note of some general conclusions resulting from the solution (9.79) for an isotropic band.

First, when the magnetic field H is directed along the disturbing force Φ_0, from (9.79) it follows that

$$\Phi(\varepsilon) = \Phi_0(\varepsilon), \tag{9.81}$$

i.e. in conductors with an isotropic dispersion relation the longitudinal magnetic field $(H||\Phi_0)$ does not change the disturbing force and, consequently, there are no longitudinal effects in such conductors.

Second, if the magnetic field H is perpendicular to the disturbing force Φ_0, then from (9.79) we have

$$\Phi(\varepsilon) = \frac{1}{1+\nu^2}\left\{\Phi_0 + \frac{e}{c}\frac{\tau}{m}[H\Phi_0]\right\}, \tag{9.82}$$

i.e. there will appear a disturbing force component perpendicular to H and Φ_0 which corresponds to the appearance of the current in this direction and, consequently, the appearance of such transverse effects asHall, Nernst–Ettingshausen effects, etc.

Third, if we consider the projection of the disturbing force in the magnetic field $\Phi(\varepsilon)$ on the direction of the vector Φ_0, then from (9.79) we obtain

$$\Phi(\varepsilon) = \frac{1+\nu^2\cos^2\alpha}{1+\nu^2}\Phi_0(\varepsilon), \tag{9.83}$$

where α is the angle between the magnetic field H and Φ_0.

For $\alpha = 0$, the first case (9.81) follows from (9.83). When $\alpha \neq 0$, i.e. when there is a transverse component of the magnetic field with respect to the direction Φ_0, the projection of the disturbing force is less than the force in the absence of the field: $\Phi(\varepsilon) < \Phi_0(\varepsilon)$. This means that in the presence of a transverse magnetic field the specimen conductivity is less than that without the field. That is, the specific resistance in the magnetic field, i.e. the magnetic resistance, must increase.

At the end of this subsection we shall emphasize the following circumstance. As seen from (9.79), the magnetic field H enters into the Boltzmann equation solution in terms of a nondimensional parameter $\nu = \Omega\tau = (eH/mc)\tau$. The question arises whether it is possible to apply the solution (9.79) to fields for which this nondimensional parameter is much greater than unity, i.e.

$$\nu = \Omega\tau \gg 1 \,\text{or}\, H \gg \frac{mc}{e\tau} \equiv H_0 \tag{9.84}$$

as we know that the Boltzmann equation itself is applicable to magnetic fields with an upper bound set by the condition (9.24). In other words, are the conditions (9.84) and (9.24) compatible? If we combine these two conditions, we shall obtain

$$\left(\frac{\tau k_0 T}{\hbar}\right) H_0 \gg H \gg H_0. \tag{9.85}$$

According to (9.56), we have $\left(\frac{\tau k_0 T}{\hbar}\right) \gg 1$; therefore the double inequality (9.85) may take place. Consequently, the conditions (9.84) and (9.24) are compatible and there is an extensive range of the magnetic field magnitude for which the condition (9.84) is fulfilled while the quantization has not yet begun.

9.3 General Expressions of Main Kinetic Coefficients

The results obtained in the previous sections allow us to construct the transport phenomenon theory in conductors with a spherically symmetric band. In this section we consider an arbitrary isotropic band and obtain the general expression for main kinetic coefficients.

All kinetic coefficients of interest measured in experiments can be expressed in terms of galvanomagnetic and thermomagnetic tensor components σ_{ik}, β_{ik} and κ_{ik}. In order to obtain an explicit form of these tensor components it is necessary to find the relation of current density and energy flux to electric field and temperature gradient. The afore-named relation is realized by means of conductivity tensors.

9.3.1 Current Density and General Form of Conductivity Tensors

Substituting the general solution of the Boltzmann equation (9.67) into (9.1) and (9.2) we shall obtain for the current density and energy flux:

$$\boldsymbol{j} = \frac{e}{4\pi^3}\int \tau(\kappa)\boldsymbol{v}(\boldsymbol{k})(\boldsymbol{v}(\boldsymbol{k})\Phi(\varepsilon))\left(\frac{\partial f_0}{\partial \varepsilon}\right)\mathrm{d}\boldsymbol{k}, \tag{9.86}$$

$$\boldsymbol{w} = -\frac{1}{4\pi^3}\int (\varepsilon-\mu)\tau(\kappa)\boldsymbol{v}(\boldsymbol{k})(\boldsymbol{v}(\boldsymbol{k})\Phi(\varepsilon))\left(\frac{\partial f_0}{\partial \varepsilon}\right)\mathrm{d}\boldsymbol{k}. \tag{9.87}$$

Proceeding to a spherical coordinate system in the $\boldsymbol{k}$-space and integrating over the angles we obtain for the current density and energy flux components:

$$j_i = \frac{e}{3\pi^2\hbar^2}\int\left(\frac{\partial \varepsilon}{\partial k}\right)^2\tau(k)\Phi_i(\varepsilon)\left(\frac{\partial f_0}{\partial \varepsilon}\right)k^2\mathrm{d}k, \tag{9.88}$$

$$w_i = -\frac{1}{3\pi^2\hbar^2}\int\left(\frac{\partial \varepsilon}{\partial k}\right)^2(\varepsilon-\mu)\tau(k)\Phi_i(\varepsilon)\left(\frac{\partial f_0}{\partial \varepsilon}\right)k^2\mathrm{d}k. \tag{9.89}$$

Passing from the integration over the wave vector magnitude to integration over energy, we have:

$$j_i = -\frac{e}{3\pi^2}\int\left(-\frac{\partial f_0}{\partial \varepsilon}\right)\frac{\tau(\varepsilon)}{m(\varepsilon)}\Phi_i(\varepsilon)k^3(\varepsilon)\mathrm{d}\varepsilon, \tag{9.90}$$

$$w_i = \frac{1}{3\pi^2}\int\left(-\frac{\partial f_0}{\partial \varepsilon}\right)(\varepsilon-\mu)\frac{\tau(\varepsilon)}{m(\varepsilon)}\Phi_i(\varepsilon)k^3(\varepsilon)\mathrm{d}\varepsilon. \tag{9.91}$$

Here such definition of the electron effective mass for the arbitrary isotropic band $1/m(\varepsilon) = (1/\hbar^2 k)(\partial\varepsilon/\partial k)$ is used.

It is seen that the current and energy flux density components are governed by the respective components of the force $\Phi(\varepsilon)$, whose explicit form is derived from the Boltzmann equation solution and was presented in (9.79). Without loss of generality assume that the magnetic field is directed along the z-axis: $H_x = H_y = 0$, $H_z = H$. Electric field and temperature gradient are in a xy-plane. Then, according to (9.79), the components of $\Phi(\varepsilon)$ will have the following form:

$$\Phi_x = \frac{1}{1+\nu^2}\left(\Phi_{0x} - \nu\Phi_{0y}\right), \tag{9.92}$$

$$\Phi_y = \frac{1}{1+\nu^2}\left(\Phi_{0y} + \nu\Phi_{0x}\right), \tag{9.93}$$

$$\Phi_z = \Phi_{0z}, \tag{9.94}$$

where Φ_{0x}, Φ_{0y} and Φ_{0z} are given by the formula (9.45).

From (9.90) and (9.94) it follows that the longitudinal current density component j_z does not depend on the magnetic field, i.e. there is no longitudinal magnetoresistance for a spherical band, as might be expected.

Using (9.92) and (9.93) in (9.90) we obtain the following expressions:

$$\begin{aligned} j_x &= \sigma_{11}E_x - \sigma_{12}E_y - \beta_{11}\nabla_x T + \beta_{12}\nabla_y T, \\ j_y &= \sigma_{12}E_x + \sigma_{11}E_y - \beta_{12}\nabla_x T - \beta_{11}\nabla_y T \end{aligned} \tag{9.95}$$

for the current density components in conductors with a spherically symmetric band in an arbitrary nonquantizing magnetic field.

From (9.92), (9.93) and (9.91) while accounting for the lattice thermal conductivity κ_{ph} we obtain for the energy flux density the following expressions:

$$\begin{aligned} w_x &= T\beta_{11}E_x - T\beta_{12}E_y - (\kappa_{11} + \kappa_{ph})\nabla_x T + \kappa_{12}\nabla_y T, \\ w_y &= T\beta_{12}E_x + T\beta_{11}E_y - \kappa_{12}\nabla_x T - (\kappa_{11} + \kappa_{ph})\nabla_y T. \end{aligned} \tag{9.96}$$

Kinetic tensor components included in expressions (9.95) and (9.96) have the following form:

$$\sigma_{11} = ne^2\left\langle \frac{\tau}{m}\frac{1}{1+\nu^2}\right\rangle, \quad \sigma_{12} = ne^2\left\langle \frac{\tau}{m}\frac{\nu}{1+\nu^2}\right\rangle, \tag{9.97}$$

$$\beta_{11} = -\frac{ne}{T}\left\langle \frac{\tau}{m}\frac{(\varepsilon-\mu)}{1+\nu^2}\right\rangle, \quad \beta_{12} = -\frac{ne}{T}\left\langle \frac{\tau}{m}\frac{(\varepsilon-\mu)\nu}{1+\nu^2}\right\rangle, \tag{9.98}$$

$$\kappa_{11} = \frac{n}{T}\left\langle \frac{\tau}{m}\frac{(\varepsilon-\mu)^2}{1+\nu^2}\right\rangle, \quad \kappa_{12} = \frac{n}{T}\left\langle \frac{\tau}{m}\frac{(\varepsilon-\mu)^2\nu}{1+\nu^2}\right\rangle. \tag{9.99}$$

The averaging symbol $\langle\ldots\rangle$ denotes integrals of the following form

$$\langle A(\varepsilon)\rangle = \frac{1}{3\pi^2 n}\int^{\infty}\left(-\frac{\partial f_0}{\partial\varepsilon}\right)k^3(\varepsilon)A(\varepsilon)\mathrm{d}\varepsilon, \tag{9.100}$$

where n is the conduction electron concentration, $\nu = \Omega\tau$.

In (9.95) and (9.96) we took into account the symmetry of the conductivity tensor components: $\sigma_{21} = -\sigma_{12}$, $\sigma_{22} = \sigma_{11}$, $\beta_{21} = -\beta_{12}$, $\beta_{22} = \beta_{11}$, $\kappa_{21} = -\kappa_{12}$, $\kappa_{22} = \kappa_{11}$, which follows from the general formula (9.4).

9.3.2 General Expressions of Main Kinetic Coefficients

Below we shall give definitions of the basic kinetic coefficients.

Galvanomagnetic Effects

Hall effect, i.e. the initiation of a transverse electric field E_y in the presence of a current j_x in the x -direction and in the absence of current j_y if the magnetic field is directed along the z-axis is determined from the conditions $j_y = 0$, $\nabla_x T = \nabla_y T = 0$ and characterized by a constant R. Then we have

$$R = E_y / j_x H. \tag{9.101}$$

Resistance in a magnetic field is determined from the conditions: $j_y = 0$, $\nabla_x T = \nabla_y T = 0$. Then we have

$$\rho(H) = E_x / j_x. \tag{9.102}$$

Nernst effect, i.e. the initiation of a longitudinal temperature gradient in the absence of heat flux in this direction: $w_x = 0$, $j_y = 0, \nabla_y T = 0$. Then we have

$$B = \nabla_x T / j_x. \tag{9.103}$$

Note that this effect also takes place in the absence of a magnetic field. The presence of a transverse magnetic field alters only the value B.

Ettingshausen effect, i.e. the initiation of transverse temperature difference in the presence of electric current in the x-direction ($j_x \neq 0$) and in the absence of current in the y-direction: $j_y = 0$, $w_y = 0$, as well as in the absence of the gradient $\nabla_x T = 0$. Then for the Ettingshausen coefficient P we have

$$P = -\nabla_y T / H j_x. \tag{9.104}$$

Thermomagnetic Effects

Transverse Nernst–Ettingshausen effect, i.e. the transverse electric field E_y associated with a temperature gradient $\nabla_x T$ is determined from the conditions $j_x = j_y = 0, \nabla_y T = 0$ and characterized by the coefficient Q:

$$Q = -E_y / H \nabla_x T. \tag{9.105}$$

Longitudinal Nernst–Ettingshausen effect, i.e. the variations of the thermopower in a transverse magnetic field is found from the conditions $j_x = j_y = 0$, $\nabla_x T \neq 0$. Then we have:

$$\alpha(H) = E_x / \nabla_x T. \tag{9.106}$$

Maggi–Righi–Leduc effect, i.e. the thermal conductivity variation in a transverse magnetic field is determined from the conditions $j_x = j_y = 0$, $\nabla_y T = 0$. Then we have:

$$\kappa(H) = -w_x/\nabla_x T. \tag{9.107}$$

Righi–Leduc effect, i.e. the transverse temperature difference $\nabla_y T$ associated with the presence of the gradient $\nabla_x T$ is determined from the conditions $j_x = j_y = 0$, $w_y = 0$ and characterized by the coefficient S:

$$S = -\nabla_y T/H\nabla_x T. \tag{9.108}$$

Proceeding from the definition of the main kinetic coefficients and the relations (9.95) and (9.96) it is easy to show that the kinetic coefficients are found in terms of the components σ_{ik}, β_{ik} and κ_{ik} in the following way:

resistance in a transverse magnetic field

$$\rho(H) = \sigma_{11}(\sigma_{11}^2 + \sigma_{12}^2)^{-1}, \tag{9.109}$$

Hall coefficient

$$R = -H^{-1}\sigma_{12}(\sigma_{11}^2 + \sigma_{12}^2)^{-1}, \tag{9.110}$$

transverse Nernst–Ettingshausen effect

$$Q = H^{-1}\left(\sigma_{12}\beta_{11} - \sigma_{11}\beta_{12}\right)\left(\sigma_{11}^2 + \sigma_{12}^2\right)^{-1}, \tag{9.111}$$

thermopower in a transverse magnetic field

$$\alpha(H) = \left(\sigma_{11}\beta_{11} + \sigma_{12}\beta_{12}\right)\left(\sigma_{11}^2 + \sigma_{12}^2\right)^{-1}, \tag{9.112}$$

thermal conductivity in a transverse magnetic field

$$\kappa(H) = \kappa_{\mathrm{ph}} + \kappa_{11} - T\beta_{11}\alpha(H) - TH\beta_{12}Q, \tag{9.113}$$

Righi–Leduc coefficient

$$S = \frac{1}{H}\frac{1}{\kappa(H)}\left[TH\beta_{11}Q - T\beta_{12}\alpha(H) + \kappa_{12}\right]. \tag{9.114}$$

Nernst coefficient B and Ettingshausen coefficient P can be expressed by $\alpha(H)$, Q and $\kappa(H)$ as follows:

$$B = T\alpha(H)/\kappa(H), \quad P = TQ/\kappa(H). \tag{9.115}$$

Substituting expressions (9.97)–(9.99) into (9.109)–(9.112), for the first four kinetic coefficients we obtain

$$\rho(H) = \frac{1}{ne^2}\frac{1}{D}\left\langle \frac{\tau}{m}\frac{1}{1+\nu^2}\right\rangle, \tag{9.116}$$

$$R = -\frac{1}{ne^2 H}\frac{1}{D}\left\langle \frac{\tau}{m}\frac{\nu}{1+\nu^2}\right\rangle, \tag{9.117}$$

$$Q = \frac{k_0}{e}\frac{1}{DH}\left[\left\langle\frac{\tau}{m}\frac{1}{1+\nu^2}\right\rangle\left\langle\frac{\tau}{m}\frac{\nu x}{1+\nu^2}\right\rangle - \left\langle\frac{\tau}{m}\frac{x}{1+\nu^2}\right\rangle\left\langle\frac{\tau}{m}\frac{\nu}{1+\nu^2}\right\rangle\right], \tag{9.118}$$

$$\alpha(H) = -\frac{k_0}{e}\left\{\frac{1}{D}\left[\left\langle\frac{\tau}{m}\frac{1}{1+\nu^2}\right\rangle\left\langle\frac{\tau}{m}\frac{x}{1+\nu^2}\right\rangle + \left\langle\frac{\tau}{m}\frac{x\nu}{1+\nu^2}\right\rangle\left\langle\frac{\tau}{m}\frac{\nu}{1+\nu^2}\right\rangle\right] - \eta\right\}, \tag{9.119}$$

where $\eta = \mu/k_0T$ is the reduced chemical potential, $x = \varepsilon/k_0T$ is the reduced energy, and

$$D = \left\langle\frac{\tau}{m}\frac{1}{1+\nu^2}\right\rangle^2 + \left\langle\frac{\tau}{m}\frac{\nu}{1+\nu^2}\right\rangle^2. \tag{9.120}$$

The formulae (9.116)–(9.119) are common for conductors with an arbitrary isotropic band. Using the above-mentioned formulae (9.116)–(9.119) it is possible to calculate other kinetic coefficients B and P.

Using the above-mentioned general formulae it is possible to consider various limiting cases: absence of a magnetic field ($\nu = 0$), weak ($\nu \ll 1$) and strong ($\nu \gg 1$) magnetic field.

From formulae (9.116)–(9.119) it follows that for determination of analytical dependences of kinetic coefficient it is necessary to know an explicit form of the dispersion law $\varepsilon(k)$, and also the dependence of effective mass m and relaxation time τ from energy. Let us note that components of kinetic tensors include only the relation τ/m.

The dispersion law is determined from the decision of Schrödinger's equation and the finding of the dependence $\tau(\varepsilon)$ is determined from the theory of scattering. The following section is devoted to calculation of the relaxation time for various scattering mechanisms.

9.4 Main Relaxation Mechanisms

In this section we shall consider some basic mechanisms of charge carrier scattering in conductors with the isotropic dispersion relation and give the respective relaxation times.

In the case where it is possible to introduce the notion of relaxation time the problem of transport phenomena theory is greatly simplified and is divided into three essentially independent problems: (1) the Boltzmann equation solution in the relaxation time approximation; (2) calculation of the current densities and energy flux, and consequently the calculation of kinetic coefficients measured experimentally; and (3) calculation of the relaxation time $\tau(\boldsymbol{k})$ for different scattering mechanisms. We have solved the first and second problems in the preceding sections/chapters. The present section/chapter deals with the third problem.

We shall dwell upon some basic mechanisms of charge carriers' scattering and derive the respective expressions for relaxation time for any spherically symmetric band. As seen from (9.52), in order to calculate the relaxation time $\tau(\boldsymbol{k})$ it is necessary to know the probability of transition of a charge carrier $W(\boldsymbol{k}, \boldsymbol{k}')$ from one state with wave vector k to another with $\boldsymbol{k}'$ due to scattering by lattice defects.

Using the method of the nonstationary disturbance theory it is possible to show that the transition probability during unit time takes the form:

$$W(\boldsymbol{k}, \boldsymbol{k}') = \frac{2\pi}{\hbar} \left| \langle \boldsymbol{k}' \left| \hat{\mathscr{H}}' \right| \boldsymbol{k} \rangle \right|^2 \delta(\varepsilon_{k'} - \varepsilon_k), \tag{9.121}$$

where $\hat{H}'$ describes the interaction of the electron with impurities (defects) or with phonons, considered as a small disturbance.

This formula underlies the theory of charge carrier scattering. Therefore, note once again the conditions under which the formula (9.121) was derived.

First, the formula (9.121) was derived in the first order disturbance theory which imposes the condition $\hat{H}' \ll \hat{H}$ on the disturbance magnitude ($\hat{H}$ is the nondisturbed Hamiltonian). This corresponds to the Born approximation in the scattering theory.

Second, the disturbance action time Δt must be much less than the time between two successive disturbance inclusions, i.e. $\Delta t \ll \tau$, where τ is the relaxation time. Only with the fulfilment of this condition is the Boltzmann equation considered fair.

Now we shall go on to a discussion of particular scattering mechanisms using the formulae (9.52) and (9.121).

9.4.1 Charge Carrier Scattering by Ionized Impurity Atoms

At low temperatures for which lattice oscillations are not intensive and phonon gas is rarefied, the basic scattering mechanism in conductors is the scattering of charge carriers by impurity atoms.

In the present problem in the plane-wave approximation the following expression for the transition probability takes place:

$$W(\boldsymbol{k}, \boldsymbol{k}') = \frac{2\pi}{\hbar} \frac{N_i}{V} \left| \int U(\boldsymbol{r}) \exp\left[i(\boldsymbol{k} - \boldsymbol{k}', \boldsymbol{r}) \right] \mathrm{d}\boldsymbol{r} \right|^2 \delta(\varepsilon_{k'} - \varepsilon_k). \tag{9.122}$$

where $N_i = N/V$ is the impurity atom concentration, and $U(r)$ is the scattering potential.

The general formula (9.122) can be used to calculate the transition probability in the case of scattering by different potentials produced by impurity atoms or lattice defects.

Impurity atoms usually form discrete energy levels in the energy gap near the edges of allowed bands. They are therefore ionized easily, and at low temperatures the main mechanism is scattering by impurity ions. A positive (donor) or a negative (acceptor) impurity ion generates a long-range Coulomb field with potential $\varphi = \pm e/æ r$ at a point $\boldsymbol{r}$ of the semiconductor lattice, where æ is the dielectric constant of the crystal, and e is the ion charge.

If in (9.122) $U(r) = e\varphi = \pm e^2/\text{æ}r$ is used as the disturbance potential and the relaxation time is calculated in accordance with (9.52), it is easy to see that τ^{-1} diverges logarithmically, and consequently the notion of mobility loses its sense. In order to obtain a finite time of relaxation (and mobility) it is necessary to make an attempt to somehow confine the sphere of action of the Coulomb impurity ion potential. The first attempt to solve this problem was made by Conwell and Weisskopf. They confined the ion action to a sphere of radius equal to half of the mean distance between adjacent impurity ions. This seemingly artificial approach provided a qualitatively correct result for the temperature dependence of mobility at low temperatures.

A more rigorous consideration of the problem of charge carrier scattering by impurity ions consists in taking account of the Coulomb potential screening by charge carriers. In this case, the potential of a singly ionized impurity atom at a point distance r from the ion location can be presented as

$$\varphi(r) = \pm(e/\text{æ}r)\exp(-r/r_0), \tag{9.123}$$

where r_0 is referred to as the radius of ion field screening and is defined generally by the formula

$$r_0^{-2} = (4\pi e^2/\text{æ}) \int (-\partial f_0/\partial\varepsilon) g(\varepsilon)\mathrm{d}\varepsilon, \tag{9.124}$$

where $f_0(\varepsilon)$ is the equilibrium electron distribution function, $g(\varepsilon)$ is the state density for an arbitrary isotropic band defined by the formula

$$g(\varepsilon) = (1/\pi^2)k^2(\varepsilon)\mathrm{d}k(\varepsilon)/\mathrm{d}\varepsilon. \tag{9.125}$$

Using the scattering potential $U(r) = e\varphi(r)$ in (9.122) and accounting for the spherical symmetry of this potential we obtain for the transition probability the expression

$$W(\boldsymbol{k}, \boldsymbol{k}') = \frac{2\pi}{\hbar}\frac{N_i}{V}\left(\frac{4\pi e^2}{\text{æ}}\right)^2 \frac{\delta(\varepsilon_{k'} - \varepsilon_k)}{\left[(\boldsymbol{k} - \boldsymbol{k}')^2 + \boldsymbol{r}_0^{-2}\right]^2}. \tag{9.126}$$

As noted above and as seen from (9.126), scattering by impurity ions is elastic. Therefore, this process can be characterized by the relaxation time and it is possible to use the formula (9.52). Substituting (9.126) into (9.52) and going on to spherical coordinates with polar axis along $\boldsymbol{k}$ integrating with respect to $\boldsymbol{k}'$ using δ-functions, and taking into account that the root of the equation $\varepsilon(\boldsymbol{k}') = \varepsilon(\boldsymbol{k})$ is $k' = k$, we finally obtain for the relaxation time the expression

$$\tau(k) = \frac{\hbar\text{æ}^2}{2\pi e^4 N_i F_{\text{imp}}(k)} k^2 \left(\frac{\partial\varepsilon}{\partial k}\right), \tag{9.127}$$

where

$$F_{\text{imp}}(k) = \ln(1+\xi) - \frac{\xi}{(1+\xi)}, \quad \xi = (2kr_0)^2. \tag{9.128}$$

In the case of a simple parabolic band accurate to a slowly varying factor $F_{\rm imp}(k)$ we have from (9.127) $\tau \sim \varepsilon^{3/2}$, or

$$\tau(\varepsilon) = \frac{æ^2 (2m_n)^{1/2} (k_0 T)^{3/2}}{\pi e^4 N_i F_{\rm imp}(\varepsilon)} \left(\frac{\varepsilon}{k_0 T} \right)^{3/2}, \tag{9.129}$$

where $F_{\rm imp}(\varepsilon)$ is given by (9.128) with $\xi = 8m_n \varepsilon r_0^2/\hbar^2$. By virtue of (9.124) the screening radius in this case is

$$r_0^2 = \frac{æ k_0 T}{6\pi e^2 n} \frac{F_{3/2}}{F_{1/2}}. \tag{9.130}$$

It is easy to obtain

$$r_0 = \left(\frac{æ k_0 T}{4\pi\, \mathrm{e}^2 n} \right)^{1/2} \tag{9.131}$$

for nondegenerate electron gas, and

$$r_0 = \left[\frac{æ \hbar^2}{4 m_n e^2} \left(\frac{\pi}{3n} \right)^{1/3} \right]^{1/2} \tag{9.132}$$

for strongly degenerate electron gas.

It is possible to consider two limiting cases depending on the value of the product kr_0.

For $kr_0 \gg 1$ we have from (9.127) and (9.128)

$$\tau(k) = \frac{\hbar æ^2}{2\pi e^4 N_i} k^2 \left(\frac{\partial \varepsilon}{\partial k} \right) \frac{1}{\ln(4k^2 r_0^2)}. \tag{9.133}$$

It is seen that for $r_0 \to \infty$, $\tau(k) \to 0$. Thus, to obtain a finite mobility it is necessary to take into account the impurity ion of Coulomb potential screening.

For $kr_0 \ll 1$ we obtain from (9.127) and (9.128) the following expression

$$\tau(k) = \frac{\hbar æ^2}{16\pi e^4 N_i r_0^4} k^{-2} \left(\frac{\partial \varepsilon}{\partial k} \right). \tag{9.134}$$

In this limiting case of strong screening a charged impurity atom behaves as a point defect with a short-range potential.

For a nondegenerate electron gas with a parabolic dispersion relation we shall write down, according to (9.129), an explicit form for the applicability condition of the Boltzmann equation in the case of scattering by a weakly screened charged impurity atom (9.133). The condition $\tau(k_0 T) \gg \hbar/k_0 T$ imposes the following restrictions on the impurity ion concentration $N_{\rm i}$:

$$N_{\rm i} \ll \frac{æ^2 (2m_n)^{1/2} (k_0 T)^{5/2}}{\pi \hbar e^4 \ln(8 m_n k_0 T r_0^2/\hbar^2)}. \tag{9.135}$$

At such impurity ion concentrations, charge carriers' scattering occurs as a separate event by different centres and the mean free time between two successive collisions becomes much larger than the time of the carrier's presence in the scattering ion field.

9.4.2 Charge Carrier Scattering by Phonons in Conductors with Arbitrary Isotropic Band

At relatively high temperatures, charge carriers' scattering by phonons becomes the dominant relaxation mechanism. This occurs because of two reasons. First, according to (9.129), the relaxation time defined for scattering by impurity ions, $\tau_i \sim \varepsilon^{3/2}$, increases as the temperature (energy) τ_i rises, so that this mechanism is not too effective; in other words, charge carriers of high energy easily go past the impurity ions without changing their motion direction, i.e. without scattering. Second, as the temperature increases, the phonon number rises, and consequently the probability of the charge carrier-phonon "collisions" increases.

Now we shall lead the relaxation time for charge carriers' scattering by phonons of various types: acoustic, nonpolar and polar optical phonons. To do this, it is necessary to find explicitly the Hamiltonian of the disturbance $\hat{\mathbf{H}}'$ describing the conduction electron interaction with the various types of phonons.

Scattering by Acoustic Phonons, Deformation Potential Method

The conduction electron–lattice interaction energy $\hat{\mathscr{H}}'$ is in essence electron energy variation due to lattice oscillations. This variation, which is due to electron motion disturbance, can be found using different methods. The first method of the determination of $\hat{\mathscr{H}}'$ was based on Bloch's assumption of deformed ions.

Here, we shall use the *method of deformation potential* which was proposed by Bardeen and Schockley. The essence of this method is as follows.

When an elastic wave propagates in a crystal, an elementary cell becomes deformed and changes its volume (the lattice constant changes), which results in a variation of the positions of the bottom of the conduction band and the top of the valence band as the bandwidth is sensitive to the lattice constant value. It is the former variation that is the energy of the interaction of conduction electrons with lattice oscillations. Obviously, it must be related to the shift value. It is known that for acoustic oscillations at $\boldsymbol{q} \to 0$ all atoms in the elementary cell oscillate in phase. Then, in this case, it is possible to write the following expression for the shift $\boldsymbol{u}_{\text{acous}}(\boldsymbol{r})$:

$$\boldsymbol{u}_{\text{acous}}(\boldsymbol{r}) = \frac{1}{\sqrt{N}} \sum_q \sum_{j=1}^{3} \boldsymbol{e}_j(\boldsymbol{q}) \left\{ b_j(\boldsymbol{q}) \exp(\mathrm{i}\boldsymbol{q}\boldsymbol{r}) + b^*{}_j(\boldsymbol{q}) \exp(-\mathrm{i}\boldsymbol{q}\boldsymbol{r}) \right\}, \quad (9.136)$$

where $\boldsymbol{e}_j(\boldsymbol{q})$ is the polarization unit vector, $\boldsymbol{q}$ is the wave vector, and $b_j(\boldsymbol{q}) \sim \exp(-\mathrm{i}\omega_j(\boldsymbol{q})t)$ are the complex normal coordinates that depend harmonically on time.

Since for long-wave acoustic oscillations the elementary cell almost does not deform (only the centre of mass oscillates), the interaction energy cannot be proportional to the shift itself, but must be a linear function of the first shift derivatives $\boldsymbol{u}_{\text{acous}}(\boldsymbol{r})$ with respect to the coordinates:

$$\hat{\mathscr{H}}'_{\text{acous}} = E_1 \operatorname{div} \boldsymbol{u}_{\text{acous}}(\boldsymbol{r}), \tag{9.137}$$

which is referred to as the deformation potential and the coefficient of proportionality E_1 as the deformation potential constant which is to be obtained experimentally ($\sim 10\, eV$).

Considering formulae (9.136) and (9.137) we shall have for the probability of electron transition from the state $\boldsymbol{k}$ to the state $\boldsymbol{k}'$ owing to the interaction with acoustic phonons the following final expression:

$$W_{\text{acous}}(\boldsymbol{k}, \boldsymbol{k'}) = \sum_q w_1(q) \left(A^+_{kk'}(\boldsymbol{q}) + A^-_{kk'}(\boldsymbol{q}) \right), \tag{9.138}$$

where

$$A^{\pm}_{kk'}(\boldsymbol{q}) = \left(N_q + \frac{1}{2} \mp \frac{1}{2} \right) \delta(\varepsilon_{k'} - \varepsilon_k \mp \hbar\omega(\boldsymbol{q}))\delta_{k',k\pm q} \tag{9.139}$$

and

$$w_1(q) = \frac{\pi E_1^2 q^2}{NM\omega(q)}, \tag{9.140}$$

where $\omega(q)$ is the longitudinal acoustic phonon frequency.

The first term in the right-hand side of the expression (9.138) is the transition probability $\boldsymbol{k} \to \boldsymbol{k}'$ due to phonon absorption, while the second is due to phonon emission; the δ -functions and δ-symbols entering into (9.138) express the laws of energy and impulse conservation respectively:

$$\varepsilon_{k'} = \varepsilon_k \pm \hbar\omega(\boldsymbol{q}), \quad \boldsymbol{k}' = \boldsymbol{k} \pm \boldsymbol{q}. \tag{9.141}$$

It is seen that scattering by acoustic phonons is, generally speaking, not elastic. However, we shall show that at not very low temperatures this process is of an elastic nature. To do this, using $\omega(q) = v_0 q$ and the impulse conservation law $q = \left| \boldsymbol{k}' - \boldsymbol{k} \right|$ we estimate that

$$\frac{\hbar\omega(q)}{\bar{\varepsilon}_k} = \frac{v_0\sqrt{mk_0T}}{k_0T} = \sqrt{\frac{T_0}{T}} \ll 1, \tag{9.142}$$

where we have assumed that $\hbar \left| \boldsymbol{k}' - \boldsymbol{k} \right| \approx \sqrt{mk_0T}, \bar{\varepsilon}_k \approx k_0T$. In the expression (9.142) $T_0 = mv_0^2/k_0$, v_0 is the sound velocity in crystal, and m is the effective mass of the charge carriers. Assuming that $m \approx 10^{-28}$g, $v_0 = 3 \times 10^5$ cm/s we obtain $T_0 \approx 1$K. Thus, at $T \gg T_0 \approx 1$K the energy of absorbed or emitted phonons is negligible compared with the charge carrier energy, and the scattering is almost elastic. Therefore, in the calculations where applicable we can assume $\hbar\omega(q) \ll \varepsilon_k$ and neglect the phonon energy.

In the case of a degenerate electron gas, we have $\bar{\varepsilon}_k \approx \mu_F$ and the ratio

$$\frac{\hbar\omega(q)}{\bar{\varepsilon}_k} \approx \sqrt{\frac{T_0}{T_{\mathrm{deg}}}} \ll 1, \tag{9.143}$$

where $T_{\mathrm{deg}} = \mu_F/k_0$ is the electron gas degeneracy temperature. Thus, the nonelasticity of scattering can be neglected for both nondegenerate and degenerate electron gases. In addition, as seen from (9.142), $\hbar\omega(q) \ll k_0T$; therefore it is possible to expand the Planck function. Then we have

$$N_q + 1 \approx N_q \approx k_0T/\hbar\omega(q). \tag{9.144}$$

As electron scattering by acoustic phonons is elastic it is possible to neglect the phonon energy in the arguments of the δ-functions in (9.139). Then accounting for (9.140) and (9.144) after a trivial summation over $\boldsymbol{q}$ using the impulse conservation law we obtain from (9.138) a simple expression for the probability of the transition $\boldsymbol{k} \to \boldsymbol{k}'$ in the case of scattering by acoustic phonons:

$$W_{\mathrm{acous}}(\boldsymbol{k}, \boldsymbol{k}') = \frac{2\pi}{MN}\frac{E_1^2}{\hbar\upsilon_0^2}k_0T\delta(\varepsilon_{k'} - \varepsilon_k). \tag{9.145}$$

Then it is possible to substitute (9.145) into (9.52) and calculate the relaxation time:

$$\frac{1}{\tau(k)} = \frac{2\pi}{MN}\frac{E_1^2}{\hbar\upsilon_0^2}k_0T\sum_{k'}\left(1 - \frac{\boldsymbol{k}\boldsymbol{k}'}{kk'}\right)\delta(\varepsilon_{k'} - \varepsilon_k). \tag{9.146}$$

Hence, it follows that in the summation over $\boldsymbol{k}'$ the second term yields zero. Consequently, in the elastic approximation and when (9.145) is fulfilled the arrival in the $\boldsymbol{k}$-state from every possible $\boldsymbol{k}'$-state during the relaxation is of no account.

Later we shall consider an arbitrary isotropic, i.e. spherically symmetric band, and turn the summation over $\boldsymbol{k}'$ in (9.146) into an integral. If in the integral obtained the volume elements in $\boldsymbol{k}'$-space are written in spherical coordinates, the integral over the angles will yield 4π while the integral over the magnitude k' can be taken using δ-functions. As a result, we shall obtain for the relaxation time the expression

$$\tau(k) = \frac{\pi\hbar\rho}{E_1^2}\frac{\upsilon_0^2}{k_0T}\frac{1}{k^2}\left(\frac{\partial\varepsilon}{\partial k}\right), \tag{9.147}$$

where $\rho = MN/V$ is the crystal density.

In the case of a standard band, we have $\varepsilon_k = \hbar^2k^2/2m_n$ and from (9.147) follows the well-known formula

$$\tau(k) = \frac{2\pi\hbar^4\rho\upsilon_0^2}{E_1^2(2m_nk_0T)^{3/2}}\left(\frac{\varepsilon}{k_0T}\right)^{-1/2}, \tag{9.148}$$

from which it is seen that the relaxation time of the charge carriers with respect to acoustic phonons is $\tau \sim T^{-1}\varepsilon^{-1/2}$.

Scattering by acoustic lattice oscillations takes place in all crystals irrespective of the elementary cell complexity. However, if the crystal has a simple lattice (one atom in each elementary cell), scattering by acoustic oscillations is the only mechanism of charge carrier–lattice interaction. In those crystals where the lattice is complex, i.e. two or more atoms or ions are in each elementary cell, alongside with scattering by acoustic oscillations there are other mechanisms of charge carrier–lattice interaction which are sometimes more important than acoustic scattering. They include scattering by nonpolar and polar optical lattice oscillations. We shall dwell upon these scattering mechanisms separately.

Scattering by Nonpolar Optical Phonons, Deformation Potential Method

It is known that for long-wave optical oscillations $q \to 0$ atoms in an elementary cell oscillate almost out of phase so that the center of gravity remains immovable. Therefore, in this case the crystal deformation and accordingly the band edge variation, i.e. the interaction energy, will be proportional to the shift of any atom in an elementary cell $\boldsymbol{u}_k(\boldsymbol{r})$:

$$\boldsymbol{u}_k(\boldsymbol{r}) = \frac{1}{\sqrt{N}} \sum_{qj} \boldsymbol{e}_{kj}(\boldsymbol{q}) \left\{ b_j(\boldsymbol{q}) \exp(\mathrm{i}\,\boldsymbol{qr}) + b^*{}_j(\boldsymbol{q}) \exp(-\mathrm{i}\,\boldsymbol{qr}) \right\}. \quad (9.149)$$

Consequently, the interaction operator of the charge carriers with optical lattice oscillations can be presented as

$$\hat{\mathscr{H}}'_{\mathrm{opt}} = \sum_{j=4}^{3s} \boldsymbol{A}_j \boldsymbol{u}_j, \quad (9.150)$$

where $\boldsymbol{u}_j$ is the shift corresponding to the jth branch, $\boldsymbol{A}_j$ is a constant vector that is governed by the symmetry of arrangement of the charge carrier energy band minima.

For the sake of simplicity, let us consider the scattering of charge carriers as occurring at a specified minimum on one optical branch, and present the vector $\boldsymbol{A}$ in the form

$$\boldsymbol{A} = E_0 \boldsymbol{b}_g, \quad (9.151)$$

where $\boldsymbol{b}_g = (\pi/a)\boldsymbol{g}$ is the reciprocal lattice vector, a is the lattice constant, $\boldsymbol{g}$ is a unit vector directed from the Brillouin zone center towards the minimum, and E_o is the optical deformation potential constant with dimensions of energy.

Taking account of the last two relations as well as (9.149), we shall find for the Hamiltonian of the deformation interaction of the electrons with long-wave optical lattice oscillations the following expression:

$$\hat{\mathscr{H}}'_{\text{opt}} = \frac{\pi}{a} \frac{E_0}{\sqrt{N}} \sum_q (\boldsymbol{ge}) \{b(\boldsymbol{q}) \exp(\mathrm{i}\, \boldsymbol{qr}) + b^*(\boldsymbol{q}) \exp(-\mathrm{i}\, \boldsymbol{qr})\}. \tag{9.152}$$

As a result, for the transition probability $\boldsymbol{k} \to \boldsymbol{k}'$ due to scattering by non-polar optical phonons we obtain, similarly to (9.138), the expression

$$W_{opt}(\boldsymbol{k}, \boldsymbol{k}') = \sum_q w_0(q) \left(A^+_{kk'}(\boldsymbol{q}) + A^-_{kk'}(\boldsymbol{q})\right), \tag{9.153}$$

where $A^+_{kk'}(\boldsymbol{q})$ is given by the formula (9.139) and

$$w_0(q) = \frac{\pi E_0^2}{NM\omega(q)} \left(\frac{\pi}{a}\right)^2 (\boldsymbol{ge})^2. \tag{9.154}$$

If optical phonon dispersion is not taken into account, i.e. we assume that at $q \to 0$

$$\omega(q) = \omega_0 \tag{9.155}$$

the expression under the sum sign in (9.153) will not depend on the phonon wave vector $\boldsymbol{q}$ and the summation over $\boldsymbol{q}$ becomes trivial. We then have

$$W_{\text{opt}}(\boldsymbol{k}, \boldsymbol{k}') = w_0 \left[N_0 \delta(\varepsilon_{k'} - \varepsilon_k - \hbar\omega(\boldsymbol{q})) + (N_0 + 1)\delta(\varepsilon_{k'} - \varepsilon_k + \hbar\omega(\boldsymbol{q}))\right], \tag{9.156}$$

where

$$w_0(q) = \frac{\pi E_0^2}{NM\omega_0} \left(\frac{\pi}{a}\right)^2 \tag{9.157}$$

and

$$N_0 = \left[\exp\left(\hbar\omega_0/k_0 T\right) - 1\right]^{-1} \tag{9.158}$$

is the number of optical phonons with the limiting frequency ω_0 at temperature T. During the transition from (9.154) to (9.157) we have assumed that $(\boldsymbol{ge}) = 1$.

The optical phonon energy $\hbar\omega_0$ is not small and scattering in this case is essentially nonelastic. Therefore, generally speaking, scattering by optical phonons cannot be described by the relaxation time. However, if the transition probability $W(\boldsymbol{k}, \boldsymbol{k}')$ is an even function of $\boldsymbol{k}'$, i.e. $W(\boldsymbol{k}, \boldsymbol{k}') = W(\boldsymbol{k}, -\boldsymbol{k}')$, the relaxation time can be introduced even in the case of nonelastic scattering. Indeed, if in (9.36) we take into account $W(\boldsymbol{k}, \boldsymbol{k}') = W(\boldsymbol{k}, -\boldsymbol{k}')$ as well as a nonequilibrium addition to the distribution function $f_1 \sim \boldsymbol{k}'$, then in the summation over $\boldsymbol{k}'$ the second term in (9.36) vanishes, and the relaxation time assumes the form

$$\frac{1}{\tau(k)} = \sum_{k'} W(\boldsymbol{k}, \boldsymbol{k}') \frac{1 - f_0(\varepsilon_{k'})}{1 - f_0(\varepsilon_k)}. \tag{9.159}$$

It is seen that in this case relaxation is governed by the process where electrons leave the $\boldsymbol{k}$-states while the process of arriving in this state is of no account.

The above condition $W(\boldsymbol{k}, \boldsymbol{k}') = W(\boldsymbol{k}, -\boldsymbol{k}')$ is satisfied by the scattering probability of the charge carriers with the isotropic dispersion relation (9.156) by nonpolar optical phonons with limiting frequency ω_0. Substituting (9.156) into (9.159) and going over to integrate over $\boldsymbol{k}'$ shall obtain for the relaxation time the following

$$\frac{1}{\tau(k)} = \frac{Vw_0}{6\pi^2}\frac{N_0}{f_0(\varepsilon)}\left[f_0(\varepsilon+\hbar\omega_0)\exp\left(\frac{\hbar\omega_0}{k_0T}\right)\frac{\partial}{\partial\varepsilon}k^3(\varepsilon+\hbar\omega_0)\right.$$
$$\left.+\,\Theta(\varepsilon-\hbar\omega_0)f_0(\varepsilon-\hbar\omega_0)\frac{\partial}{\partial\varepsilon}k^3(\varepsilon-\hbar\omega_0)\right], \qquad (9.160)$$

where $\Theta(x)$ is the step function: $\Theta(x) = 0$ at $x < 0$, $\Theta(x) = 1$ at $x > 0$, and $k(\varepsilon)$ is an arbitrary function of energy which establishes the dispersion relation of charge carriers.

Note that (9.160) is the general expression for the relaxation time in the case of scattering by nonpolar phonons from which all particular cases follow. At high temperatures when $k_0T \gg \hbar\omega_0$ and it is possible to neglect the nonelasticity, τ does not depend on the degree of degeneracy of the carriers, and the following simple expression for the relaxation time of an arbitrary isotropic band follows from (9.160), taking into account (9.158):

$$\tau(k) = \frac{1}{\pi\hbar}\left(\frac{\hbar\omega_0}{E_0}\right)^3\frac{\rho a^3}{k_0T}\frac{1}{k^2}\left(\frac{\partial\varepsilon}{\partial k}\right). \qquad (9.161)$$

A comparison of this formula with (9.147) shows that the function $\tau(\varepsilon, T)$ at high temperatures in the cases of scattering by acoustic phonons and by nonpolar optical phonons is the same: $\tau \sim T^{-1}k^{-2}(\varepsilon)\frac{\partial\varepsilon}{\partial k}$.

For a parabolic band we have $k(\varepsilon) = (1/\hbar)(2m_n\varepsilon)^{1/2}$, and it follows from (9.161) that

$$\tau(\varepsilon) = \frac{2}{\pi}\left(\frac{\hbar\omega_0}{E_0}\right)^2\frac{\hbar^2\rho a^2}{(2m_nk_0T)^{3/2}}\left(\frac{\varepsilon}{k_0T}\right)^{-1/2}. \qquad (9.162)$$

In the case of a nondegenerate electron gas, (9.160) assumes the form (9.162) at high temperatures $k_0T \gg \hbar\omega_0, \varepsilon > \hbar\omega_0$, while at low temperatures $k_0T \ll \hbar\omega_0, \varepsilon < \hbar\omega_0$ the phonon emission process is not possible and

$$\tau = \frac{\sqrt{2}}{\pi}\frac{(\hbar\omega_0)^{1/2}}{E_0^2}\frac{\hbar^2\rho a^2}{m_n^{3/2}}\exp\left(\frac{\hbar\omega_0}{k_0T}\right), \qquad (9.163)$$

which depends only on the absorption of phonons with energy $\varepsilon < \hbar\omega_0$. It is seen that at low temperatures the relaxation time for scattering by nonpolar optical phonons does not depend on energy but rather on temperature exponentially.

Scattering by Polar Optical Phonons

The mechanism of scattering by nonpolar optical phonons discussed in the preceding subsection is the only mechanism of interaction of the charge carriers with optical oscillations in crystals with a covalent bond, i.e. homopolar crystals. Among these are semiconductors of the n-Ge type. The elementary cell of these crystals consists of neutral atoms, and therefore in case of optical oscillations only deformation potential occurs. However, in heteropolar crystals, i.e. ion crystals (NaCl, KCl, etc.) or crystals with an ionic fraction of the chemical bond (compounds of the $A^{III}B^{V}$ type such as InSb, GaAs etc.), for optical lattice oscillations electric polarization occurs alongside with deformation potential, which causes additional interaction of these oscillations with charge carriers. This mechanism, which is referred to as *scattering by polar optical phonons*, is more significant in many crystals than scattering by acoustic and optical deformation potentials. The present point is devoted to this scattering mechanism.

Let us consider a cubic crystal with an ionic fraction of the chemical bond with two ions in an elementary cell (semiconductors of the $A^{III}B^{V}$ and $A^{II}B^{VI}$ types). Long-wave optical oscillations are of interest to us when ions of different signs, shifting towards opposite sides, cause lattice polarization. Such polarization propagates in the crystal and forms a polarization wave. The polarization vector which occurs in this case at the point $\boldsymbol{r}$ has the form

$$\boldsymbol{P}(\boldsymbol{r}) = \left(\frac{NM_0\omega^2(q)}{4\pi V æ^*}\right)^{1/2}(\boldsymbol{u}_1 - \boldsymbol{u}_2), \tag{9.164}$$

where $M_0 = M_1M_2/(M_1+M_2)$ is the reduced elementary cell mass, M_1, M_2 are the ion masses, $\omega(q)$ is the frequency of the longitudinal optical phonons,

$$1/æ^* = 1/æ_\infty - 1/æ_0, \tag{9.165}$$

where $æ_\infty$ and $æ_0$ are, respectively, the high-frequency and static dielectric constants of the crystal, and N is the number of elementary cells in the crystal of volume V.

Accounting for (9.149) the expression (9.164) assumes the following form:

$$\boldsymbol{P}(\boldsymbol{r}) = \left(\frac{M_0\omega^2}{4\pi V æ^*}\right)^{1/2}\sum_q \sum_{j=4,5,6} \boldsymbol{e}_j \left\{b_j(\boldsymbol{q})\exp(\mathrm{i}\,\boldsymbol{qr}) + b_j^*(\boldsymbol{q})\exp(-\mathrm{i}\,\boldsymbol{qr})\right\}, \tag{9.166}$$

where $M = M_1 + M_2$ is the elementary cell mass.

The polarization vector $\boldsymbol{P}(\boldsymbol{r})$ is equivalent to the presence of a bound charge with density $\rho_g(\boldsymbol{r}) = -\mathrm{div}\,\boldsymbol{P}(\boldsymbol{r})$, which corresponds to the scalar potential φ satisfying the Poisson equation

$$\nabla\varphi = -4\pi\rho_g = 4\pi\,\mathrm{div}\,\boldsymbol{P}(\boldsymbol{r}). \tag{9.167}$$

Substituting (9.166) into this equation it is possible to find φ and thereby find the disturbance energy $\hat{\mathscr{H}}'_{\text{pol}} = \pm e\varphi$ related to optical polarization lattice oscillations. As a result, we have

$$\hat{\mathscr{H}}'_{\text{pol}} = \mp \mathrm{i}\, e \left(\frac{4\pi M\omega^2(q)}{V æ^*}\right)^{1/2} \sum_{q} \sum_{j=4,5,6} \frac{1}{q^2} (\boldsymbol{e}_j \boldsymbol{q}) \left\{ b_j(\boldsymbol{q}) \exp(\mathrm{i}\, \boldsymbol{q}\boldsymbol{r}) \right.$$
$$\left. + b^*{}_j(\boldsymbol{q}) \exp(-\mathrm{i}\, \boldsymbol{q}\boldsymbol{r}) \right\} . \tag{9.168}$$

It is seen from this expression that the charge carriers interact only with the longitudinal optical polarization oscillations, for which $\boldsymbol{e}_j || \boldsymbol{q}$ and $(\boldsymbol{e}_j \boldsymbol{q}) = q$. Therefore, only one term remains in the sum over j, and in further discussions we shall omit the index showing the branch number.

Following the same procedure as in the case of scattering by acoustic phonons, it is easy to obtain from (9.168) the transition probability $\boldsymbol{k} \to \boldsymbol{k}'$ caused by interaction of the charge carriers with polar optical phonons:

$$W_{\text{pol}}(\boldsymbol{k}, \boldsymbol{k}') = \sum_{q} w(q) \left(A^{+}_{kk'}(\boldsymbol{q}) + A^{-}_{kk'}(\boldsymbol{q}) \right), \tag{9.169}$$

where $A^{\pm}_{kk'}(\boldsymbol{q})$ is given by the formula (9.139) and

$$w(q) = \frac{4\pi^2 e^2 \omega(q)}{V æ^* q^2}, \tag{9.170}$$

where $\omega(q)$ is the longitudinal optical oscillation frequency.

If the optical phonon dispersion $\omega(q) = \omega_0$ is not taken into account, the summation over $\boldsymbol{q}$ in (9.168) using the impulse conservation law yields the closed expression for the transition probability in the case of scattering by polar optical phonons

$$W_{\text{pol}}(\boldsymbol{k}, \boldsymbol{k}') = \frac{4\pi^2 e^2}{V æ^*} \frac{\omega}{(\boldsymbol{k}' - \boldsymbol{k})^2} \left[N_0 \delta(\varepsilon_{\boldsymbol{k}'} - \varepsilon_{\boldsymbol{k}} - \hbar\omega_0) \right.$$
$$\left. + (N_0 + 1) \delta(\varepsilon_{\boldsymbol{k}'} - \varepsilon_{\boldsymbol{k}} + \hbar\omega_0) \right], \tag{9.171}$$

where N_o is the number of optical phonons with limiting frequency ω_0.

It is seen from (9.171) that, in contrast to nonpolar phonons, the transition probability for scattering by polar optical phonons depends significantly on the directions of $\boldsymbol{k}$ and $\boldsymbol{k}'$. Therefore, the formula (9.159) for polar optical phonons is not valid, and in the general case it is impossible to introduce a relaxation time for scattering by polar optical oscillations.

The process of scattering by polar optical phonons can be described in terms of a relaxation time only in two limiting cases: high temperatures ($k_0 T \gg \hbar\omega_0$) and low temperatures ($k_0 T \ll \hbar\omega_0$). Let us consider these cases separately.

In the region of high temperatures ($k_0 T \gg \hbar\omega_0$) the nonelasticity can be neglected and we have $\varepsilon_{k'} \approx \varepsilon_k$ and $N_0 + 1 \approx N_0 \approx k_0 T / \hbar\omega$. Then (9.171) assumes the form

$$W_{\mathrm{pol}}(\boldsymbol{k},\boldsymbol{k}') = \frac{8\pi^2 e^2}{V æ^*}\frac{k_0 T}{\hbar}\frac{1}{(\boldsymbol{k}'-\boldsymbol{k})^2}\delta(\varepsilon_{k'}-\varepsilon_k). \qquad (9.172)$$

Substituting (9.172) into (9.52), changing the summation to integrate over $\mathrm{d}\boldsymbol{k}'$ in spherical coordinates and using a δ-function in the integral over k' it is easy to obtain the expression $\tau(k)$ below which is valid for an arbitrary isotropic band

$$\tau(k) = \frac{æ^*\hbar}{2e^2 k_0 T}\left(\frac{\partial\varepsilon}{\partial k}\right). \qquad (9.173)$$

Note that, in contrast to nonpolar optics, $\tau(k)$ does not depend on the phonon frequency (the formulae (9.173) and (9.161)) for scattering by polar optical phonons at high temperatures. As for a standard band $\varepsilon = \hbar^2 k^2/2m$ (9.173) assumes the known form

$$\tau(\varepsilon) = \frac{1}{2\alpha}\left(\frac{\hbar}{\omega_0 k_0 T}\right)^{1/2}\left(\frac{\varepsilon}{k_0 T}\right)^{1/2}, \qquad (9.174)$$

where

$$\alpha = \frac{\mathrm{e}^2}{æ^*\hbar}\left(\frac{m_n}{2\hbar\omega_0}\right)^{1/2} \qquad (9.175)$$

is the nondimensional parameter introduced by Frohlich which characterizes the interaction of a conduction electron with longitudinal long-wave optical phonons in ionic crystals.

The relaxation time calculation presented here is based on the disturbance theory and is valid for $\alpha < 1$.

Note that $\tau_{\mathrm{pol}} \sim T^{-1}\varepsilon^{1/2}$, while $\tau_{\mathrm{nonpol}} \sim T^{-1}\varepsilon^{-1/2}$ (9.162). Consequently, as the energy ε increases, scattering by nonpolar phonons can become dominant ($\tau_{\mathrm{nonpol}} < \tau_{\mathrm{pol}}$).

In the region of low temperatures ($k_0 T \ll \hbar\omega_0$) the scattering process by optical phonons is essentially nonelastic, and it is impossible to introduce relaxation time as we have done for elastic interaction (see the formula (9.52)). However, other circumstances make it possible here to describe the relaxation process in terms of a relaxation time. We know that in this case the interaction mechanism consists of the transitions $\varepsilon(\boldsymbol{k}) \to \varepsilon(\boldsymbol{k}') = \varepsilon(\boldsymbol{k}) \pm \hbar\omega_0$ (leaving the state $\boldsymbol{k}$) and $\varepsilon(\boldsymbol{k}') \pm \hbar\omega_0 \to \varepsilon(\boldsymbol{k})$ (arrival in the $\boldsymbol{k}$-state) due to phonon absorption and emission. It is obvious that at low temperatures ($k_0 T \ll \hbar\omega_0$) only one of these four processes can take place: $\varepsilon(\boldsymbol{k}) \to \varepsilon(\boldsymbol{k}') = \varepsilon(\boldsymbol{k}) + \hbar\omega_0$, i.e. only leaving the $\boldsymbol{k}$-state for other $\boldsymbol{k}'$- states is possible with phonon absorption. Having absorbed the phonon, the same electrons instantaneously (spontaneously) emit a phonon with the same energy $\hbar\omega_0$ since the ratio of the emission probability to the absorption probability is $(N_0+1)/N_0 \approx \exp(\hbar\omega_0/k_0 T) \gg 1$. Owing to this process there occurs a relaxation with respect to impulse, and the electron energy does not change. Since these processes of phonon absorption and emission occur instantaneously the distribution function $f(\boldsymbol{k}')$ has no time to change. Therefore, it is possible to neglect in (9.36) the term proportional to $f_1(\boldsymbol{k}')$. Then the relaxation time will assume the form of (9.159) but with the difference that (9.159) will now include the portion of the probability (9.171) that is associated with phonon absorption, i.e. in this case we have

$$\frac{1}{\tau(k)} = \frac{4\pi^2 e^2 \omega_0 N_0}{V æ^*} \sum_{k'} \frac{1 - f_0(\varepsilon_{k'})}{1 - f_0(\varepsilon_k)} \frac{1}{(k' - k)^2} \delta(\varepsilon_{k'} - \varepsilon_k - \hbar\omega_0). \tag{9.176}$$

Converting the summation into integration over $\mathrm{d}\boldsymbol{k}'$ and using spherical coordinates we first calculate the integration over angles. We then go from integration over $\mathrm{d}k'$ to that over $\mathrm{d}\varepsilon'$ and apply the energy conservation law. As a result, for an arbitrary isotropic band at an arbitrary degree of degeneracy of charge carriers we obtain the following expression:

$$\frac{1}{\tau(\varepsilon)} = \frac{e^2\omega_0}{æ^*} \frac{f_0(\varepsilon + \hbar\omega_0)}{f_0(\varepsilon)} \frac{k(\varepsilon + \hbar\omega_0)}{k(\varepsilon)} \frac{\partial k(\varepsilon + \hbar\omega_0)}{\partial \varepsilon} \ln\left[\frac{k(\varepsilon + \hbar\omega_0) + k(\varepsilon)}{k(\varepsilon + \hbar\omega_0) - k(\varepsilon)}\right]. \tag{9.177}$$

In this formula it is assumed that $\varepsilon \ll \hbar\omega_0$ and therefore $N_0 \approx \exp(-\hbar\omega_0/k_0T)$.

For a parabolic band $k(\varepsilon) = (1/\hbar)(2m_n\varepsilon)^{1/2}$. If the logarithmic term is expanded in terms of the small quantity $\varepsilon/\hbar\omega_0$ then the simple expression follows from (9.177)

$$\frac{1}{\tau(\varepsilon)} = \frac{2\alpha\omega_0 f_0(\varepsilon + \hbar\omega_0)}{f_0(\varepsilon)}, \tag{9.178}$$

where α is given by the formula (9.175).

In the absence of degeneracy we have $f_0(\varepsilon) = \exp((\mu - \varepsilon)/k_0T)$ and from (9.178) we obtain the well-known result

$$\tau = (2\alpha\omega_0)^{-1} \exp(\hbar\omega_0/k_0T). \tag{9.179}$$

It is seen that for scattering by polar optical phonons τ does not depend only on ε in the case of a nondegenerate electron gas with a simple parabolic band. In the general case, as it follows from (9.177) and (9.178), it does depend on energy, but not as a power law. On the other hand, τ depends on temperature exponentially.

9.4.3 Generalized Formula for Relaxation Time

Now we shall sum up the obtained results from the preceding sections.

In the simplest case where the band has a parabolic shape and the charge carriers' motion is described by a plane-wave, the relaxation time for all elastic scattering mechanisms can be reduced to the formula

$$\tau(\varepsilon) = \tau_{0r}(T)\,(\varepsilon/k_0T)^{r-1/2}. \tag{9.180}$$

Table 9.1. Values of parameters entering into the expressions (9.181) and (9.183) for relaxation time for different mechanisms of elastic scattering

Scattering mechanism	Scattering parameter r	$\tau_{0r}(T)$	A_{0r}
Acoustic phonons (deformation potential)	0	$\dfrac{2\pi\hbar^4\rho v_0^2}{E_1^2(2m_n k_0 T)^{3/2}}$	$\dfrac{\pi E_1^2 k_0 T}{\hbar\rho v_0^2}$
Nonpolar optical phonons at high temperatures $(k_0T \gg \hbar\omega_0)$[a]	0	$\dfrac{2}{\pi}\left(\dfrac{\hbar\omega_0}{E_0}\right)^2 \dfrac{\hbar^2\rho\alpha^2}{(2m_n k_0 T)^{3/2}}$	$\pi^3\hbar\left(\dfrac{E_0}{\hbar\omega_0}\right)^2\dfrac{k_0T}{\rho\alpha^2}$
Polar optical phonons at high temperatures $(k_0T \gg \hbar\omega_0)$[a]	1	$\dfrac{1}{2\alpha}\left(\dfrac{\hbar}{\omega_0 k_0 T}\right)^{1/2}$	$\dfrac{2\pi^2 e^2 k_0 T}{æ^*\hbar}$
Impurity ions	2	$\dfrac{æ^2(2m_n)^{1/2}(k_0T)^{3/2}}{\pi e^4 N_i F_{\mathrm{imp}}(\varepsilon)}$	$\dfrac{2\pi^3 N_i F_{\mathrm{imp}}(k)}{\hbar æ^2}$

[a] At low temperatures $(k_0T \ll \hbar\omega_0)$ in the case of scattering by polar or nonpolar optical phonons τ does not depend on energy ($r = 1/2$), and for a parabolic band it is given by the formulae (9.162) and (9.179) respectively.

The values of the scattering parameter r and the mean relaxation time $\tau_{0r}(T)$ which are obtained as a result of a comparison of (9.180) with the formulae (9.129), (9.148), (9.162) and (9.174) for different elastic scattering mechanisms are given in Table 9.1.

It is seen that for almost all scattering mechanisms $\tau_{0r}(T)$ does not depend on energy, and $\tau(\varepsilon)$ is a power function of energy. An exception is scattering by impurity ions. In this case, the function $F_{\mathrm{imp}}(\varepsilon)$ occurs in $\tau_{0r}(T)$ due to screening, the explicit form of which is given by (9.128). However, this function has a rather weak dependence on energy; therefore in the integration it is usually considered as constant and factored outside the integral sign, assuming that $\varepsilon = \bar{\varepsilon}$, $\bar{\varepsilon}$ is the mean charge carriers' energy.

If the band is parabolic, it is convenient to present the $\tau^{-1}(\varepsilon)$ as follows:

$$\tau^{-1} = a_{0r}(T)\varepsilon^r g(\varepsilon), \tag{9.181}$$

where $g(\varepsilon) = \frac{(2m_n)^{3/2}\varepsilon^{1/2}}{2\pi^2\hbar^3}$ is the state density for the parabolic band and

$$a_{0r}(T) = \frac{(k_0T)^{r-1/2}}{\tau_{0r}(T)}\frac{2\pi^2\hbar^3}{(2m_n)^{3/2}}. \tag{9.182}$$

Assume that the band is nonparabolic but spherically symmetric, i.e. the energy is an arbitrary function of the wave number $\varepsilon = \varepsilon(k)$ or is the same k

in an arbitrary energy function $k = k(\varepsilon)$. However, we shall assume that charge carriers are described by the plane wave. In this case, according to the formulae (9.128), (9.147), (9.161), (9.173) and (9.127), $\tau^{-1}(\varepsilon)$ for all mechanisms has the form

$$\tau^{-1} = A_{0r} k^{-2r}(\varepsilon) g(k(\varepsilon)), \tag{9.183}$$

where $g(k(\varepsilon)) = (1/3\pi^2)\mathrm{d}k^3(\varepsilon)/\mathrm{d}\varepsilon$ is the state density of an arbitrary spherically symmetric band, and the values of A_{0r} are given in Table 9.1.

From (9.181) and (9.183) it follows that τ^{-1} has a common structure

$$\tau^{-1} \sim W(\varepsilon) g(\varepsilon), \tag{9.184}$$

where $W(\varepsilon)$ is the scattering probability. It is seen that for different scattering mechanisms, for a parabolic band we have

$$W(\varepsilon) \sim \varepsilon^{-r}, \tag{9.185}$$

while for a nonparabolic band

$$W(\varepsilon) \sim k^{-2r}(\varepsilon). \tag{9.186}$$

Note, however, that for deriving (9.183) and consequently (9.186) the nonparabolicity was taken into account inconsistently. It was assumed that the dispersion relation of charge carriers was nonparabolic, but when the matrix element, i.e. the transition probability $\boldsymbol{k} \to \boldsymbol{k}'$, was calculated the plane wave was used as the wave functions of the initial and final states.

9.5 Boltzmann Equation Solution for Anisotropic Band in Relaxation Time Tensor Approximation

9.5.1 Current Density

The present section/chapter briefly discusses the Boltzmann equation solution for conductors with anisotropic band in relaxation time tensor approximation by which the current density is found. The anisotropic band nonparabolicity is taken into account and the degree of charge carriers' degeneracy is taken to be arbitrary. The results obtained are applicable to n–Ge, n–Si and lead chalcogenides: PbTe, PbSe and PbS.

For accuracy, we shall consider an electron semiconductor with many minima in the energy spectrum. The isoenergetic surface near each minimum is an ellipsoid of the general form described by the following equation:

$$B(\varepsilon) = \frac{\hbar^2}{2}\left(\frac{k_x^2}{m_{01}} + \frac{k_y^2}{m_{02}} + \frac{k_z^2}{m_{03}}\right), \tag{9.187}$$

where m_{0i} are the effective mass components at the conduction band bottom, and $B(\varepsilon)$ is any smooth function of energy ε. In particular, for n–Ge and n–Si,

$B(\varepsilon) = \varepsilon$; for lead chalcogenides (PbTe, PbSe,PbS), $B(\varepsilon) = \varepsilon(1 + \varepsilon/\varepsilon_g)$; and for all the above-mentioned semiconductors, $m_{01} = m_{02} = m_{0\perp}$, $m_{03} = m_{0||}$ is an ellipsoid of revolution.

The effective mass tensor relative to the major ellipsoid axes will be, naturally, diagonal:

$$m_{ik}^{-1} = m_i^{-1}\delta_{ik}, \tag{9.188}$$

where m_i are the diagonal components of the effective mass relating the velocity and impulse components:

$$v_i = \hbar^{-1}(\partial\varepsilon/\partial k_i) = \hbar k_i/m. \tag{9.189}$$

In the general case, $m_i^{-1} = (1/\hbar^2 k_i)(\partial\varepsilon/\partial k_i)$ depends on energy and this dependence is given by the function $B(\varepsilon)$:

$$m_i(\varepsilon) = m_{0i}(\partial B(\varepsilon)/\partial\varepsilon). \tag{9.190}$$

9.5.2 The Boltzmann Equation Solution

If we confine our discussion to elastic scattering ($\varepsilon = \varepsilon'$), then in accordance with the principle of detailed equilibrium $W(\boldsymbol{k}, \boldsymbol{k}') = W(\boldsymbol{k}', \boldsymbol{k})$, and the Boltzmann equation can be rewritten as

$$v\nabla_r f - \frac{e}{\hbar}\left(E_0 + \frac{1}{c}[\boldsymbol{v}\mathscr{H}]\right)\nabla_k f = \sum_{\boldsymbol{k}'} W(\boldsymbol{k}, \boldsymbol{k}')(f(\boldsymbol{k}') - f(\boldsymbol{k})), \tag{9.191}$$

whose solution can be presented as

$$f(\boldsymbol{k}) = f_0(\boldsymbol{k}) - (\partial f_0/\partial\varepsilon)(\boldsymbol{v}\boldsymbol{P}(\varepsilon)), \tag{9.192}$$

where $\boldsymbol{P}(\varepsilon)$ is an unknown vector to be defined and has the meaning of a generalized force impulse.

Let us substitute (9.192) into the right-hand side of (9.191) and introduce the inverse relaxation time tensor $\hat{\tau}^{-1}$ as follows:

$$\sum_{k'} W(\boldsymbol{k}, \boldsymbol{k}')(f(\boldsymbol{k}') - f(\boldsymbol{k})) \\ = -\left(\frac{\partial f_0}{\partial\varepsilon}\right)\boldsymbol{P}(\varepsilon)\sum_{k'} W(\boldsymbol{k}, \boldsymbol{k}')(\boldsymbol{v}' - \boldsymbol{v}) \Rightarrow \left(\frac{\partial f_0}{\partial\varepsilon}\right)\boldsymbol{P}(\hat{\tau}^{-1}\boldsymbol{v}). \tag{9.193}$$

Then, in a linear approximation the equation (9.191) assumes the form

$$\boldsymbol{v}\nabla_r f_0 - \frac{e}{\hbar}\boldsymbol{E}_0\nabla_k f_0 + \frac{e}{c\hbar}[\boldsymbol{v}\mathscr{H}]\nabla_k(\boldsymbol{v}\boldsymbol{P}(\varepsilon)) = \left(\frac{\partial f_0}{\partial\varepsilon}\right)\boldsymbol{P}(\hat{\tau}^{-1}\boldsymbol{v}). \tag{9.194}$$

In a coordinate system with axes along the major ellipsoid axes (9.187), the relaxation time tensor $\hat{\tau}^{-1}$, in accordance with (9.188), is diagonal:

$$\tau_{ik} = \tau_i \delta_{ik}. \tag{9.195}$$

To transform the factor $\nabla_k(\upsilon \boldsymbol{P})$ in (9.194) we introduce the unit vector $\boldsymbol{n}$ in the $\boldsymbol{k}$-space. Then

$$\nabla_k(\boldsymbol{\upsilon P}) = \sum_{\alpha\beta} n_\alpha \frac{\partial}{\partial k_\alpha}(\upsilon_\beta P_\beta) = \sum_{\alpha\beta} n_\alpha \left(P_\beta \frac{\partial \upsilon_\beta}{\partial k_\alpha} + \upsilon_\beta \frac{\partial P_\beta}{\partial k_\alpha} \right), \tag{9.196}$$

where n_α are the unit vector components in the $\boldsymbol{k}$-space. If (9.189) and $\frac{\partial \upsilon_\beta}{\partial k_\alpha} = \frac{\hbar}{m_\beta}\delta_{\alpha\beta}$ are taken into account, (9.196) can be presented as follows:

$$\nabla_k(\boldsymbol{\upsilon P}) = \hbar(\hat{m}^{-1}\boldsymbol{P}) + \hbar\boldsymbol{\upsilon}\sum_\beta \upsilon_\beta \frac{\partial P_\beta}{\partial \varepsilon}, \tag{9.197}$$

where $\hat{m}^{-1}$ is the diagonal tensor of the inverse effective mass (9.188).

Substituting (9.197) into (9.194) and accounting for (9.42) and (9.43), we obtain the following expression for the vector $\boldsymbol{P}(\varepsilon)$:

$$(\boldsymbol{\upsilon}\Phi_0) + \frac{e}{c}\,\boldsymbol{\upsilon}\left[\mathscr{H}(\hat{m}^{-1}\boldsymbol{P})\right] = (\boldsymbol{P}(\hat{\tau}^{-1}\boldsymbol{\upsilon})), \tag{9.198}$$

where Φ_0 is the generalized disturbing force (9.45). To solve the vector equation (9.198) it is convenient to present $\boldsymbol{P}(\varepsilon)$ in the form

$$\boldsymbol{P}(\varepsilon) = (\hat{\tau}^{-1}\Phi(\varepsilon)). \tag{9.199}$$

Then, accounting for the diagonality of the tensor $\hat{\tau}^{-1}$ (9.195), the right-hand side of (9.198) will assume the form

$$(\boldsymbol{P}(\hat{\tau}^{-1}\boldsymbol{\upsilon})) = (\Phi\boldsymbol{\upsilon}). \tag{9.200}$$

Taking account of the last two relations and noting that $\upsilon \neq 0$, we obtain an equation for $\Phi(\varepsilon)$ from (9.198),

$$\Phi_0 + \frac{e}{c}\left[H\hat{m}^{-1}(\hat{\tau}\Phi)\right] = \Phi. \tag{9.201}$$

Solving this equation while accounting for the tensor diagonalities $\hat{m}$ and $\hat{\tau}$, (9.188) and (9.195), it is possible to find the vector components $\Phi(\varepsilon)$. As a result, we have the following compact solution for the required vector $\boldsymbol{P}(\varepsilon)$ (9.199)

$$\boldsymbol{P}(\varepsilon) = \frac{1}{1+\nu_0^2}\left\{\hat{\tau}\Phi_0 + \frac{e}{c}\hat{\tau}\left[H\hat{m}^{-1}(\hat{\tau}\Phi_0)\right] + \frac{e^2}{c^2}\frac{|\hat{\tau}|}{|\hat{m}|}(H\Phi_0)(\hat{m}H)\right\}, \tag{9.202}$$

where

$$\nu_0^2 = \frac{e^2}{c^2}\frac{|\hat{\tau}|}{|\hat{m}|}(\hat{m}H)(\hat{\tau}^{-1}H) \tag{9.203}$$

and $|\hat{\tau}| = \tau_1\tau_2\tau_3$ $|\hat{m}| = m_1 m_2 m_3$ are the determinants of the diagonal relaxation time tensors $\hat{\tau}$ and effective mass tensor $\hat{m}$.

Note that in the case of scalar effective mass and relaxation time, (9.202) is reduced to the well-known solution (9.69) with (9.79).

9.5.3 Current Density

The solution (9.202) provides the possibility of calculating current density in the most general case of anisotropic spectrum and scattering in an arbitrary magnetic field.

Based on the general expression (9.1) it is possible to reduce the ith current density component for the spectrum (9.187) to the form

$$j_i = -\frac{e(8m_{01}m_{02}m_{03})^{1/2}}{3\pi^2\hbar^3}\int\left(-\frac{\partial f_0}{\partial\varepsilon}\right)B^{3/2}(\varepsilon)\frac{P_i(\varepsilon)}{m_i(\varepsilon)}\mathrm{d}\varepsilon. \tag{9.204}$$

In this case, we deform the ellipsoid (9.187) by the substitution $k_i = \sqrt{m_{0i}}k'_i$ and go on to integrate over energy, accounting for the fact that P_i and m_i depend on ε. Let us introduce an averaging formula for anisotropic spectrum (9.187):

$$\langle A(\varepsilon)\rangle = \frac{(8m_{01}m_{02}m_{03})^{1/2}}{3\pi^2\hbar^3 n_0}\int\left(-\frac{\partial f_0}{\partial\varepsilon}\right)B^{3/2}(\varepsilon)A(\varepsilon)\mathrm{d}\varepsilon, \tag{9.205}$$

where

$$n_0 = \frac{(8m_{01}m_{02}m_{03})^{1/2}}{3\pi^2\hbar^3}\int\left(-\frac{\partial f_0}{\partial\varepsilon}\right)B^{3/2}(\varepsilon)\mathrm{d}\varepsilon \tag{9.206}$$

is the charge carriers' concentration in a valley with the spectrum (9.187). Then (9.204) has the symbolic form

$$j_i = -en_0\left\langle\frac{P_i(\varepsilon)}{m_i(\varepsilon)}\right\rangle. \tag{9.207}$$

This formula has an obvious physical sense: current density components are defined by the average of the corresponding drift velocity components, since P_i are the impulse components of the force that gives rise to drift, and P_i/m_i is the drift velocity.

A

Definite Integrals Frequently Met in Statistical Physics

A.1 Gamma-Function or Euler Integral of Second Kind

$$\Gamma(n) = \int_0^\infty x^{n-1} \mathrm{e}^{-x} \mathrm{d}x; \qquad n > 0. \tag{A.1}$$

Write down this integral for $\Gamma(n+1)$ and integrate up by parts once. As a result, we get the following recurrent formula

$$\Gamma(n+1) = n\Gamma(n). \tag{A.2}$$

For any integer number $(n > 0)$ from (A.2) it follows that

$$\Gamma(n+1) = n \cdot (n-1) \cdot (n-2) \ldots 1\Gamma(1). \tag{A.3}$$

Inasmuch as the integral entering here is

$$\Gamma(1) = \int_0^\infty e^{-x} \mathrm{d}x = 1, \tag{A.4}$$

then (A.3) takes the form

$$\Gamma(n+1) = n! \tag{A.5}$$

For a semi-integer $n = \frac{2k+1}{2}$ from (A.2) it follows that

$$\begin{aligned} &\Gamma\left(\frac{3}{2}\right) = \frac{1}{2}\Gamma\left(\frac{1}{2}\right); \Gamma\left(\frac{5}{2}\right) = \frac{1}{2} \cdot \frac{3}{2}\Gamma\left(\frac{1}{2}\right); \\ &\Gamma\left(\frac{7}{2}\right) = \frac{1}{2} \cdot \frac{3}{2} \cdot \frac{5}{2}\Gamma\left(\frac{1}{2}\right). \end{aligned} \tag{A.6}$$

In the general case

$$\Gamma\left(\frac{2k+1}{2}\right)=\frac{(2k-1)!!}{2^k}\Gamma(1/2); k=1,2,3,\ldots \tag{A.7}$$

Here the double factorial $(2k-1)!!$ is the product of subsequent odd numbers from 1 to $(2k-1)$, i.e. $(2k-1)!!=1\cdot 3\cdot 5\cdot\ldots\cdot(2k-1)$. According to the definition

$$\Gamma(1/2)=\int\limits_0^\infty x^{-1/2}\mathrm{e}^{-x}\mathrm{d}x. \tag{A.8}$$

To calculate this integral introduce the replacement $x^{1/2}=t$. Then we get

$$\Gamma(1/2)=2\int\limits_0^\infty \mathrm{e}^{-t^2}\mathrm{d}t. \tag{A.9}$$

In as much as the latter integral equals $\sqrt{\pi}/2$, then

$$\Gamma(1/2)=\sqrt{\pi}. \tag{A.10}$$

As a result, for an arbitrary semi-integer argument (A.7) takes the form

$$\Gamma\left(\frac{2k+1}{2}\right)=\frac{(2k-1)!!}{2^k}\sqrt{\pi};\quad k=0,1,2,3,\ldots \tag{A.11}$$

A.2 Integral of Type

$$I_n=\int\limits_0^\infty x^n\mathrm{e}^{-ax^2}\mathrm{d}x; a>0, n\geq 0 \text{ is an integer number.} \tag{A.12}$$

Carry out the replacement $ax^2=y$. Then we have

$$I_n=\frac{1}{2}a^{-\frac{n+1}{2}}\int\limits_0^\infty \mathrm{e}^{-y}y^{\frac{n-1}{2}}\mathrm{d}y=\frac{1}{2}a^{-\frac{n+1}{2}}\Gamma\left(\frac{n+1}{2}\right). \tag{A.13}$$

Hence for $n=0$

$$I_0=\frac{1}{2}a^{-1/2}\Gamma(1/2)=\frac{1}{2}\sqrt{\frac{\pi}{a}}, \tag{A.14}$$

for even n, i.e. $n=2k$

$$I_{2k}=\frac{1}{2}a^{-\frac{2k+1}{2}}\Gamma\left(\frac{2k+1}{2}\right) \tag{A.15}$$

or [(see A.11)]

$$I_{2k} = \frac{(2k-1)!!}{2^{k+1}}\sqrt{\frac{\pi}{a^{2k+1}}}, \quad k \geq 1 \tag{A.16}$$

for odd n, i.e. $n = (2k+1)$

$$I_{2k+1} = \frac{1}{2}\frac{1}{a^{k+1}}\Gamma(k+1) = \frac{k!}{2a^{k+1}}, \quad k \geq 0. \tag{A.17}$$

Adduce values of the integral I_n in particular cases:

$$I_1 = \frac{1}{2a}; I_2 = \frac{1}{4}\frac{\sqrt{\pi}}{a^{3/2}}; I_3 = \frac{1}{2a^2}; I_1 = \frac{3}{8}\frac{\sqrt{\pi}}{a^{5/2}}. \tag{A.18}$$

If in an integral (A.12) the boundaries change in the limits from $-\infty$ to $+\infty$, for odd n we get

$$\int\limits_{-\infty}^{\infty} x^n \mathrm{e}^{-ax^2}\mathrm{d}x = 0, \tag{A.19}$$

and for even n

$$\int\limits_{-\infty}^{\infty} x^n \mathrm{e}^{-ax^2}\mathrm{d}x = 2\int\limits_{0}^{\infty} x^n \mathrm{e}^{-ax^2}\mathrm{d}x. \tag{A.20}$$

A.3 Integral of Type

$$K_n = \int\limits_0^{\infty} \frac{x^n \mathrm{d}x}{\mathrm{e}^x - 1}, \tag{A.21}$$

n is an integer or a semi-integer positive number. Transpose a certain part of the integrand:

$$\frac{1}{\mathrm{e}^x - 1} = \frac{\mathrm{e}^{-x}}{1-\mathrm{e}^{-x}} = \mathrm{e}^{-x}(1 + \mathrm{e}^{-x} + \mathrm{e}^{-2x} + \cdots) = \mathrm{e}^{-x}\sum_{k=0}^{\infty}\mathrm{e}^{-kx} \tag{A.22}$$

and substitute this series into (A.21). As a result we have

$$K_n = \int\limits_0^{\infty} x^n \sum_{k=0}^{\infty} \mathrm{e}^{-(k+1)x}\mathrm{d}x. \tag{A.23}$$

If we introduce the replacement of variables $(k+1)x = t$, then

$$K_n = \sum_{k=0}^{\infty}(k+1)^{-(n+1)x}\int\limits_0^{\infty} t^n \mathrm{e}^{-t}\mathrm{d}t \tag{A.24}$$

and we get

$$K_n = \Gamma(n+1)\zeta(n+1), \tag{A.25}$$

where $\Gamma(n)$ is the gamma-function [see (A.1)], and

$$\zeta(n) = \sum_{k=1}^{\infty} \frac{1}{k^n} \tag{A.26}$$

is the Riemann function.

All information about the $\Gamma(n)$- function are given in Appendix A.

Adduce some values of $\zeta(n)$ function:

$$\begin{aligned} &\zeta(2) = \tfrac{\pi^2}{6}; \zeta(3) = 1,202; \zeta(4) = \tfrac{\pi^4}{90}; \\ &\zeta(5) = 1,037; \zeta(3/2) = 2,612; \zeta(5/2) = 1,341. \end{aligned} \tag{A.27}$$

If we take into account these values, we can find necessary values of the integral of the K_n type

$$K_{1/2} = \Gamma(3/2)\zeta(3/2) = 2,33; K_1 = \Gamma(2)\zeta(2) = \frac{\pi^2}{6}; \tag{A.28}$$

$$K_{3/2} = \Gamma(5/2)\zeta(5/2) = 1,78; K_2 = \Gamma(3)\zeta(3) = 2,4; \tag{A.29}$$

$$K_3 = \Gamma(4)\zeta(4) = \frac{\pi^4}{15}. \tag{A.30}$$

A.4 Integral of Type

$$M_n = \int_0^{\infty} \frac{x^n \mathrm{e}^x \mathrm{d}x}{(\mathrm{e}^x - 1)^2}; \quad n > 1. \tag{A.31}$$

In order to calculate this integral, we expand a certain part of the integrand function into an infinitive series

$$(\mathrm{e}^x - 1)^{-2} = \mathrm{e}^{-2x}(1 - \mathrm{e}^{-x})^{-2} = \mathrm{e}^{-2x}(1 + 2\mathrm{e}^{-x} + 3\mathrm{e}^{-2x} + \cdots)$$

or

$$(\mathrm{e}^x - 1)^{-2} = \mathrm{e}^{-2x} \sum_{m=0}^{\infty} (m+1)\mathrm{e}^{-mx}. \tag{A.32}$$

Substituting (A.32) into (A.31), we get

$$M_n = \sum_{m=0}^{\infty} (m+1) \int_0^{\infty} x^n \mathrm{e}^{-(m+1)x} \mathrm{d}x. \tag{A.33}$$

Introducing the replacement of variables $(m+1)x = t$, we get

$$M_n = \sum_{m=0}^{\infty} (m+1)^{-n} \int_0^{\infty} t^n \mathrm{e}^{-t} \mathrm{d}t; n > 1. \tag{A.34}$$

If we take into account definitions of $\Gamma(n)$ (A.1) and $\zeta(n)$ (A.26) functions, the integral takes the form

$$M_n = \Gamma(n+1)\zeta(n). \tag{A.35}$$

In particular cases

$$M_{3/2} = \Gamma(5/2)\zeta(3/2) = 3,48; M_2 = \Gamma(3)\zeta(2) = \frac{\pi^2}{3}; \tag{A.36}$$

$$M_{5/2} = \Gamma(7/2)\zeta(5/2) = 4,45; M_3 = \Gamma(4)\zeta(3) = 7,21; \tag{A.37}$$

$$M_4 = \Gamma(5)\zeta(4) = \frac{4\pi^4}{15}. \tag{A.38}$$

A.5 Integral of Type

$$L_n = \int_0^{\infty} \frac{x^n \mathrm{d}x}{\mathrm{e}^x + 1}. \tag{A.39}$$

To compute this integral, expand the expression $(\mathrm{e}^x + 1)^{-1}$ into a series in powers of e^{-x}:

$$(\mathrm{e}^x + 1)^{-1} = \mathrm{e}^{-x}(1 + \mathrm{e}^{-x})^{-1} = \mathrm{e}^{-x} \sum_{k=0}^{\infty} (-1)^k \mathrm{e}^{-kx}. \tag{A.40}$$

Having substituted the latter expression into (A.39), we get

$$L_n = \sum_{k=0}^{\infty} (-1)^k \int_0^{\infty} x^n \mathrm{e}^{-(k+1)x} \mathrm{d}x. \tag{A.41}$$

If we introduce the replacement of variables $(k+1)x = t$, the integral takes the form

$$L_n = \sum_{k=0}^{\infty} \frac{(-1)^k}{(k+1)^{n+1}} \int_0^{\infty} t^n \mathrm{e}^{-t} \mathrm{d}t = \Gamma(n+1) \sum_{k=0}^{\infty} \frac{(-1)^k}{(k+1)^{n+1}}. \tag{A.42}$$

Transpose the series entering into this expression. To do this in (A.42) add and subtract a series consisting of k. As a result we get:

$$\sum_{k=0}^{\infty} \frac{(-1)^k}{(k+1)^{n+1}} = \sum_{k=0}^{\infty} \frac{1}{(k+1)^{n+1}} - 2\sum_{m=0}^{\infty} \frac{1}{(2m+2)^{n+1}} \tag{A.43}$$

or

$$\sum_{k=0}^{\infty} \frac{(-1)^k}{(k+1)^{n+1}} = (1-2^{-n})\sum_{k=0}^{\infty} \frac{1}{(k+1)^{n+1}} = (1-2^{-n})\zeta(n+1). \tag{A.44}$$

Substituting (A.44) into (A.42), we finally get

$$L_n = (1-2^{-n})\Gamma(n+1)\zeta(n+1). \tag{A.45}$$

Note that at $n = 0$ expression (A.45) turns into uncertainty, inasmuch as $\zeta(1) = \infty$.

However, at $n = 0$ integral (A.39) can be calculated immediately. Indeed, at $n = 0$

$$L_0 = \int\limits_0^{\infty} \frac{\mathrm{d}x}{\mathrm{e}^x + 1} = \int\limits_1^{\infty} \frac{\mathrm{d}y}{y(y+1)} = -\ln\left(\frac{1+y}{y}\right)\Bigg|_1^{\infty} = \ln 2. \tag{A.46}$$

In particular cases from (A.45) we find

$$L_{1/2} = (1-2^{-1/2})\Gamma(3/2)\zeta(3/2) = 0.673, \tag{A.47}$$

$$L_1 = \frac{1}{2}\Gamma(2)\zeta(2) = \frac{\pi^2}{12}, \tag{A.48}$$

$$L_2 = (1-2^{-2})\Gamma(3)\zeta(3) = 1.8, \tag{A.49}$$

$$L_3 = (1-2^{-3})\Gamma(4)\zeta(4) = \frac{7\pi^4}{120}. \tag{A.50}$$

B

Jacobian and Its Properties

The Jacobian, defined as

$$\frac{\partial(u,v)}{\partial(x,y)} = \begin{vmatrix} \frac{\partial u}{\partial x} & \frac{\partial u}{\partial y} \\ \frac{\partial v}{\partial x} & \frac{\partial v}{\partial y} \end{vmatrix}, \tag{B.1}$$

possesses the following properties:

1. $$\frac{\partial(u,v)}{\partial(x,y)} = \left(\frac{\partial u}{\partial x}\right)_y \left(\frac{\partial v}{\partial y}\right)_x - \left(\frac{\partial u}{\partial y}\right)_x \left(\frac{\partial v}{\partial x}\right)_y . \tag{B.2}$$

2. $$\frac{\partial(u,v)}{\partial(x,y)} = -\frac{\partial(v,u)}{\partial(x,y)}; \ \frac{\partial(u,v)}{\partial(x,y)} = -\frac{\partial(u,v)}{\partial(y,x)}. \tag{B.3}$$

3. $$\frac{\partial(u,y)}{\partial(x,y)} = \left(\frac{\partial u}{\partial x}\right)_y . \tag{B.4}$$

4. $$\frac{\partial(u,v)}{\partial(x,y)} = \frac{\partial(u,v)}{\partial(t,s)} \frac{\partial(t,s)}{\partial(x,y)}. \tag{B.5}$$

5. $$\frac{\mathrm{d}}{\mathrm{d}t} \frac{\partial(u,v)}{\partial(x,y)} = \frac{\partial\left(\frac{\mathrm{d}u}{\mathrm{d}t}, v\right)}{\partial(x,y)} + \frac{\partial\left(u, \frac{\mathrm{d}v}{\mathrm{d}t}\right)}{\partial(x,y)}. \tag{B.6}$$

These properties of the Jacobian are frequently used when finding thermodynamic relationships and calculating thermodynamic coefficients.

Bibliograpy

1. L.D. Landau, E.M. Lifshitz, *Statistical Physics of A Course of Theoretical Physics*, vol 5 (Elsevier, Butterworth-Heinemann, 1980) [L.D. Landau, E.M. Lifshitz, *Statistical Physics*, vol 5 (Moscow, Fizmatlit, 2002) (in Russian)]
2. L.D. Landau, E.M. Lifshitz, *Mechanics of A Course of Theoretical Physics*, vol 1 (Elsevier, Oxford, 1996)
3. L.D. Landau, E.M. Lifshitz, *Quantum Mechanics: Non-Relativistic Theory*, of A Course of Theoretical Physics, vol. 3 (Butterworth-Heinemann, Oxford, 2003) [L.D. Landau, E.M. Lifshitz, *Quantum Mechanics: Non-Relativistic Theory*, vol 3 (Fizmatlit, Moscow, 2002) (in Russian)]
4. R. Kubo, *Thermodynamics* (North-Holland Publishing Company, Amsterdam, 1968) [R. Kubo, *Thermodynamics* (Mir, Moscow, 1970) (in Russian)]
5. R. Kubo, *Statistical Mechanics* (North-Holland, Amsterdam, 1965) [R. Kubo, *Statistical Mechanics* (Mir, Moscow, 1970) (in Russian)]
6. I.P. Bazarov, *Thermodynamics* (Pergamon Press, Oxford, 1964) [I.P. Bazarov, *Thermodynamics* (Vishaya School, Moscow, 1991) (in Russian)]
7. K. Huang, *Statistical Mechanics* (Wiley, New York, 1987) [K. Huang, *Statistical Mechanics* (Mir, Moscow, 1966) (in Russian)]
8. E. Fermi, *Thermodynamics.* PH, 1973. [E. Fermi, *Thermodynamics.* Kharkov (1973) (in Russian)]
9. J.W. Gibbs, *Elementary Principles in Statistical Mechanics* (Yale Unimpressed, New Heaven, 1902) [J.W. Gibbs, *Elementary Principles in Statistical Mechanics* (Moscow, Nauka, 1946) (in Russian)]
10. J.W. Gibbs, *Thermodynamics. Statistical Mechanics* (Nauka, Moscow, 1982) (in Russian)]
11. F. Reif, *Statistical Physics* (McGraw-Hill, Moscow, 1970) [F. Reif, *Statistical Physics* (Nauka, Moscow, 1972) (in Russian)]
12. J.S. Blakemore, *Semiconductor Statistics* (Pergamon Press, Oxford, 1962) [J.S. Blakemore, *Semiconductor Statistics* (IL, Moscow, 1964) (in Russian)]
13. A.I. Anselm, *Fundamentals of Statistical Physics and Thermodynamics* (Nauka, Moscow, 1973) (in Russian)]
14. A.I. Anselm, *Introduction to Semiconductor Theory* (Prentice Hall, Englewood Cliffs, NJ, 1982) [A.I. Anselm, *Introduction to Semiconductor Theory* (Nauka, Moscow, 1978) (in Russian)]

15. Yu.B. Rumer, M.Sh. Ryvkin, *Thermodynamics, Statistical Physics and Kinetics* (Mir Pub, Moscow, 1980) [Yu.B. Rumer, M.Sh. Ryvkin. *Thermodynamics, Statistical Physics and Kinetics* (Nauka, Moscow, 1977) (in Russian)]
16. B.M. Askerov, *Electron Transport Phenomena in Semiconductors* (World Scientific, Singapore, 1994) [B.M. Askerov, *Electron Transport Phenomena in Semiconductors* (Nauka, Moscow, 1985) (in Russian)]
17. Smirnov, *A Course of Higher Mathematics*, vol 3 (Pergamon Press, Oxford, 1964), part 2. [Smirnov,*A Course of Higher Mathematics* (Nauka, Moscow, 1974) (in Russian)]
18. O. Madelung, *Festkörpertheorie.* I, II. (Springer, Heidelberg, 1988) [O. Madelung, *Solid State Theory* (Nauka, Moscow, 1980) (in Russian)]
19. K. Seeger, *Semiconductor Physics* (Springer, New York, 2004), 9th ed. [K. Seeger, *Semiconductor Physics* (Mir, Moscow, 1977) (in Russian)]
20. L.I. Schiff, *Quantum Mechanics* (McGraw-Hill, New York, 1968) [L.I. Schiff, *Quantum Mechanics* (IL, Moscow, 1957) (in Russian)]
21. C. Kittel, *Quantum Theory of Solids.* (Wiley, New York, 1987), 2nd ed. [C. Kittel, *Quantum Theory of Solids* (Nauka, Moscow, 1980) (in Russian)]
22. R.E. Peierls, *Quantum Theory of Solids* (Clarendon Press, Oxford, 2001) [R.E. Peierls, *Quantum Theory of Solids* (IL, Moscow, 2002) (in Russian)]
23. A.I. Anselm, B.M. Askerov, Thermomagnetic phenomena in semimetals in strong magnetic field. Sov. Phys.-Solid State **2** 2060 (1960) Chemical potential and criterion of degeneracy. Soviet Phys.-Solid State **2** 2571 (1960). Fizika Tverdogo Tela (ru) **2**(9), 2310 (1960); (11), 2821 (1960)
24. A.I. Anselm, B.M. Askerov, Quantum theory of Nernst-Ettingshausen in semiconductors. Fizika Tverdogo Tela (ru) **9** 31 (1967)
25. B.M. Askerov, M. Cankurtaran, Isobaric specific heat and thermal expansion of solids in the debye approximation. Physica Status Solidi (b) **185** 341–348 (1994)
26. M. Cankurtaran, B.M. Askerov, Equation of state, isobaric specific heat and thermal expansion of solids with polyatomic basis in the Einstein–Debye approximation. Physica Status Solidi (b) **194** 499–507 (1996)
27. B.M. Askerov, M.M. Machmudov, Kh.A. Gasanov, Equation of state of an electron gas and theory of the thermal voltage in a quantizing magnetic field. J Semicond. **32**(3), 261 (1998). Fizika i Tekhnika Poluprovodnikov (ru) **32** 290 (1998)

Index

MIX
Papier aus verantwortungsvollen Quellen
Paper from responsible sources
FSC® C105338

Printed by Books on Demand, Germany